AQA Human Biology

H 30908

A2

Exclusively endorsed by AQA

£20·99

Pauline Lowrie

Bev Goodger

Nelson Thornes

Published in 2009 by:
Nelson Thornes Ltd
Delta Place
27 Bath Road
CHELTENHAM
GL53 7TH
United Kingdom

09 10 11 12 13 / 10 9 8 7 6 5 4 3 2 1

A catalogue record for this book is available from the British Library

ISBN 978 0 7487 8278 9

Cover photograph by Photolibrary/Image Source Ltd

Illustrations include artwork drawn by Barking Dog Art and GreenGate Publishing

Page make-up by GreenGate Publishing, Tonbridge, Kent

Printed and bound in Croatia by Zrinski

Photograph Acknowledgements

The authors and publisher are grateful to the following for permission to reproduce photographs and other copyright material in this book.

Alamy/Arco Images GmbH: p171 (top); **/Chris Mattison:** p115 (bottom); **/David Robertson:** p159; **/Janine Wiedel Photolibrary:** p3; **/Kevin Britland:** p175 (bottom); **/MedicalRF.com:** p26; **/Picture Partners:** p40; **/Sandy Young:** p115 (top); **Corbis/Eric Fougére/Kipa:** p36; **Fotolia:** p59 (left), p59 (right), p101, p175 (middle), p176, p188; **Getty/LWA/Dann Tardif:** pviii; **iStockphoto.com:** p22, p145, p150 (bottom), p167, p173, p175 (top), p211, p226; **Martyn Chillmaid:** p186 (right), p194, p196, p215, p253; **Oxford Scientific/ Chris Barry:** p119 (bottom); **Pauline Lowrie:** p96 , p123; **Science Photo Library/Agema Infrared Systems:** p230; **/Alex Bartel:** p197; **/Astrid & Hanns-Frieder Michler:** p141; **/Bill Barksdale/AGstockUSA:** p181 (top); **/Biophoto Associates:** p50, p54 (top); **/Chris Knapton:** p95, p179, p239; **/Chris Martin Mahr:** p174; **/Colin Cuthbert:** p213; **/Cordellia Molloy:** p193 ; **/Cristina Pedrazzini:** p66; **/David Parker:** p98; **/Don Fawcett:** p13; **/Dr Gary Gaugler:** p252; **/Dr P. Marazzi:** p249; **/Duncan Shaw:** p234; **/Eddy Gray:** p187; **/Emmeline Watkins:** p186 (left); **/Eye of Science:** p45; **/Georgette Douwma:** p177; **/Innerspace Imaging:** p138; **/James King-Holmes:** p192 (left); **/Jean-Claude Revy-A. Goujeon, ISM:** p6; **/Jean-Loup Charmet:** p141; **/Klaus Guldbrandsem:** p88; **/ Leslie J Borg:** p178, p208; **/Mauro Fermariello:** p246; **/Michael Abbey:** p202(bottom); **/Moredun Scientific Ltd:** p202(top); **/NIAID/ CDC:** p199 (bottom); **/Pascal Goetgheluck:** p54 (bottom); **/Paul Rapson:** p225; **/Peter Menzel:** p86; **/R.Maisonneuve, Publiophoto Diffusion:** p92; **/Robert Brook:** p207, p209, p227; **/Robert Longuehaye, NIBSC:** p84; **/Saturn Stills:** p192 (right); **/Scott Camazine/ Sue Trainor:** p21; **/Sheila Terry:** p171 (bottom), p180; **/Simon Fraser:** p80, p199 (top); **/Sinclair Stammers:** p181 (bottom), p240; **/Steve Gschmeissner:** p160, p247; **/Sue Ford:** p119 (top); **/Thomas Deerinck, NCMIR:** p41, p49, p149; **Wellcome/Dr David Becker:** p24; **Wellcome:** p150 (top).

Every effort has been made to trace and contact all copyright holders and we apologise if any have been overlooked. The publisher will be pleased to make the necessary arrangements at the first opportunity.

The authors would like to thank Dr Julian Rayner for his help with parts of this book.

Contents

AQA introduction

Nelson Thornes has worked in partnership with AQA to ensure this book and the accompanying online resources offer you the best support for your A2 course.

All resources have been approved by senior AQA examiners so you can feel assured that they closely match the specification for this subject and provide you with everything you need to prepare successfully for your exams.

These print and online resources together **unlock blended learning**; this means that the links between the activities in the book and the activities online blend together to maximise your understanding of a topic and help you achieve your potential.

These online resources are available on *kerboodle!* which can be accessed via the internet at **http://www.kerboodle.com/live**, anytime, anywhere. If your school or college subscribes to this service you will be provided with your own personal login details. Once logged in, access your course and locate the required activity.

For more information and help visit **http://www.kerboodle.com**

Icons in this book indicate where there is material online related to that topic. The following icons are used:

💡 Learning activity

These resources include a variety of interactive and non-interactive activities to support your learning:

- Animations
- Simulations
- Maths skills
- Key diagrams
- Glossary

✔ Progress tracking

These resources include a variety of tests which you can use to check your knowledge on particular topics (Test yourself) and a range of resources that enable you to analyse and understand examination questions (On your marks...).

You will also find the answers to the examination-style questions online.

🔎 Research support

These resources include WebQuests, in which you are assigned a task and provided with a range of web links to use as source material for research.

These are designed as Extension resources to stretch you and broaden your learning, in order for you to attain the highest possible marks in your exams.

Web links

Our online resources feature a list of recommended weblinks, split by chapter. This will give you a head start, helping you to navigate to the best websites that will aid your learning and understanding of the topics in your course.

🔬 How science works

These resources are a mixture of interactive and non-interactive activities to help you learn the skills required for success in this new area of the specification.

🔺 Practical

This icon signals where there is a relevant practical activity to be undertaken, and support is provided online.

How to use this book

This book covers the specification for your course and is arranged in a sequence approved by AQA.

The textbook will cover all three of the Assessment Objectives required in your AQA A Level Human Biology course.

The main text of the book will cover AO1 – Knowledge and understanding. This consists of the main factual content of the specification. The other Assessment Objectives (AO2 – Application of knowledge and understanding and AO3 – How science works) make up around 50% of the assessment weighting of the specification, and as such will be covered in the textbook in the form of the feature 'Applications and How science works' (see below). You will **not** be asked to recall the information given under these headings for the purpose of examinations.

The book content is divided into the two theory units of the AQA Human Biology A2 specification; Unit 4 – Bodies and cells in and out of control and Unit 5 – The air we breathe, the water we drink, the food we eat. Units are then further divided into chapters, and then topics, making the content clear and easy to use.

Unit openers give you a summary of the content you will be covering, and a recap of ideas from GCSE and AS Level that you will need.

The features in this book include:

Learning objectives

At the beginning of each section you will find a list of learning objectives that contain targets linked to the requirements of the specification. The relevant specification reference is also provided.

Key terms

Terms that you will need to be able to define and understand are highlighted in bold blue type within the text, e.g. **allele**. You can look up these terms in the glossary on page 260.

Hint

Hints to aid your understanding of the content.

Link

Synoptic links are highlighted in the margin near the relevant text using the link icon accompanied by brief notes which include references to where the linked topics are to be found in this book or the AS book.

Applications and How science works

These features may cover either or both of the assessment objectives AO2 – Application of knowledge and understanding and AO3 – How science works, both key parts of the new specification.

As with the specification, these objectives are integrated throughout the content of the book. This feature highlights opportunities to apply your knowledge and understanding and draws out aspects of 'How science works' as they occur within topics, so that it is always relevant to what you are studying. The ideas provided in these features intend to teach you the skills you will need to tackle this part of the course, and give you experience that you can draw upon in the examination. You will not be examined on the exact information provided in 'Application and How science works' features.

For more information, see 'How science works' on page 1 for more detail.

Summary questions

Short questions that test your understanding of the subject and allow you to apply the knowledge and skills you have acquired to different scenarios. Answers are supplied at the back of the book. These answers are more than just a mark scheme. They often include explanations of the answers to aid learning and understanding. These answers are not exhaustive and there may be acceptable alternatives.

AQA Examiner's tip

Hints from AQA examiners to help you with your studies and to prepare for your exam.

AQA Examination-style questions

Questions from past AQA papers that are in the general style that you can expect in your exam. These occur at the end of each chapter to give practice in examination-style questions for a particular topic. They also occur at the end of each unit; the questions here may cover any of the content of the unit. These questions relate to earlier specifications but have been chosen because they are relevant to the new specification. Despite careful selection there may be certain terms that do not exactly match the new requirements. They should therefore be treated in the same way as Applications and used for examination practice and application of knowledge, rather than learning their content. When you answer the examination-style questions in this book, remember that quality of written communication (QWC) will be assessed in any question or part-question in the Unit 4 and 5 papers where extended descriptive answers are required.

Synopticity

Synoptic questions or part-questions are a key feature of your A2 examination papers. Such a question may require you to draw on knowledge, understanding and skills from AS Level that underpin the A2 topic which the question is about. They link knowledge, understanding and skills from topics in the A2 theory unit on which the question is set, perhaps in a new context, which is described in the question.

Stretch and challenge

Some of the questions in the papers for Units 4 and 5 are designed to test the depth of your knowledge and understanding of the subject. Such questions may require you to solve a problem where you have to decide on a suitable strategy and appropriate methods, possibly linking different ideas from within the unit, discuss in an extended written answer a controversial issue involving biology, perhaps in terms of advantages and disadvantages, that affects people or society at large.

The questions test your ability to think deeply and clearly about biology and to provide solutions and answers that are coherent and clear.

Answers to these questions are supplied online.

AQA examination questions are reproduced by permission of the Assessment and Qualifications Alliance. Nelson Thornes is responsible for the solution(s) given and they may not constitute the only possible solution(s).

Studying A2 Human Biology

Welcome to Human Biology at A2 level.

This book aims to make your study of human biology successful and interesting.

The book is written to cover the content of the A2 course for the AQA specification. Each chapter in the book corresponds exactly to the subdivisions of each unit of the specification.

A2 course structure (% of total A2 marks is shown in brackets)

Unit 4 Bodies and cells in and out of control
            (40%)   Chapters 1–7

Unit 5 The air we breathe, the water we drink, the food we eat
            (40%)   Chapters 8–13

Unit 6 Investigative and practical skills in A2 Human Biology
            (20%)   Online resources

Using the book

You will find that the A2 course builds on the skills and understanding you developed in your AS course. The topics in the A2 course will deepen your knowledge and understanding of human biology. The course provides a solid biology foundation for those who intend to proceed to university to study biology or a related biological or medical science such as medicine, dentistry or veterinary science, where a good understanding of biology is necessary to fully appreciate key ideas, concepts and techniques.

New ideas are presented in the book in a careful step-by-step manner to enable you to develop a firm understanding of concepts and ideas. Human Biology at A2 Level will require you to describe and explain facts and processes in detail and with accuracy. However, the course is also about developing skills so that you can apply what you have learned. Examination papers will test skills such as interpreting new information, analysing experimental data, and evaluating information. In the AQA specification you will see sections which begin 'candidates should be able to ....' These are the sections which set out the skills you will need to develop to achieve success. You will find 'Application and How science works' features in the book. These features present relevant and challenging information which will enable you to develop these skills. The factual content of these sections is **not** required for examination purposes.

The AQA specification also emphasises how scientists work and how their work affects people in their everyday lives. For example, information is often presented in newspapers and on TV on science issues such as the possible side-effects of vaccines or drugs. Such reports may even contain conflicting evidence. The validity of evidence and the accuracy of conclusions is constantly questioned by scientists. Information in the text and in the accompanying resources will enable you to analyse evidence and data and to evaluate the way scientists obtain new evidence.

Checking your progress

You will find questions at the end of each chapter so that you can check your progress as you complete each section. Each chapter represents a manageable amount of learning so that you do not try to achieve too much too quickly. At the end of each unit there are questions written by AQA examiners in the same style that you will meet in examinations.

Investigative and practical skills

There are two routes for the assessment of Investigative and Practical Skills:

Either, **Route T**: Practical Skills Assessment (PSA) + Investigative Skills Assignment (ISA), which will be marked by your teacher.

Or, **Route X**: Practical Skills Verification (PSV) (assessed by your teacher) + Externally Marked Practical Assessment (EMPA), which is set and marked by an external AQA appointed examiner.

Both routes form 20% of the total A2 assessment and will involve carrying out practical work, collecting and processing data, and then using the data to answer questions in a written test. The resources which accompany the book provide examples of investigations so that you can develop your practical and investigative skills as you progress through the topics in Units 4 and 5.

The book and accompanying resources provides a wealth of material specifically written for your A2 Human Biology course. As well as helping you to achieve success, you should find the resources interesting and challenging.

How science works

You have already gained some skills through the 'How science works' component of your AS course. As you progress through your A2 Human Biology course, you will develop your scientific skills further and learn about important new ideas and applications through the 'How science works' component of your A level course. These skills are a key part of how every scientist works. Scientists use them to probe and test new theories and applications in whatever field of work they are working in. Now you will develop them further and gain new skills as you progress through the course.

'How science works' is developed in this book through relevant features in the main content of the book and are highlighted accordingly. The 'How science works' features in this book will help you to develop the relevant 'How science works' skills necessary for examination purposes but more importantly these features should give you a thorough grasp of how scientists work, as well as a deeper awareness of how science is used to improve the quality of life for everyone.

When carrying out their work scientists:

- Use theories, models and ideas to develop scientific explanations and make progress when validated evidence is found that supports a new theory or model.

- Use their knowledge and understanding when observing objects or events, in defining a problem and when questioning the explanations of themselves or of other scientists

- Make observations that lead to explanations in the form of hypotheses. In turn hypotheses lead to predictions that can be tested experimentally.

- Carry out experimental and investigative activities that involve making accurate measurements, and recording measurements methodically.

- Analyse and interpret data to look for patterns and trends, to provide evidence and identify relationships.

- Evaluate methodology, evidence and data, and resolve conflicting evidence.

- Appreciate that if evidence is reliable and reproducible and does not support a theory, the theory must be modified or replaced with a different theory.

- Communicate the findings of their research to provide opportunity for other scientists to replicate and further test their work.

- Evaluate, and report on the risks associated with new technology and developments.

- Consider ethical issues in the treatment of humans, other organisms and effects on the environment.

- Appreciate the role of the scientific community in validating findings and developments .

- Appreciate how society in general uses science to inform decision making.

UNIT 1

Bodies and cells in and out of control

Chapters in this unit

Human bodies consist of complex organ systems carrying out different functions.

In this unit, you will build on what you learned at AS Level to understand how the different organ systems of the body are coordinated so that the body works efficiently as a whole. You will learn in Chapter 1 how the reproductive system works, and how knowledge of how it works enables humans to control it. This is important in birth control methods, so that pregnancy may be planned. It is also important in enabling people with infertility problems to have children. In Chapter 2 you will go on to learn how the human body develops from infancy to adulthood and about the changes that occur with ageing.

Chapter 3 looks at the cellular level, where genetic information is stored in the form of DNA. DNA codes for the production of proteins, many of which are enzymes. You will learn how genes are passed on from generation to generation, producing variation in the offspring. You will also learn how DNA carries the coded information to make proteins, and how defects in DNA can lead to inherited conditions. Cancer may also result from changes in DNA. You will learn how environmental factors can increase the incidence of cancer, and how enzymes coded for by DNA control cellular reactions.

Scientists can use their knowledge of DNA to insert new genes into organisms. In Chapter 4 you will learn how this is done and how these methods can be used to produce new organisms with novel features. You will go onto evaluate the ethical and moral issues that arise from the use of gene technology. Chapter 4 also looks at how genomes, such as that of the human, have been sequenced and what this has taught us.

Coordinating the different organ systems of the body involves both nervous and hormonal communication. Chapter 5 looks at how the nervous system works to detect stimuli, and how this information is processed and interpreted. You will learn how the glands and skeletal muscle can be stimulated by the nervous system, and how drugs can affect the way in which the nervous system works.

Chapter 6 looks at how the body reacts to different situations. You will learn how the nervous and hormonal systems work together to produce coordinated responses to stimuli which we perceive as threatening or frightening.

Finally, in Chapter 7 you will learn how the human body operates control systems to maintain the body's internal environment within restricted limits. This process is called homeostasis.

What you already know

Whilst the material in this unit is intended to be self-explanatory, there is certain information from GCSE and the AS Human Biology course that will prove very helpful to the understanding of this unit. Knowledge of the following elements of GCSE and the AS Human Biology course will be of assistance:

- The body consists of cells, tissues, organs and systems.
- Polysaccharides are carbohydrate polymers made of monomers such as glucose.
- Proteins are made of amino acids. The overall shape of a protein (its tertiary structure) is held together by hydrogen and ionic bonds, as well as disulphide bridges.
- Fats are made of glycerol and fatty acids.
- There are bacteria in the human gut that may be beneficial, including those that produce vitamin K.
- Enzymes are proteins that lower the activation energy for a reaction to occur. Their action can be explained by the lock and key and induced fit models.
- Enzyme activity is affected by factors such as temperature and pH.
- Cell membranes are composed of a phospholipid bilayer with proteins embedded in it.
- Substances can cross cell membranes by diffusion, active transport and osmosis.
- Cystic fibrosis is an inherited disorder, in which the CFTR protein in plasma membranes does not function properly.
- The heart is composed of cardiac muscle, and it pumps blood round the body. It is myogenic, meaning that it can contract without being stimulated by the nervous system.
- Over large distances, the efficient supply of materials is provided by mass transport in the blood.
- Some diseases and conditions are the result of lifestyle decisions.
- Mitosis is a type of cell division that results in identical copies of the parent cell.
- Chemical carcinogens and radiation may damage DNA and cause mutations in the genes controlling growth. This can lead to uncontrolled cell division and cancer.
- Meiosis is a form of cell division that results in haploid daughter cells called gametes.
- Organisms can be classified into a taxonomic hierarchy: kingdom, phylum, class, order, family, genus and species.
- During exercise, the heart rate and the ventilation increase. The stimulus for these changes is an increase in carbon dioxide concentration in the blood.
- Parasites are organisms that live in or on another organism, called its host, from which it feeds. It generally causes harm. Parasites are highly adapted to their way of life.

1 IVF – babies for those who cannot conceive naturally

1.1 Male and female reproductive systems

Learning objectives:

■ How do the male and female reproductive systems work?

Specification reference: 3.4.1

■ The male reproductive system

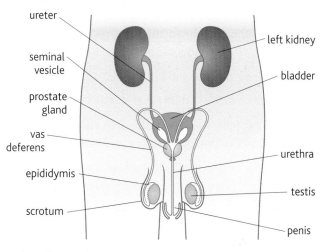

Figure 1 *The male reproductive system*

Male **gametes**, or **sperm** are produced in the **testes**. Figure 1 shows that the testes are in a sac called the **scrotum**. This keeps the testes at about 35 °C, 2 °C below normal body temperature. This is the optimum temperature for sperm development.

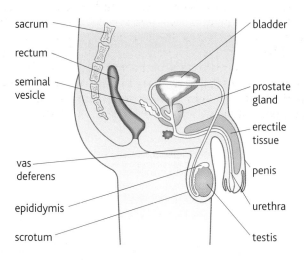

Figure 2 *A section through the testis*

The testis is divided internally into several lobules. Each lobule contains many long tubules called **seminiferous tubules**. This is where sperm are produced. You will learn more about this in Topic 1.3. Between the seminiferous tubules are **interstitial cells**, which secrete **testosterone**. The seminiferous tubules join together to form larger tubes called the vasa efferentia. In turn, these join together to form a long, coiled tube called the epididymis. This is where sperm are stored. The epididymis leads into another tube called the vas deferens. This carries the sperm towards the urethra. As sperm pass along the vasa deferentia, secretions are added from the seminal vesicles. These produce mucus which makes the sperm more motile. The prostate gland produces an alkaline fluid that also helps sperm motility, as well as neutralising any acid urine that remains in the urethra. The mixture of sperm and secretions is called semen. The semen is carried along the urethra and is ejaculated through the penis.

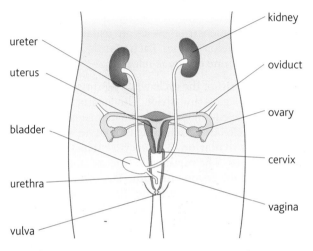

Figure 3 *The female reproductive system*

The female reproductive system

Figure 3 shows the female reproductive system. The female gametes (eggs or **ova**) are produced in the **ovaries**. A woman releases one ovum approximately every month from the age of puberty (at about 12 years of age) until the menopause (at about 50 years of age). The ovum is fertilised by a sperm in the **oviduct** which is lined with mucus-secreting ciliated epithelial cells. The fertilised ovum is called a **zygote**. As the zygote travels down the oviduct, it divides to form a hollow ball of cells called the **blastocyst**. You will learn more about the blastocyst in Topic 1.4. The blastocyst reaches the uterus after about a week. Here, it implants into the lining of the uterus, called the **endometrium**. Some of the cells in the blastocyst form the embryo and the rest form the placenta. The embryo develops into a fetus within the uterus.

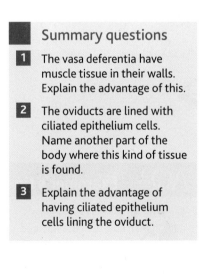

Summary questions

1. The vasa deferentia have muscle tissue in their walls. Explain the advantage of this.

2. The oviducts are lined with ciliated epithelium cells. Name another part of the body where this kind of tissue is found.

3. Explain the advantage of having ciliated epithelium cells lining the oviduct.

1.2 The menstrual cycle

Figure 1 *The internal structure of an ovary. The pink structure at the top is the corpus luteum and the white ovals are developing follicles*

The structure of the ovaries

Figure 1 shows the structure of a human ovary. It contains many **Graafian follicles** at different stages of development. Each follicle is a ball of cells surrounding an immature ovum, or **oocyte**.

A female is born with about a million immature follicles in her ovaries. By the time a female enters puberty, aged about 12 to 13 years, only about 400 000 of these follicles remain.

In human females, the time of the first menstrual cycle is called the **menarche**. The first few menstrual cycles are usually infertile, but after that, follicles begin to develop further. Each month, a few follicles start to develop, but usually only one oocyte is released.

Figure 2 shows the first part of the cycle when a primary follicle starts to develop. The follicle cells grow and divide. Fluid starts to build up in the follicle. The mature follicle is about 15 mm in diameter. On about day 14 of the cycle, the follicle bursts, releasing the oocyte, surrounded by a few follicle cells. This is **ovulation.** The oocyte enters the oviduct and is moved towards the uterus. The ruptured follicle continues to grow and forms a structure called the **corpus luteum.** The corpus luteum produces hormones. If the woman becomes pregnant, the corpus luteum remains in the ovary, inhibiting further menstrual cycles. If the woman does not become pregnant, the corpus luteum degenerates and another cycle begins.

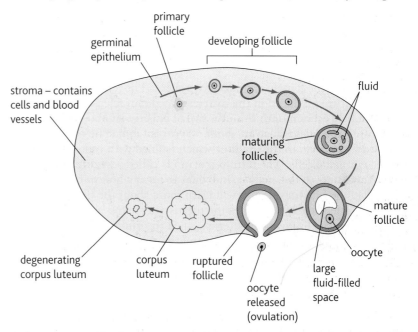

Figure 2 *Stages in development of a Graafian follicle in the ovary*

AQA Examiner's tip

Be careful to get the terminology right. It is an oocyte that is released from the follicle at ovulation – the remains of the follicle remain in the ovary. A common exam error is to say that a follicle or an ovum is released at ovulation.

The menstrual cycle

The menstrual cycle continues until the woman reaches the **menopause** at about the age of 50. The events of the menstrual cycle are controlled by hormones. Figure 3 summarises the events of the menstrual cycle.

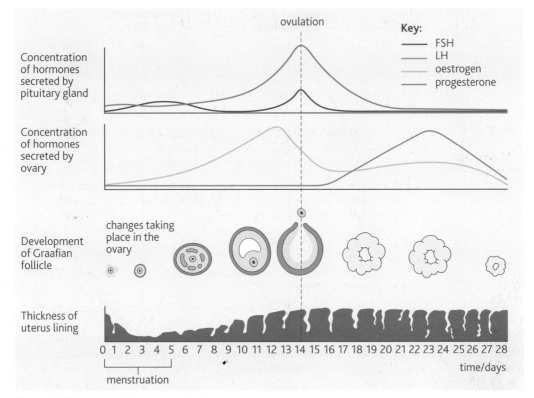

Figure 3 *The events of the menstrual cycle*

The **pituitary gland** is at the base of the brain. The menstrual cycle starts when the pituitary gland secretes a hormone called **follicle stimulating hormone (FSH)**. FSH travels in the blood to the ovary, where it causes a follicle to develop. As the follicle grows and develops, it releases the hormone **oestrogen**. Oestrogen stimulates the thickening of the endometrium. It also inhibits secretion of FSH, and stimulates the production of another hormone, **luteinising hormone (LH)** from the pituitary gland. However, most of this LH is temporarily stored in the pituitary gland.

By about day 11, the oestrogen concentration in the blood is very high, because of the size of the developing follicle. This high level of oestrogen causes the LH stored in the pituitary gland to be released in a surge, and it also causes a surge in FSH. This surge in LH brings about ovulation. The mature follicle bursts, releasing the oocyte into the oviduct.

LH stimulates the ruptured follicle to develop into a corpus luteum. The corpus luteum continues to secrete some oestrogen, but it now secretes large amounts of another hormone, **progesterone**. Progesterone inhibits FSH and LH, and stimulates the growth of blood vessels in the endometrium. As LH is now inhibited, the corpus luteum starts to degenerate. It produces less progesterone. This fall in progesterone means that FSH is no longer inhibited, so FSH is released again and a new cycle starts.

The fall in progesterone and oestrogen leads to **menstruation**. This is when the outer layer of the endometrium is shed and lost from the body. The interactions of the hormones involved in the menstrual cycle can be seen in Figure 4, on the next page.

AQA Examiner's tip

Remember that follicles and oocytes are present in the ovary from birth. So be sure to say that FSH causes a follicle to *develop*. A common error is to say that FSH causes the *production* of a follicle.

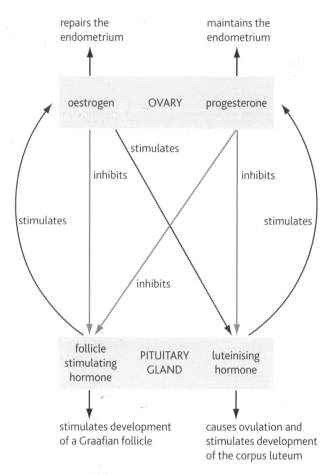

Figure 4 *Interactions of the hormones involved in the menstrual cycle*

How science works

Differences between the menstrual and the oestrus cycle

Humans are unique in having a menstrual cycle. Other mammals have an oestrous cycle. There are three main differences between an oestrous cycle and a menstrual cycle:

■ In an oestrous cycle, the outer layer of the endometrium is not shed from the body but is reabsorbed if pregnancy does not occur. (Some menstrual flow does occur in species closely related to humans, such as chimpanzees.)

■ In an oestrous cycle, the time of ovulation is called oestrus. The female's behaviour is very different at this time, and it is the only time of the cycle when she is sexually receptive to the male.

■ Oestrous cycles generally continue until death, while menstrual cycles stop at the menopause.

You may remember from AS level that scientists believe humans evolved from ancestors who had oestrous cycles. The change to having a menstrual cycle may have encouraged pair-bonding and a change to family groups from the larger social groups that are seen in animals such as apes.

Summary questions

1 Explain why it is important that oestrogen inhibits FSH during the first stage of the menstrual cycle.

2 a Suggest how the development of a menstrual cycle in human ancestors, rather than an oestrous cycle, may have encouraged the development of pair-bonding.

  b Explain why scientists think pair-bonding was important for human survival.

3 Explain what causes some cells to respond to a hormone such as FSH, while others do not.

1.3 Gametogenesis

Learning objectives:

■ How are gametes produced?

■ What are the differences between male and female gametes?

Specification reference: 3.4.1

Gametogenesis is the formation of gametes. 'Genesis' is a Greek word meaning 'the coming into being of something'. **Spermatogenesis** is the formation of sperm and **oogenesis** is the formation of ova.

In both males and females, gametogenesis can be divided up into the same three stages:

1 **multiplication**, in which the cells divide by mitosis

2 **growth phase**, in which the cells produced by mitosis grow in size

3 **maturation**, in which these cells divide by meiosis to form haploid cells. These then differentiate into gametes.

■ Spermatogenesis and oogenesis

You will see that the processes of spermatogenesis and oogenesis are very similar. However, there are two important differences:

■ The first division of meiosis occurs in the female before she is born. A secondary oocyte is released from the follicle at ovulation. Meiosis does not finish until after fertilisation.

■ In spermatogenesis, all the haploid cells formed as a result of meiosis develop into sperm. However, in oogenesis the haploid cells divide unequally, so tiny **polar bodies** are formed, which have no function. Only one ovum is produced.

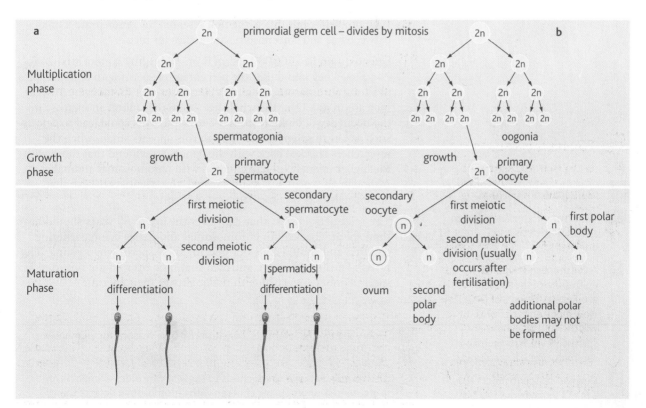

Figure 1 a *Spermatogenesis – formation of sperm (or spermatozoa),* **b** *Oogenesis – formation of ova*

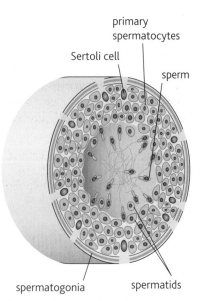

primary
spermatocytes

Sertoli cell

sperm

spermatogonia

spermatids

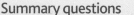

Figure 2 *A section through a seminiferous tubule*

■ The structure of the testes

You learned about the structure of the testes in Topic 1.1. You will recall that the testes contain seminiferous tubules. The total length of seminiferous tubules in the testes of one man is about 1 km. Sperm are produced inside these seminiferous tubules.

The outer layer of the seminiferous tubule is made up of **germinal epithelium**. This is where the cells divide by mitosis during the multiplication stage. As the cells divide, they move towards the centre of the tubule. These cells are called **spermatogonia**. As they move nearer the centre, the growth stage occurs and the spermatogonia develop into primary **spermatocytes**. Finally meiosis occurs. After the first meiotic division, the cells are called secondary spermatocytes. The haploid cells formed at the end of meiosis are called **spermatids**. These spermatids embed themselves into large **Sertoli cells** which supply the developing sperm with nutrients. They also protect the sperm from the immune system, as they are genetically different from the body cells.

The sperm released into the lumen of the seminiferous tubules are still not fully mature and cannot swim. They pass to the epididymis where they become fully motile, but they still need to undergo further changes before they are capable of fertilising an egg. You will learn more about this in Topic 1.4.

■ How science works

Older fathers run higher risk of fetal defects

Recent research suggests that older fathers are more likely to produce fetuses with chromosomal defects. This can lead to miscarriages or birth defects. It was known that older women had a higher risk of such problems, but this is the first research to demonstrate a link between age and chromosome anomalies for men.

Scientists studied 200 000 sperm from 18 healthy donors aged 24 to 74 years. They found that the percentage of sperm with double copies of all the chromosomes (diploidy) increases by 17% for every 10-year increase in age. Over the whole age group they found an increase in the frequency of diploidy from 0.2 to 0.4%. This could lead to defects such as Down's Syndrome. Though this appears small, when the researchers checked back with the donors they found that those with the higher frequencies had children with chromosomal anomalies. Although all sperm are freshly created, the scientists believe that Sertoli cells may become damaged over time.

The scientists suggest that men over the age of 55 years should have their sperm checked for chromosomal anomalies before fathering children and should consider freezing their sperm when young. The British Fertility Society confirmed that men over 55 are not used as sperm donors anyway due to the quality of their sperm.

© *New Scientist Magazine*

1 How valid are the scientists' conclusions? Give reasons for your answer.

2 Explain how a chromosome anomaly can lead to Down's syndrome.

3 Use your knowledge of oogenesis to suggest why older women have a greater risk of producing children with chromosomal defects than younger women.

4 Suggest how damage to Sertoli cells might result in defective sperm.

Summary questions

1 Give **two** differences between spermatogenesis and oogenesis.

2 A cell in the germinal epithelium of the seminiferous tubules has 12 units of DNA. How many units of DNA will there be in: a a primary spermatocyte, b a secondary spermatocyte, c a spermatid?

3 Explain why sperm produced in a man's seminiferous tubules need to be protected from the man's immune system.

1.4 Fertilisation

Learning objectives:

- How does fertilisation take place?

- How does an embryo start to develop?

Specification reference: 3.4.1

During **copulation**, or sexual intercourse, the spongy tissues in the penis become swollen with blood making it hard and erect. This makes it easier to enter the vagina of the female. Semen is ejaculated from the penis and released at the end of the vagina, close to the cervix. Semen is alkaline to neutralise the acidity of the vagina.

The sperm use their flagella to swim through the mucus in the cervix and towards the oviduct. This usually takes between 3 and 5 hours, but can be much less. Contractions of the uterine muscles are thought to help the sperm reach the oviduct.

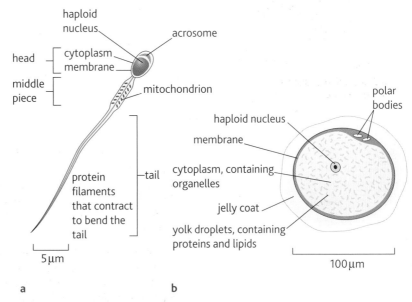

Figure 1 **a** *Structure of a human sperm,* **b** *structure of a human oocyte*

Capacitation

When sperm are first ejaculated, they are not able to fertilise the oocyte. As they travel through the female reproductive tract, they undergo a process called **capacitation**. This process, which takes a few hours, results in changes that make the sperm capable of fertilising the oocyte. Capacitation involves the removal of cholesterol and some glycoproteins from the membrane above the acrosome. This makes the membrane more fluid and more permeable to calcium ions.

The acrosome reaction

About 200 million sperm are ejaculated into the vagina, but only a few hundred reach the oocyte and only one sperm will fertilise it. Before the sperm can fertilise the oocyte, the **acrosome reaction** must take place. This involves the acrosome, a membrane-bound sac of enzymes in the head of the sperm, releasing enzymes that help the sperm to digest its way through the jelly coat, also known as the zona pellucida.

Fertilisation

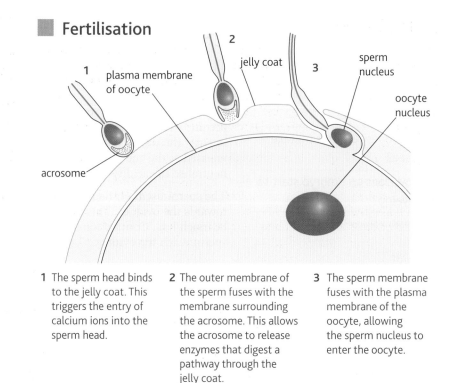

1 The sperm head binds to the jelly coat. This triggers the entry of calcium ions into the sperm head.

2 The outer membrane of the sperm fuses with the membrane surrounding the acrosome. This allows the acrosome to release enzymes that digest a pathway through the jelly coat.

3 The sperm membrane fuses with the plasma membrane of the oocyte, allowing the sperm nucleus to enter the oocyte.

Figure 2 *The acrosome reaction*

Figure 3 shows the events of fertilisation. Only one sperm is shown in Figure 3. There would actually be hundreds of sperm surrounding the oocyte. When the successful sperm digests its way through the jelly coat, the sperm membrane fuses with the oocyte membrane, allowing the nucleus of the sperm to enter the oocyte. As the sperm membrane fuses with the oocyte membrane, calcium ions are released from storage sites in the oocyte. This causes vesicles in the oocyte, called cortical granules, to fuse with the oocyte membrane and release their contents into the space between the oocyte membrane and the jelly coat. The process is called the **cortical reaction**. It is not known exactly what happens, but it causes changes in the jelly coat which forms a **fertilisation membrane** that prevents the entry of more sperm.

1 The acrosome reaction occurs: the sperm binds to the jelly coat and the acrosome enzymes are released.

2 The enzymes have enabled the sperm to penetrate the jelly coat. The sperm membrane fuses with the plasma membrane of the oocyte.

3 The sperm nucleus enters the oocyte and the cortical reaction occurs. Cortical granules are released by the oocyte that change the structure of the jelly coat. This means that no more sperm can enter the oocyte. The oocyte nucleus undergoes the second division of meiosis.

4 The male and female nuclei fuse, forming a diploid zygote.

5 A spindle forms, and the zygote divides by mitosis.

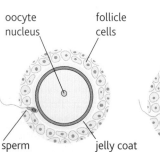

Figure 3 *Fertilisation*

Implantation

The fertilised oocyte becomes a **zygote**. This is a diploid cell. As it travels along the oviduct, it divides by mitosis. After 3–4 days, it has formed a ball of cells called a **morula** which looks rather like a blackberry. By about 5–6 days, it has formed a **blastocyst**. You can see this in Figure 5. The blastocyst is a hollow ball of cells. The outer layer of cells, the **trophoblast**, will develop into the membranes around the embryo and the beginnings of the placenta. The inner cell mass is the part of the blastocyst from which the embryo will develop.

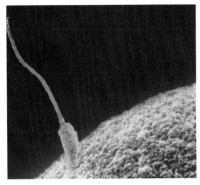

Figure 4 *The moment of fertilisation*

The blastocyst reaches the uterus about a week after fertilisation. It takes a few days for it to implant into the endometrium. The jelly coat surrounding the blastocyst lyses (breaks open). The trophoblast cells at one end of the embryo form branched extensions called villi that secrete enzymes. These enzymes digest the outer layer of the endometrium, releasing nutrients for the developing embryo.

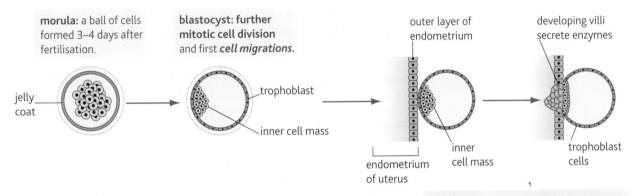

Figure 5 *From fertilisation to implantation*

1.5 The placenta

Learning objectives:

- How is the placenta structured?

- What is the function of the placenta?

- How are materials transferred between the embryo and the mother?

Specification reference: 3.4.1

We saw in Topic 1.4 that the **blastocyst** implants in the endometrium of the uterus. Some of the cells in the blastocyst form **chorionic villi** that secrete enzymes to digest the outer layer of the endometrium. As the chorionic villi develop, they become surrounded by maternal blood spaces. This provides a large surface area over which the fetus can obtain nutrients and exchange gases with the mother's blood. These chorionic villi are the beginning of the placenta.

The developing embryo

The cells in the inner mass of the blastocyst differentiate into the many different kinds of cell that make up tissues and organs. Twenty-five days after ovulation, the embryo has a primitive beating heart. Four weeks after ovulation the brain is growing rapidly, the eyes and ears have started to form, and the gut is beginning to develop. At 10 weeks after fertilisation, the main body organs have formed and the embryo becomes a **fetus**. All the main body organs have now formed, although the fetus is only about 35 mm long from its rump to the top of its head. The fetus is protected inside a sac called the **amnion**, which secretes **amniotic fluid**. The **umbilical cord** connects the fetus to the placenta and carries blood between the two. This contains two umbilical arteries, carrying blood from the fetus to the placenta, and an umbilical vein, carrying blood from the placenta to the fetus.

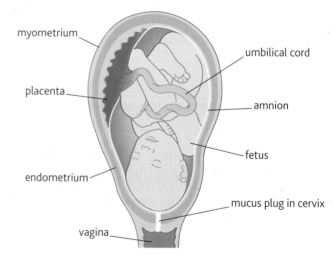

Figure 1 *A fetus in the later stages of pregnancy*

The function of the placenta

The fetus depends on the placenta for its development. Nutrients, such as glucose, amino acids and mineral ions, as well as water and oxygen, pass from the mother's blood into the fetus's blood in the placenta. Antibodies also pass from the mother to the fetus so that the fetus is protected from the same diseases as the mother. This immunity lasts for a short time after birth. Carbon dioxide and urea from the fetus pass in the opposite direction to the nutrients and antibodies.

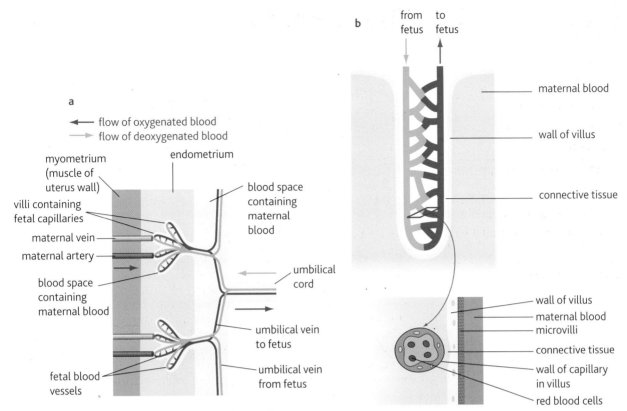

Figure 2 *The structure of the placenta*

Figure 2a shows the structure of the placenta. It contains a large number of chorionic villi. Figure 2b shows a close-up view of a villus. The capillaries are separated from the maternal blood by a thin membrane, between 2–6 μm thick. The membrane has three layers:

■ The wall of the villus, which consists of a thin layer of fused cells. The cells in the wall of the villus have microvilli, which considerably increase the surface area.

■ A thin layer of connective tissue.

■ Endothelium cells making up the capillary wall.

The mother's blood and the fetal blood flow in opposite directions. This is called **countercurrent flow**. It ensures that there is a concentration gradient all the way along the capillary (Figure 3). Blood from the fetus arriving in the placenta has a very low oxygen concentration (0) but it receives oxygen by diffusion from the mother's blood which has a higher oxygen concentration (2).

■ Hormones and the placenta

The blastocyst and the developing placenta produce the hormone **human chorionic gonadotrophin (hCG)**. This hormone is secreted in large amounts in the first 2 months of pregnancy and is present in the urine of pregnant women. This factor forms the basis of pregnancy tests.

In the first 3 months of pregnancy, the corpus luteum continues to secrete progesterone and a small amount of oestrogen. This inhibits contraction of the myometrium (the muscles in the wall of the uterus) and maintains the endometrium. After the first 3 months of pregnancy, the corpus luteum begins to degenerate and the placenta secretes

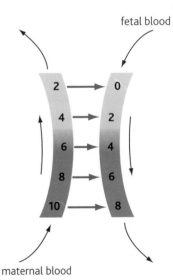

Figure 3 *Countercurrent flow in the placenta*

■ **Hint**

Note that the mother's blood and the fetus's blood do not mix. They are separated from each other by a thin membrane which, to some extent, controls the materials that cross it.

increasing amounts of progesterone, together with some oestrogen. Progesterone also inhibits the production of FSH from the pituitary gland.

Application and How science works

Screening in pregnancy

Pregnant women are usually offered screening if there is an inherited disorder in the family, or if the mother is over 35 years of age and more likely to give birth to a child with Down's Syndrome. In amniocentesis a hypodermic needle is inserted through the mother's abdomen and a small sample of amniotic fluid is extracted. This will contain some fetal cells. The needle is guided into a suitable position by using an ultrasound scan as shown in the diagram.

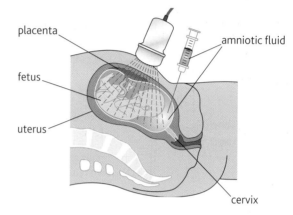

Figure 4 *The process of amniocentesis*

The cells are extracted and stimulated so that they divide by mitosis. A spindle inhibitor is then added so that the cells do not complete mitosis. The cells are placed in a dilute salt solution so that they swell up, and then the chromosomes are photographed. The images of the chromosomes are then manipulated on a computer screen and arranged in homologous pairs. This enables chromosome defects to be detected.

Another screening technique involves taking a sample of the chorionic villi cells using a plastic catheter inserted through the vagina.

1   Explain why the fetal cells are stimulated to divide by mitosis.

2   At what stage of mitosis will the cells be? Explain your answer.

3   a Why is a dilute salt solution used to make the cells swell up?

    b Explain why a dilute salt solution will cause the cells to swell up.

4   What information could be obtained by examining the chromosomes in this way?

Application and How science works

Pregnancy testing

You will recall from AS that antibodies can be used to test for specific substances in urine or blood. A dipstick-type pregnancy test can be used to test for hCG in urine. Figure 5 shows how this works.

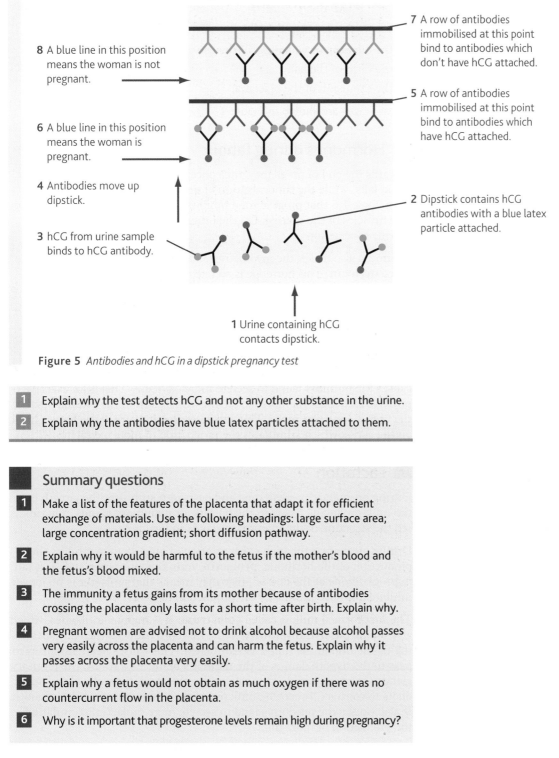

8 A blue line in this position means the woman is not pregnant.

6 A blue line in this position means the woman is pregnant.

4 Antibodies move up dipstick.

3 hCG from urine sample binds to hCG antibody.

7 A row of antibodies immobilised at this point bind to antibodies which don't have hCG attached.

5 A row of antibodies immobilised at this point bind to antibodies which have hCG attached.

2 Dipstick contains hCG antibodies with a blue latex particle attached.

1 Urine containing hCG contacts dipstick.

Figure 5 *Antibodies and hCG in a dipstick pregnancy test*

1 Explain why the test detects hCG and not any other substance in the urine.

2 Explain why the antibodies have blue latex particles attached to them.

Summary questions

1 Make a list of the features of the placenta that adapt it for efficient exchange of materials. Use the following headings: large surface area; large concentration gradient; short diffusion pathway.

2 Explain why it would be harmful to the fetus if the mother's blood and the fetus's blood mixed.

3 The immunity a fetus gains from its mother because of antibodies crossing the placenta only lasts for a short time after birth. Explain why.

4 Pregnant women are advised not to drink alcohol because alcohol passes very easily across the placenta and can harm the fetus. Explain why it passes across the placenta very easily.

5 Explain why a fetus would not obtain as much oxygen if there was no countercurrent flow in the placenta.

6 Why is it important that progesterone levels remain high during pregnancy?

1.6 Labour and lactation

Learning objectives:

- What happens during birth?
- What causes the mother to produce milk?

Specification reference: 3.4.1

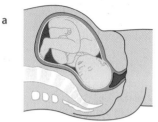

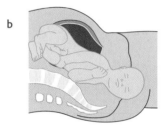

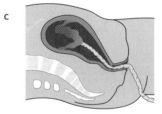

Figure 1 *The stages of labour:* **a** *dilation of the cervix,* **b** *birth of the baby,* **c** *expulsion of the placenta*

Birth (parturition)

There are three stages of birth:

- the dilation, or widening, of the cervix (the neck of the uterus)
- the birth of the baby
- the expulsion of the placenta.

These stages are shown in Figure 1.

Hormones during labour

Towards the end of pregnancy, the concentration of **progesterone** in the blood falls, while the concentration of **oestrogen** rises. You will recall from Topic 1.5 that progesterone inhibits contraction of the myometrium (the muscles of the uterus). On the other hand, oestrogen stimulates uterine contractions.

Oestrogen also makes the myometrium more sensitive to another hormone called **oxytocin**. This hormone is secreted by the pituitary gland. Oxytocin stimulates contraction of the myometrium. The term 'labour' is used to describe the regular occurrence of these uterine contractions. Another hormone produced late in pregnancy is called **relaxin**. This softens the cervix and makes the pelvis more flexible, making the birth easier.

The pressure of the fetus's head against the cervix stimulates stretch receptors. These send impulses to the hypothalamus in the brain which causes the pituitary gland to secrete more oxytocin. This is an example of **positive feedback**. As more oxytocin is released from the pituitary gland, the uterus contracts more. This increases the pressure of the fetus's head on the cervix which stimulates the production of more oxytocin.

Lactation

During pregnancy, progesterone and oestrogen stimulate the growth of the milk-producing tissue in the breasts, or **mammary glands**. You can see the structure of the mammary glands in Figure 2.

During pregnancy, the high concentration of progesterone inhibits the production of the hormone, **prolactin**, from the pituitary gland. The fall in progesterone at the end of pregnancy means that prolactin is no longer inhibited. It now stimulates the production of milk.

The first-formed milk is called **colostrum**. It is high in antibodies, which gives some **passive immunity** to the newborn baby. As the baby suckles, the tissue around the nipple is stimulated. This causes nerve impulses to pass to the hypothalamus of the mother's brain. As a result, the pituitary gland secretes oxytocin. This causes the muscle in the walls of the milk ducts to contract, squeezing milk out of the nipple into the baby's mouth. It also causes the pituitary gland to secrete more prolactin, stimulating the production of more milk.

Prolactin also inhibits the production of FSH and LH. As a result, ovulation is unlikely while a mother is breastfeeding a baby full-time.

How science works

Wet nursing

Some women are unable to, or choose not to breastfeed their baby. This can be for many different reasons. The majority of women that do not breastfeed choose to bottle feed their baby. There are a few women, however, who choose to let another woman breastfeed their baby. This is called wet nursing. Wet nursing used to be common among the aristocracy in the 19th century as women wanted to conserve their figures. Nowadays, wet nursing is considered controversial. There are concerns that wet nurses may pass on infections through breastfeeding.

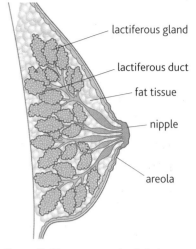

Figure 2 *The mammary gland of a human*

1. Explain why a wet nurse is able to produce enough milk to feed more than one baby.

2. Explain how a wet nurse in the 19th century might be passing on infections to babies.

3. Do you think that it is a good idea for a busy working mother to employ a wet nurse for her baby? Give reasons for your answer.

Application

Breast milk

Breast milk is made with the appropriate nutrients for a baby, in the appropriate quantities. The first breast milk produced is called colostrum. It's made in pregnancy and for the first few days after birth it varies in colour, but it may be thick and yellowish. Its constituents are important for the health of a very new baby.

Later on, the milk produced will be greater in volume. The constituents of it are appropriate for a baby's needs from shortly after birth to as long as the mother wants to breastfeed. Milk varies in its precise composition from mother to mother, and from feed to feed. For instance, the fat content of breast milk changes according to the amount of milk in the breast – breasts that have a lot of milk at any one time will have milk that is proportionately lower in fat compared to breasts that have less milk in them at any one time. All a mother must do is feed in response to the cues a baby gives.

Milk is made in response to its removal from the breast, and this is why mothers of twins (or more) can make sufficient milk to nourish their babies – the more milk is removed, the more milk is made. (Adapted from advice given to mothers by the National Childbirth Trust.)

1. Explain why the constituents of colostrum are 'important for the health of a very new baby'.

2. Explain how the volume of milk that a mother produces increases over a period of weeks.

3. Explain how a mother with twins produces more breast milk than a mother with a single baby.

How science works

Breastfeeding could slash breast cancer risk

Adapted from an article in *New Scientist*, 19 July 2002, by Danny Pearman

Scientists have concluded that women in the West could reduce their breast cancer risk by nearly 60% if they returned to having larger families and breastfeeding more, like their ancestors did. These findings help to explain why breast cancer, virtually unknown 200 years ago, is now a major killer.

Cancer Research UK announced that having an average of six children and breastfeeding them for 2 years would reduce the incidence of breast cancer to Third World levels. Scientists believe that even breastfeeding each child for an extra 6 months could prevent 5% of breast cancers each year.

The new study pools data from 47 studies, involving over 150 000 women in 30 countries. A third of the women studied had breast cancer and factors considered included number of children, duration of breastfeeding, socio-economic class and level of education.

Scientists believe that returning to pre-industrial levels of fertility and breastfeeding would slash the current UK rate of breast cancer from 6.3% to 2.7%. For each child a woman has, her risk of the disease declines by 7.0%. On top of this, for every year that she breastfeeds, her risk declines by 4.3%. However, scientists still do not know how this protects against breast cancer.

© *New Scientist Magazine*

1 Suggest why the scientists considered factors such as socio-economic class and level of education in their study.

2 Evaluate the reliability of this study.

3 Suggest one way in which it is possible that breastfeeding or having children might protect a woman from breast cancer.

Summary questions

1 Newborn babies are given some immunity against disease when they feed on colostrum. Explain why this immunity does not last for long.

2 When a woman breastfeeds her baby in the first few days after birth, she may experience uterine contractions. Explain why.

1.7 Birth control

Learning objectives:

- What methods of birth control are there?

- How do these methods of birth control work?

Specification reference: 3.4.1

Contraception is a method of birth control that prevents conception. Methods of birth control can be divided up broadly into barrier methods and hormonal methods. Not all birth control methods prevent conception. Some methods of birth control allow fertilisation to occur.

Barrier methods

Barrier methods of contraception are designed to prevent sperm meeting up with an oocyte. The **condom** is the most widely used barrier method. It consists of a thin latex sheath that is placed over the erect penis. It is the smaller item shown in Figure 1. Sperm are released into the tip of the sheath, which prevents them being released into the vagina. A condom also reduces the likelihood of passing on sexually transmitted diseases.

An alternative to this is the female condom. This is the larger of the two items in Figure 1. The female condom is made of polyurethane, and is inserted into the vagina before sexual intercourse takes place. Like the male condom, this also offers some protection against sexually transmitted diseases. Both types of condom should be used only once.

Another barrier method of contraception is the **cap**. This is a latex or silicone cap designed to fit over the cervix. A woman should spread spermicidal cream into the cap and insert it before sexual intercourse. It is held in place by suction and should be left in place for at least 8 hours following sexual intercourse. It is important that the cap is the right size for the woman. If she has a baby, she may need a new cap of a different size afterwards. The cap is very reliable when fitted correctly but it can be difficult to fit securely in place. In addition, it does not protect against sexually transmitted diseases. A rare infection called toxic shock syndrome (TSS) is associated with tampon use. If a woman has ever had TSS she will be advised not to use a cap.

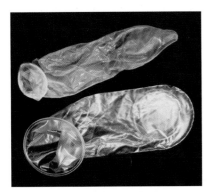

Figure 1 *Male and female condoms*

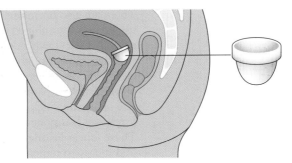

Figure 2 *The cap*

Intrauterine device (IUD)

Another form of birth control is the **intrauterine device**, or IUD. It is made of plastic, often with copper wire wound around it. It is placed inside the uterus, with two small strings that pass through the cervix. The copper is mildly spermicidal so it tends to kill sperm. It is not clear how the IUD works, but there are several theories:

- It may increase the production of chemicals called prostaglandins by the uterus, which stops a fertilised oocyte implanting

cross-section of uterus

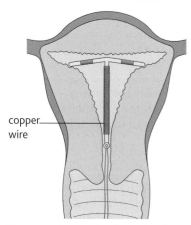

copper wire

Figure 3 *Intrauterine device (IUD)*

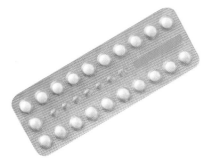

Figure 4 *The combined pill*

- It may cause changes in the endometrium so that the fertilised oocyte cannot implant.
- It may make it more difficult for the sperm and oocyte to pass through the uterus and oviducts.

The IUD should be fitted and removed by a health professional, but it can stay in place for years. In most women it causes no problems at all. However, for a few women, it may cause heavier periods and increase menstrual cramps. A few IUDs are expelled by the uterus within a few months of being fitted. In a very few cases, fitting an IUD may cause an infection. If a woman does become pregnant while using an IUD there is an increased risk of ectopic pregnancy (that is, a pregnancy resulting from the embryo implanting in the oviduct). The IUD does not protect against sexually transmitted diseases. Some people do not find this kind of birth control acceptable, since it does not stop fertilisation occurring. These people think that preventing the implantation of a very early embryo is destroying a potential human being.

Hormonal methods

Hormonal methods are based on oestrogen and/or progesterone. There are two kinds of pill – the combined pill containing both oestrogen and progesterone, and the progesterone-only pill. Progesterone inhibits the production of LH by the pituitary gland, so ovulation does not occur. Oestrogen inhibits FSH production, so follicles do not develop.

The combined pill is usually taken for 21 days. After this there is a period of 7 days when no pill is taken, or a placebo pill containing no hormones is taken. This results in the endometrium breaking down and a menstrual period occurs.

Although hormonal methods of contraception do not protect against sexually transmitted diseases they are highly effective if used properly. They can reduce menstrual cramps and make periods less heavy. They decrease a woman's risk of ovarian or endometrial cancer. There is a slightly increased risk of thrombosis for women taking the pill, but this is mainly in women who smoke or have high blood pressure.

There are also hormonal methods of contraception that do not need to be used daily. A combined hormonal contraceptive is available as an adhesive patch, which slowly releases hormones into the body. However, this is less effective on overweight women. A single injection of progesterone may be given into the upper arm. This lasts for up to 3 months. Alternatively, a progesterone implant may be used. This is a small flexible rod, about the size of a matchstick, that is inserted underneath the skin of the upper arm. It releases hormones slowly and can remain effective for up to 3 years. These longer lasting methods may be an advantage for women who do not find it easy to remember to take a pill every day.

The 'morning after' pill contains a high dose of a synthetic form of progesterone. It is not a contraceptive and is not intended as a regular method of birth control, but may be used occasionally, for example, if a condom breaks. Scientists are uncertain about exactly how it works, but they think it may work in one of three ways:

- by making the endometrium less receptive to the fertilised oocyte, preventing implantation
- by preventing the sperm and oocyte meeting in the oviduct
- by preventing ovulation.

The 'morning after' pill needs to be taken as soon as possible after unprotected sexual intercourse has taken place. It is less effective after ovulation has occurred, and has no effect once implantation has occurred. Like the IUD, this pill does not always prevent fertilisation, so some people find its use unacceptable.

How science works

Relative effectiveness of different birth control methods

The table shows the relative effectiveness of some birth control methods.

Method	Use effectiveness (actual use)/%	Theoretical effectiveness (perfect use)/%
combined pill	92	99.7
progesterone-only pill	92	99.7
IUD with copper	99.2	99.4
cervical cap – woman who has had children	68	74
cervical cap – woman who has not had children	84	91
male condom	85	98
female condom	79	95

These figures are based on 100 couples using the method for 1 year. They show the percentage of women who do not get pregnant while using the method for 1 year.

1. Explain why it is difficult to compare methods of birth control in this way.

1. Suggest how the figures for theoretical effectiveness are calculated.

2. Suggest why a cervical cap is more effective in preventing pregnancy in women who have not had children.

3. Hormonal pills are more effective in preventing pregnancy than condoms. However, young people who request contraception are usually advised to use condoms. Explain why.

4. Suggest why the pill is not 100% effective.

Summary questions

1. Explain why a woman may need a different cervical cap after she has had a baby.

2. Copy and complete the table on the right.

3. Use your knowledge of the hormones in pregnancy to explain why the 'morning after' pill is less effective after ovulation, and not effective after implantation.

Method of birth control	Advantages	Disadvantages
combined pill		
male condom		
IUD		
cap		

1.8 *In vitro* fertilisation 1

Learning objectives:

■ What are the causes of infertility?

■ How can infertility be treated?

Specification reference: 3.4.1

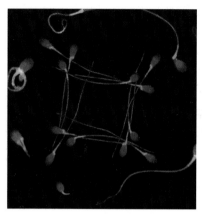

Figure 1 *Abnormal sperm cells*

■ Male infertility

Some men are infertile because they produce semen containing a large number of abnormal sperm. These sperm may have structural changes or may not be able to swim properly. Figure 1 shows a sample of abnormal sperm.

Other infertile men may produce semen containing very few sperm cells. This can be caused by many factors, including:

■ blockage of the vasa deferentia. This may result from a number of causes, including injury and disease

■ infectious diseases including STDs like chlamydia, or diseases from childhood such as mumps

■ hormonal dysfunction

■ chronic diseases such as diabetes or hypertension

■ varicocele, a swelling rather like a 'varicose vein' that appears on one or both testicles. Varicoceles account for almost 40% of the causes of male infertility.

Erectile Dysfunction (ED), the inability to obtain or maintain an erection, can also cause male infertility.

Male infertility may be treated by surgery if blockage of the vasa deferentia is suspected. If a man produces some healthy sperm, *in vitro* fertilisation may be carried out using selected sperm. If necessary, a sperm may be injected into an egg during IVF treatment.

■ How science works

Youthful 'nurse' cells could restore male fertility

A shortage of healthy Sertoli cells is thought to cause between 5 and 15% of infertility cases in men. It was assumed that the number of Sertoli cells a man has is decided during puberty, and that they then no longer divide to produce new cells. Now studies in hamsters and humans suggest these cells can be returned to a state in which they begin dividing again.

The discovery was made by scientists working in Australia. When they suppressed the production of sex hormones in adult hamsters, they found that the animals' Sertoli cells turned back into immature cells, and started dividing. The scientists then gave the hamsters follicle stimulating hormone (FSH), and found that this caused the Sertoli cells to mature.

Ironically, the investigation was helped by the fact that sex hormone suppression is being tested as a contraceptive for men. Scientists took tissue samples from the testes of six men before and after they took an experimental contraceptive, and examined its effect on their Sertoli cells. Before the treatment they found no immature Sertoli cells, but after 12 weeks, 2% had reverted to an immature state and seemed to be dividing.

The scientists believe that if men's hormones were more profoundly suppressed, or suppressed for longer, more of the Sertoli cells would revert to an immature state. They hope to test this theory, and also whether treating men with FSH causes the cells to mature as it does in hamsters. FSH doesn't appear to have any adverse effects on men's bodies.

Both hamsters and humans depend on the same hormones for sperm production, so the scientists are optimistic that FSH may have a similar effect in humans. However, some scientists are concerned about applying hamster results to humans. Hamsters are seasonal breeders, so their reproductive system may be more flexible than ours.

Some scientists are also concerned that stimulating division of Sertoli cells in humans might interfere with the cells' second function: maintaining a barrier in the testes between blood and developing sperm. Breaching this barrier could damage sperm or the tubules they develop in. However, hamsters regained fully functional testes, and were fertile after hormone suppression treatment, according to the scientists carrying out this work.

© *New Scientist Magazine*

1 Explain how a lack of healthy Sertoli cells might cause infertility in men.

2 Explain why mature Sertoli cells do not divide.

3 Suggest how these scientists could test whether this method does stimulate the production of Sertoli cells in humans.

4 Explain how stimulating the division of Sertoli cells could damage the sperm or the tubules they develop in (paragraph 6).

■ Female infertility

There are several reasons why women may be infertile. Some women have abnormal menstrual cycles, in which they do not produce enough hormones. This may mean that they do not ovulate, or ovulate only rarely. This kind of infertility may be treated using hormones.

Another cause of infertility in women is blockage of the oviducts. This may be caused by bacterial infections, such as gonorrhoea or chlamydia. Sometimes surgery may be used to unblock the oviducts, but often *in vitro* fertilisation (IVF) is used.

■ *In vitro* fertilisation (IVF)

In vitro means 'in glass'. *In vitro* fertilisation is sometimes called 'test-tube baby' treatment, but in fact fertilisation usually occurs in flat glass dishes. In IVF, sperm and oocytes are mixed together in a glass dish, and embryos are then returned to the woman's uterus. The stages involved in IVF treatment are shown in Figure 2.

Sometimes a man produces very few healthy sperm, or the sperm are not capable of fertilising the oocyte by themselves. In this case, a sperm may be injected into the oocyte using a very tiny needle. This technique is called intra-cytoplasmic sperm injection, or ICSI. You can see this in Figure 3.

The woman is given hormone injections that cause several follicles to develop in her ovary. Hormone injections are usually given for about 10 days.

↓

When the follicles have matured adequately, another hormone injection is given. This hormone is similar to LH.

↓

Semen is collected. It is washed and then placed in a liquid containing nutrients and substances that allow the sperm to become capacitated.

The mature oocytes are collected from the ovary using a needle. This is inserted into the woman's vagina and is guided into place using ultrasound.

↓

The sperm and oocytes are placed together in the culture medium for about 18 hours. During this time fertilisation should take place.

↓

The fertilised egg is passed to a special growth medium and left for about 48 hours until the egg has reached the 6–8 cell stage.

↓

The embryos judged to be the 'best' are transferred to the woman's uterus through a thin, plastic catheter, which goes through her vagina and cervix. Usually two embryos are transferred. Any extra embryos may be preserved by freezing in liquid nitrogen.

Figure 2 *Stages involved in IVF treatment*

Figure 3 *Intra-cytoplasmic sperm injection*

Summary questions

1 Name a hormone that would cause follicles to develop in a woman's ovaries.

2 Explain why a woman being given IVF treatment needs to be given an injection of a hormone similar to LH before the mature oocytes are collected.

3 Suggest why, during IVF treatment:

a sperm cells are washed

b sperm cells are placed in a medium that allows them to become capacitated.

4 Suggest why:

a a large number of oocytes are collected at one time

b only two embryos are usually replaced in the uterus.

1.9 *In vitro* fertilisation 2

Learning objectives:

- What happens to embryos that are not implanted?

- To what extent should IVF be available?

Specification reference: 3.4.1

Unused embryos

A major ethical issue involved in IVF treatment is the fate of unused embryos. You saw in Topic 1.8 that several embryos may be produced at one time. It is common for a couple to have some unused, frozen embryos left over after IVF treatment has been successful.

Some people think that that destroying these embryos is killing a potential human being. Other people do not see it like this. When babies are conceived naturally, many early embryos fail to develop and are lost naturally from the mother's body. They believe that discarding embryos after IVF is not very different from this natural process.

One possible use of these surplus embryos is in medical research. Early embryos, at the 8–16 cell stage, can be used as a source of stem cells. This could help scientists to develop cures for some serious diseases such as Parkinson's disease, Alzheimer's disease or diabetes.

Availability of IVF

The number of people requesting IVF treatment is increasing every year. One reason for this is that many professional women are leaving it until they are older to try for a baby, and sometimes they find that they cannot conceive a baby naturally. As you learned in Topic 1.8, IVF treatment is expensive and there is not enough money for everybody who would like IVF treatment to be given unlimited treatment on the NHS. Some people think that IVF treatment should be freely available. Other people think that having a baby is not a right, and that only people with a good chance of conceiving, and who are likely to make good parents, should be allowed IVF. The case studies in this topic are all about people using IVF. Perhaps they will help you to decide what you think about this.

How science works

The Diane Blood story

Diane Blood and her husband Stephen were trying for a baby when Stephen suddenly died of meningitis in February 1995. He went into a coma and never regained consciousness. While Stephen was in a coma, some of his semen was removed and frozen.

After Stephen died, Diane decided she wanted to use the semen sample to have a baby using IVF treatment. However, she found that she would not be allowed to do this. The Human Fertilisation and Embryology Authority (HFEA), who regulate IVF treatment in the UK, said that she could not do this as she did not have Stephen's written permission.

Diane Blood challenged this decision in the High Court, but she did not win. However, the Court of Appeal allowed her to take the frozen semen to another country where the law did not prevent her using the semen in IVF treatment. She finally managed to obtain the IVF treatment she wanted in a Belgian clinic. Eventually she gave birth to a son, Liam, in December 1998, followed by another son, Joel, in July 2002.

When the boys were first born, Diane was not allowed to have Stephen's name as the children's father on their birth certificates. However, after another legal battle in December 2003, a new law came into power that allowed her to re-register her two son's births, with Stephen named as their father.

1 Do you think doctors should have taken semen from Stephen Blood while he was in a coma?

2 Was it right for the Human Fertilisation and Embryology Authority to refuse IVF treatment to Diane Blood in the UK?

How science works

Who should get IVF on the NHS?

Some people believe that only those who have a good chance of success should be allowed IVF on the NHS. Obese women are less likely to conceive using IVF, and if they do there is more chance of health problems for the mother and baby. The British Fertility Society (BFS) is a national multidisciplinary organisation representing professionals practising in the field of reproductive medicine. They recommend that NHS IVF treatment should only be given to women of a healthy weight, with a body mass index under 36. They also recommend that IVF should be offered to single women and lesbian couples, but not women over 40. The BFS recommend that smokers should still be offered NHS IVF treatment but encouraged to quit smoking. Also, they recommend that couples with children from previous relationships should be offered the treatment.

1 Should the following people be allowed IVF treatment on the NHS? Give reasons for your answer in each case.
- women who are obese
- women who are very underweight
- single women
- lesbian couples
- women who smoke
- women over 40
- couples who have a child already from a previous relationship

How science works

Couple want right to choose IVF embryo with deaf gene

A deaf couple want new embryology legislation to allow them to deliberately select a deaf child from embryos created by IVF. The Government's Human Fertilisation and Embryology Bill could allow parents to screen embryos and choose to discard any with a serious disability. But Paula Garfield and Tomato Lichy argue that the rules should apply the other way, permitting them to choose to have a deaf child. They say that the proposals as they stand discriminate against deaf people. The couple said: 'If hearing people were to have the right to throw away a deaf embryo, then we as deaf people should also have the right to throw away a hearing embryo'.

The Royal National Institute for Deaf People does not support the choice of deaf embryos over those who would not be born with hearing problems. Its chief executive Jackie Ballard said: 'Deafness is a disability and we have spent a long time campaigning to improve the lives of people who live with it. But it is certainly not a slight to the deaf to say it is better to bring a child who will face the least difficulty into the world'. Opponents of embryo selection say destroying an embryo, whether it is deaf or not, is wrong. The Christian Institute's Mike Judge said: 'The dignity of human life does not depend on the ability to hear'.

© *The Christian Institute*

1 Should this couple be allowed to use IVF to have a deaf baby? Give reasons for your answer.

Summary questions

1 What do you think should happen to surplus embryos created during IVF treatment? Give reasons for your answer.

2 You have been asked by NICE (the National Institute for Clinical Excellence) to advise the NHS on who should or should not be allowed IVF treatment on the NHS. Write a set of guidelines.

1   **Figure 1** shows changes in the concentration of two female sex hormones involved in controlling the menstrual cycle.

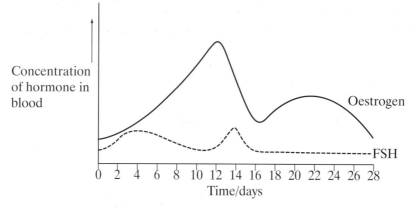

Figure 1

(a) (i)   What is the main function of FSH?

(ii) Describe the effect of oestrogen on the concentration of FSH during days 4 to 10 of this cycle.

(iii) How might the information in **Figure 1** be used to work out the most likely time for conception?   *(3 marks)*

(b) Use the information in **Figure 1** to explain one way in which oral contraceptives might act to prevent fertilisation taking place.   *(2 marks)*

2   (a) Copy and complete **Table 1**, which gives information about some of the hormones which control the menstrual cycle.   *(4 marks)*

Table 1

Hormone	Site of secretion	Target organ	One effect
Luteinising hormone (LH)			
Oestrogen	ovary	uterus	stimulates growth of uterine lining
Progesterone	ovary	uterus	

(b) **Figure 2** shows the way in which hormones are involved in controlling part of the menstrual cycle.

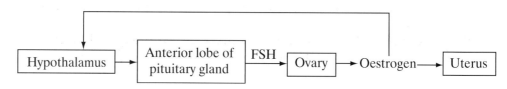

Figure 2

(i) Some women only produce very small amounts of FSH. Explain why these women are infertile.

(ii) Clomiphene is a drug used to treat this type of infertility. It blocks the action of oestrogen. Explain how treatment with clomiphene could be used to stimulate production of FSH.

(3 marks)

3   (a) Describe how the structure of the placenta is adapted to its role in supplying substances to, and removing waste products from, the fetus.

(4 marks)

(b) Explain the advantage of each of the following changes which occur in the mother during pregnancy.

(i)   an increase in the number of red blood cells

(ii) an increase in cardiac output

(2 marks)

4   **Figure 3** shows a sperm about to fertilise an ovum (oocyte).

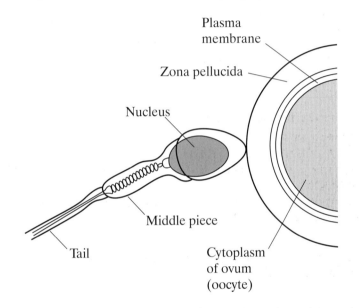

Figure 3

(a) The tip of the sperm contains enzymes which assist in fertilisation.

(i)   Name the part that contains the enzymes.

(ii) Describe the function of these enzymes.

(3 marks)

(b) Sperms may be stored for a few weeks in the tubes coming from the testes. Here, the sperms remain immobile. They only become mobile in the semen when mixed with seminal fluid.

Semen contains a mean of 70×10^6 sperms per cm^3. On average, 70% of the sperms are of normal shape and able to be capacitated.

(i)   Describe what happens during capacitation of a sperm.

(ii) A single ejaculation of semen has a volume of $3\,cm^3$. Calculate the average number of sperms in this ejaculation that will be capacitated. Show your working.

(3 marks)

(c) The seminal fluid contains the monosaccharide, fructose. Suggest how this fructose enables the sperms to move after ejaculation.

(3 marks)

Growing up, growing old and passing on your genes

2.1 Patterns of human growth

You will recall from AS level that humans have evolved a longer childhood than other primates. This means that the development of sexual maturity is delayed.

Figure 1 shows the relative growth rates of different human organ systems. You can see that the nervous system develops very quickly, both before the baby is born and afterwards. A human baby is born with a large brain that develops quickly into its adult size. The reproductive organs do not develop much until **puberty**. This means that a human child has an extended childhood when it is dependent on its parents. This pattern of growth allows a human child to learn advanced skills, such as tool use and language. It also ensures that a human does not become a parent until it has learned advanced social skills, since bringing up a highly dependent infant requires considerable skills in the parent. Figure 2 shows the mean growth rate for males and females. In this case, growth is measured by the height grown each year in centimetres.

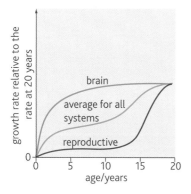

Figure 1 *Relative growth rates of different organ systems in humans*

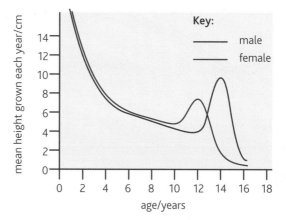

Figure 2 *The mean growth rate of males and females*

Puberty in males

Puberty is the stage in life when the sex organs and secondary sex characteristics develop. In males, this occurs between the ages of 12 and 16. The hypothalamus in the brain releases a hormone called **gonadotrophin releasing factor (GnRF)**. This stimulates the release of two different hormones from the pituitary gland, **follicle stimulating hormone (FSH)** and **luteinising hormone (LH)**. Luteinising hormone is sometimes called interstitial cell stimulating hormone (ICSH) in males. These hormones stimulate the release of another hormone, **testosterone**, from the testes. The changes that occur are mainly the result of testosterone and can be seen in Figure 3.

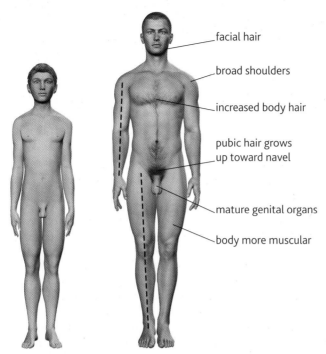

facial hair

broad shoulders

increased body hair

pubic hair grows
up toward navel

mature genital organs

body more muscular

Figure 3 *The development of male secondary sex characteristics*

The main changes that occur in the male at puberty are:

- growth of cartilage in the thorax (chest) and pectoral girdle (shoulders) so that the shoulders and chest become broader and more muscular
- growth of penis and testes
- growth of pubic hair
- growth of axillary (underarm) and facial hair, and often chest and body hair
- growth of the larynx, causing the voice to deepen
- general body growth and development of muscles
- increased activity of sweat glands and sebaceous glands.

How science works

Acne

Our skin has lots of grease-producing glands called sebaceous glands. At puberty, they start secreting a greasy liquid called sebum. This sebum can cause acne when the ducts of the glands become blocked. Blackheads are signs of a partial blockage. In bad cases of acne, sebum breaks through the wall of the ducts, causing inflammation of the surrounding cells. Sebum contains lots of triglycerides. The skin breaks these down into fatty acids which in turn break down cell membranes and so cause inflammation.

Scientists have tried to find out what the function of the glands is but have failed. Some biologists believe the sebaceous glands have evolved to cause acne. Others believe that the spots are there to signal the start of puberty, and so sexual maturity.

Other theories are that acne is a sign that a young man or woman is not yet old enough to reproduce, allowing them to explore and learn without becoming parents until they are in their late teens or twenties. Another suggestion is that acne encourages social bonding through grooming.

1 Skin bacteria break down triglycerides in sebum.
   a Name the type of reaction that is involved in this reaction.
   b Name another product of this reaction in addition to fatty acids.

2 Evaluate the different theories about the development of acne mentioned in the passage. Do you think acne has a function in humans? Give reasons for your answer.

Puberty in females

Puberty in females occurs between the ages of 10 and 15. The release of GnRF by the hypothalamus triggers the release of FSH and LH from the pituitary gland. These hormones interact to cause the events of the menstrual cycle. The time of the first menstrual cycle in a female is called the **menarche**. However, a female is not usually fertile until about 2 years after the menarche. The development of secondary sex characteristics in females is the result of female sex hormones, especially oestrogen.

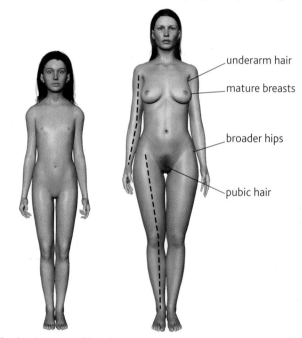

underarm hair

mature breasts

broader hips

pubic hair

Figure 4 *The development of female secondary sex characteristics*

The main changes that occur in the female at puberty are:

- growth of the female sex organs, i.e. ovaries, oviducts, uterus and vagina
- growth of breasts
- growth of pubic hair
- growth of axillary (underarm) hair
- increase of cartilage in the pelvic girdle (hips) which causes the diameter of the pelvic girdle to increase
- the 'carrying angle' of the arms and legs changes, so they are at a slight angle
- accumulation of fat, especially on the hips, breasts and thighs
- increased activity of sweat glands and sebaceous glands.

Although we know that GnRF triggers puberty, we do not know the stimulus that causes this hormone to be released. One factor that seems to stimulate its release is diet. Over the past 150 years, the age at which females begin menstruating has dramatically decreased. This could be due to a combination of different factors.

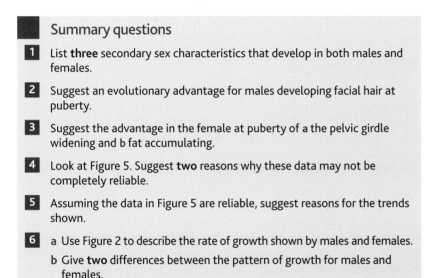

Figure 5 *The mean age at menarche of females from four different countries from 1840 to the present*

Summary questions

1 List **three** secondary sex characteristics that develop in both males and females.

2 Suggest an evolutionary advantage for males developing facial hair at puberty.

3 Suggest the advantage in the female at puberty of a the pelvic girdle widening and b fat accumulating.

4 Look at Figure 5. Suggest **two** reasons why these data may not be completely reliable.

5 Assuming the data in Figure 5 are reliable, suggest reasons for the trends shown.

6 a Use Figure 2 to describe the rate of growth shown by males and females.

 b Give **two** differences between the pattern of growth for males and females.

Learning objectives:

■ What changes happen in our body as we age?

Specification reference: 3.4.2

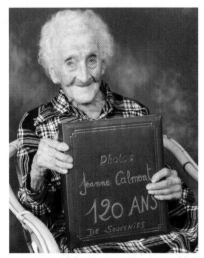

Figure 1 *Jeanne Calment, a French woman who reached the longest confirmed age in history*

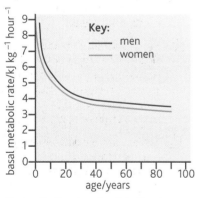

Figure 2 *How basal metabolic rate changes with age*

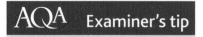

AQA **Examiner's tip**

You will need to know specific details about how BMR varies according to your age, sex and body size, as well as the factors that affect BMR. This is covered in Topic 1.1 of *AS Human Biology*.

Figure 1 is a photograph of a lady called Jeanne Calment, who died in August 1997 at the age of 122 years, 164 days. Many of her relatives lived to a considerable age, and she had a comfortable lifestyle, being born into a well-to-do family. However, she smoked until she was 117. We do not understand what causes some people, like Jeanne Calment, to live to such an old age, but we do understand some of the changes that occur as we age.

Basal metabolic rate

Basal metabolic rate is the rate at which we use energy to keep ourselves alive while we are at rest, but not asleep. The conditions under which BMR is measured are very specific. The subject must be completely at rest, must be lying down in the warm and must not have eaten for 12 hours. Our BMR drops as we age, mainly because people have fewer metabolically active cells as they get older. This explains why older people can become obese if they do not reduce their food intake. Figure 2 shows how BMR changes with age.

Cardiac output

You will recall from AS level that **cardiac output = stroke volume × heart rate**. Cardiac muscle cells become smaller and may be lost during ageing. This is especially likely if the person does not exercise regularly. Figure 3 shows how cardiac output changes with age.

Nerve conduction velocity

As people age, nerve cells (called neurones) gradually die. These are not replaced, so as people get older they have fewer neurones in their brain and the rest of the nervous system. Secondly, the speed at which impulses pass along neurones becomes slower. You will learn more about this in Topic 5.3. This happens because the myelin sheath that insulates neurones becomes thinner, so more ions leak out of the membrane surrounding the neurones. Also, synapses (the places where one neurone communicates with another) produce less transmitter substance as people age. This also makes the overall rate at which impulses are transmitted along nerves slower. You will learn more about synapses in Topic 5.4. Figure 4 shows how nerve conduction velocity decreases with age.

Female reproductive capacity

The **menopause** is the time of the last menstrual cycle of a woman. It usually occurs between the ages of 50 and 54, but may occur as early as 35 or as late as 59. The cause of the menopause is the lack of follicles in the ovary. In Topic 1.2 you learned that, from puberty, one follicle in a woman's ovaries matures every month. However, many follicles do not ever develop in this way, but gradually disappear. By the time the woman reaches menopause, there are no follicles left. This means that she is infertile.

You will also recall from Topic 1.2 that when a follicle develops it secretes **oestrogen**. Therefore, once a woman has reached the age of menopause, she no longer has oestrogen circulating in her blood. As the concentration of oestrogen in the woman's blood decreases, the concentration of **FSH** and **LH** increase. These reach their maximum concentration 1–3 years after the woman's last menstruation, but then their concentration gradually decreases. The lack of oestrogen may cause 'hot flushes' when the woman experiences sudden bouts of sweating and feeling very hot, vaginal dryness and sometimes psychological problems.

You will recall from AS level that oestrogen reduces the chances of a woman developing coronary heart disease (CHD) but after the menopause, women have the same chance of developing CHD as men. The lack of oestrogen also causes a reduction in bone density. In some women, this becomes so severe that they suffer from osteoporosis, a condition in which the bones fracture very easily. Men can also suffer from osteoporosis but it is much less common.

Some women take hormone replacement therapy (HRT) following the menopause. This consists of low doses of oestrogen and progesterone. Many women find that this greatly reduces the symptoms associated with the menopause.

How science works

Caring grandmas explain evolutionary role of menopause

Menopause in humans occurs around age 50. In other mammals, female reproduction stops only because of ageing. Menopause is genetically controlled, so the genes responsible were selected by evolution. However scientists have puzzled over what the advantage of the menopause might be.

Two hypotheses have been proposed: the first is that human childbirth is more likely to kill older women, so a woman who stops getting pregnant at 50 will still have time to raise her last child. The second is that the process allows a woman to help take care of her grandchildren – who she knows are carrying her genes.

Scientists analysed the births and deaths of 5500 people in Gambia between 1950 and 1975 – before a modern medical clinic arrived. They believe this is similar to the situation experienced during the evolution of humans. The data revealed that a child was over 10 times less likely to survive if its mother died before it was 2 years old, but that children between one and two had twice the chance of surviving if their maternal grandmother was still alive. No other relatives had any effect.

To test the first hypothesis above, the researchers used a mathematical model to calculate the growth rate of a population if the maximum age at which women could give birth was increased from 50 to 65 – an age when a mother would be more likely to die before her child reaches two. They found that this affected too few children to change the growth, or fitness, of the population significantly.

To test the second hypothesis – that a grandmother with her own young children would be less able to help a daughter's children – the team modelled a population in which grandmothers without young children of their own doubled their grandchildren's survival. They then factored menopause to arrive at a variety of ages. Making menopause arrive later in life made grandmothers without young children scarcer.

The scientists found that, when menopause happened at 50, about 60% of children had a surviving maternal grandmother with no young children of her own, but with menopause at 65 that fell to 10%. Although grandmothers had a small effect on each child's survival, in the population as a whole the effect was very important. The researchers found that, if the menopause happened later, the population grew more slowly, or even declined. The optimum age for menopause was – as in real life – 50 years.

© *New Scientist Magazine*

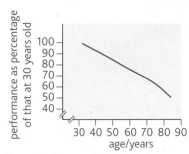

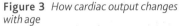

Figure 3 *How cardiac output changes with age*

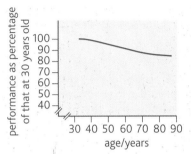

Figure 4 *How nerve conduction velocity changes with age*

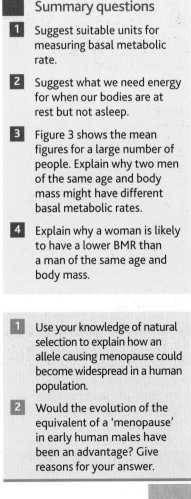

Summary questions

1 Suggest suitable units for measuring basal metabolic rate.

2 Suggest what we need energy for when our bodies are at rest but not asleep.

3 Figure 3 shows the mean figures for a large number of people. Explain why two men of the same age and body mass might have different basal metabolic rates.

4 Explain why a woman is likely to have a lower BMR than a man of the same age and body mass.

1 Use your knowledge of natural selection to explain how an allele causing menopause could become widespread in a human population.

2 Would the evolution of the equivalent of a 'menopause' in early human males have been an advantage? Give reasons for your answer.

2.3 Cancer

Learning objectives:

- Why are people more likely to get cancer when they are older?

- What groups are most at risk of developing cancer?

Specification reference: 3.4.2

Link

Look back at Topic 8.3 in *AS Human Biology* to remind yourself how cancer develops.

You will recall from AS level that cancer is a mass of abnormal cells that usually grow rapidly, invading surrounding tissues. Some cells break off and spread elsewhere in the body, forming secondary tumours or metastases. Figure 1 shows you how cell division in a normal cell is controlled by growth factors.

Genes called proto-oncogenes code for receptor proteins in a cell membrane or protein growth factors. These genes may mutate to form oncogenes. Oncogenes may produce receptor proteins that do not need growth factors to stimulate them, or they may produce unlimited amounts of growth factors. Cells also contain tumour suppressor genes that slow down cell division and cause cells with damaged DNA to die. Mutations in tumour suppressor genes may result in cells with damaged DNA replicating rapidly.

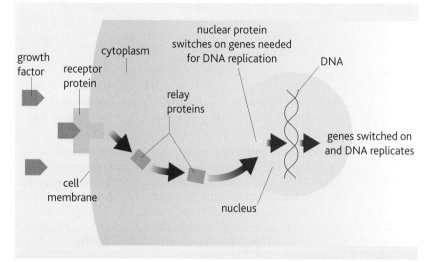

Figure 1 *Normal cell receiving signals from growth factors that tell it when to divide*

Older people are more likely to develop cancer because they have had more years of exposure to environmental factors that cause damage to DNA. Also, several mutations in proto-oncogenes and tumour suppressor genes are normally necessary before cancer develops. Older people have had more chance to accumulate this damage to their DNA.

Prostate cancer

Prostate cancer is the most common cancer of men in the UK. Figure 1 in Topic 1.1 shows that the prostate gland is a gland about the size of a walnut that lies just below the urinary bladder and surrounds the upper part of the urethra. The gland often enlarges as men get older, making the urethra narrower so it is more difficult for them to pass urine. This is called benign prostate hyperplasia. It can be treated by surgically removing some of the prostate gland, making it easier for urine to pass. However, sometimes the prostate gland enlarges as a result of prostate cancer. The symptoms are similar to those for benign prostate hyperplasia, as the man has difficulty in passing urine. Often the cancer grows very slowly so there may be no need to treat the disease.

AQA Examiner's tip

You do not need to know details about specific cancers such as prostate and breast cancer. However, you may be expected to analyse data concerning the incidence of different kinds of cancer and suggest why cancer is more common among older people.

Very few cases of prostate cancer are found in men under the age of 50, and most new cases are diagnosed in men in their 70s. You can see this in Figure 2.

You will see that the incidence of prostate cancer increases steeply as men get older, and that the oldest age groups have the highest incidence. It is estimated that 15–30% of men over 50 have some evidence of prostate cancer, rising to 60–70% of men aged 80. However, only 4% of men will die from prostate cancer. This means that men are more likely to die of some other cause before the prostate cancer kills them. If the cancer is slow-growing, it may be better not to treat it, as surgery could cause damage to the urethra or rectum or nerve supply to the bladder or penis.

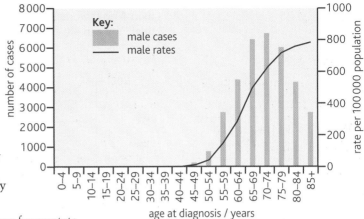

One way to detect prostate cancer is to screen men for prostate specific antigen (PSA). This test uses monoclonal antibodies which you studied at AS level. The test is very widely used in the USA, but is not currently offered routinely in the UK. This is because it could cause anxiety in a man who is otherwise well, and whose cancer may be very slow-growing.

Figure 2 *Number of new cases of prostate cancer and incidence rates, UK 2005*
Source: Cancer Research UK

▪ Application and How science works

Prostate cancer screening

The rate of prostate cancer screening is markedly different in the USA from that in the UK. In the USA it is common to test men for prostate specific antigen (PSA), which indicates that a man may have prostate cancer. In 2001, 57% of men in the USA aged 50 or older had taken a PSA test within the last 12 months. By contrast, in the UK, PSA tests are only carried out when a man reports symptoms to his GP. In the UK, only about 6% of men aged 45 to 84 had a PSA test for each year between 1999 and 2002. Despite this, scientists are not convinced that screening for PSA reduces deaths from prostate cancer. Since the late 1990s, prostate cancer deaths have fallen more rapidly in the USA than in the UK, but there is no evidence that PSA screening is responsible for this difference.

Scientists studied patterns in prostate cancer death rates in the USA and the UK between 1975 and 2004 and also compared the different methods of screening and treatment. In both countries, deaths from prostate cancer were at their highest in the early 1990s. Both countries had a similar death rate from prostate cancer at this time. After this, the death rate decreased in the USA by over 4% a year, four times faster than the decrease in the UK. The USA showed the greatest fall in death rates among people aged 75 or older, while in the UK the death rate in this age group remained steady after 2000.

The scientists carrying out this study are not convinced that screening for PSA is responsible for these differences. One explanation could be that, in the USA, gonadotrophin-releasing hormone treatment is used more often with older men. Secondly, men who were shown to have a fast-growing prostate cancer following PSA screening were offered more aggressive treatment.

1 Explain why it is important to study the differences in patterns of disease between different countries.

2 Evaluate whether PSA tests should be offered to all men over 50 years of age in the UK.

■ Breast cancer

Breast cancer is now the most common cancer in the UK. Every year, more than 44 000 women are diagnosed with breast cancer, as well as about 300 men. You can see data concerning the incidence of breast cancer in Figure 3.

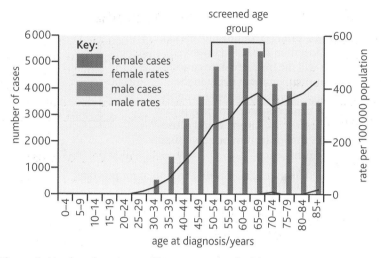

Figure 3 *Number of new cases of breast cancer and incidence rates, UK 2005*
Source: Cancer Research UK

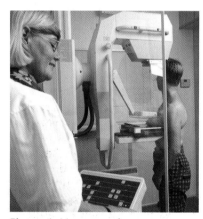

Figure 4 *Mammography*

You will notice that breast cancer is strongly related to age. More than 80% of cases occur in women over 50 years of age, with the highest incidence in the 50–64 age group.

We know that genetic factors play a part, because women with a mother, sister or daughter diagnosed with breast cancer have an 80% higher risk of developing breast cancer themselves. Despite this, 88% of breast cancers occur in women without any family history of the disease. Factors that seem to increase the risk of developing breast cancer include:

- obesity
- using hormone replacement therapy for 5 years or longer
- using oral contraceptives
- a sedentary lifestyle
- consuming alcohol.

You will also recall from Topic 1.6 that women who have smaller families, do not breastfeed, and delay childbearing until they are older, also appear to have an increased risk of breast cancer.

In the UK, screening for breast cancer is offered to all women aged 50–70 on the NHS. This consists of a breast X-ray, called a **mammogram**. You can see this in Figure 4.

Screening is offered every 3 years. A tumour shows up as a dense patch. However, a dense area of tissue may be detected that is not cancer. A woman whose mammogram shows cause for concern is offered further tests. In 2004–5, nearly 14 000 breast cancers were detected in this way. Younger women, unless they are at high risk of developing breast cancer, are not screened. One reason for this is that it is harder to spot a tumour in the breast tissue of a woman who has not yet reached the menopause.

Summary questions

1. Name two environmental factors that increase the chances of developing cancer.

2. How is it possible that some men develop breast cancer?

3. Explain why cancer is more likely to be cured if it is detected early.

4. Evaluate whether screening for breast cancer should be offered to all women over the age of 25.

2.4 Alzheimer's disease

Learning objectives:

- What causes Alzheimer's disease?

- What are the issues raised by having an increasing number of elderly people in the population?

Specification reference: 3.4.2

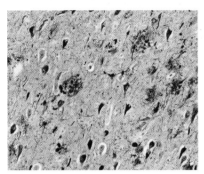

Figure 1 *Brain tissue from a person with Alzheimer's disease showing neurofibrillary tangles (dark teardrop shapes) and amyloid plaques (rounded brown masses)*

Dementia

Dementia is a general term for a loss of memory, language and intellectual ability. This condition is more likely to occur in older people. Currently, about 700 000 people in the UK have dementia and it is estimated that there will be over a million people with dementia by 2025. It causes 60 000 deaths a year directly, and costs the UK over 17 billion pounds a year, mainly in caring for these people. The personal cost to families and carers may be even greater.

Alzheimer's disease

Alzheimer's disease is the most common type of dementia in the UK, accounting for over half of all cases of dementia. During the course of this disease, physical changes occur in the brain. Neurones and synapses are lost from parts of the cerebral cortex, the part of the brain concerned with learning and memory. Some of the neurones develop twisted fibres inside them, called 'tangles'. Plaques of amyloid protein build up between the neurones. You can see this in Figure 1.

Alzheimer's disease is a progressive disease, in which the brain becomes more damaged over time. As a result the symptoms become more severe. In the early stages of the disease, people with Alzheimer's will sometimes be forgetful, or may have difficulty finding the right words. Later, they may become confused and often forget the names of people around them, forget where they are or important appointments, and forget recent events. Many people with Alzheimer's develop mood swings, so they may become depressed or angry. They will experience frustration because of their memory loss, and they may be frightened by it. As the disease worsens, a person with Alzheimer's may become increasingly withdrawn, partly because of an inability to communicate, but also because they become less confident. Sufferers will need increasing support from other people as the disease progresses, until eventually they will need help for even the most basic daily activities, such as washing, dressing or eating.

Causes of Alzheimer's disease

The cause of Alzheimer's disease is not known. The plaques that build up in the brain are made of a small protein called beta-amyloid. Beta-amyloid is part of a larger transmembrane protein called amyloid precursor protein, or APP. APP is believed to be important in helping neurones to grow. An enzyme cuts off beta-amyloid from APP. In people with Alzheimer's disease, it appears that there is too much beta-amyloid protein, or that it has a slightly different structure from usual.

Healthy neurones are held in shape by microtubules, and a protein called tau is associated with these microtubules. In Alzheimer's disease, something goes wrong and tau protein builds up in the neurones, causing them to die.

Preventing Alzheimer's disease

Dementia affects one in 14 people over the age of 65 and one in 6 over the age of 80. However, younger people may also be affected. Currently, it is estimated that there are 15 000 people under the age of 65 with dementia in the UK. There are several factors that increase the risk of developing Alzheimer's disease, although it seems that there is no single factor that actually causes Alzheimer's. Instead, it appears that a combination of factors is responsible.

■ Genetic factors. There are a few families where there is clear evidence that the disease has been passed on from one generation to the next. Often, this involves a kind of dementia that appears in relatively young people. In most cases, however, there is no evidence that you are more likely to develop Alzheimer's yourself if you have a relative with the disease.

■ People who have had severe head or whiplash injuries, especially if this happens when they are aged 50 years or more, appear to be at increased risk of developing dementia. Boxers who receive continual blows to the head are also at risk.

■ Some research suggests that people who smoke, have high blood pressure or high cholesterol levels are also more likely to develop Alzheimer's disease.

Treatment of Alzheimer's disease

There is no cure available for Alzheimer's disease at the moment. However, there are some drugs that can help with the symptoms and slow down the progress of the disease. People with Alzheimer's disease have a shortage of the chemical acetylcholine in their brains. This is a chemical that operates at synapses, and you will learn more about this in Topic 5.4.

■ An ageing population

In the UK, as in most developed countries of the world, we have an ageing population. People are living longer, but the proportion of young people in the population has reduced. The proportion of people aged 65 and over was 16% in 2006 but this is expected to rise to 22% by 2031. The cost of supporting the elderly, in paying their pensions and their health and social-care costs, comes from taxes. These taxes have to be paid by the working population. In 2006 there were 3.3 people of working age for every person of state pension age. This ratio is expected to fall to 2.9 by 2031, and this takes into account the increase in state pension age to 68 that will be taking place.

Summary questions

1 Alzheimer's disease cannot be diagnosed reliably until after a person has died. Suggest why.

2 Suggest some of the problems that will be faced by a person caring for a family member with Alzheimer's in their own home.

How science works

Warning over social care funding

Health experts are worried that the ageing population means there will not be enough state funding for the care of the elderly and the disabled. If the number of elderly people in the population continues to rise at the rate it is, there will be very little money available for social care. The government will need to reconsider how social care is funded. As part of this process they will ask the public how they think it should be funded. Currently state support is means tested. This means that many people must pay for their own care. Many people think that this is unfair, and that everybody who needs care should receive help from the government.

1 Is it fair to make elderly and disabled people pay towards their own care costs?

2 Some people think that all elderly and disabled people should receive free care, without means testing. Do you agree with this? Give reasons for your answer.

3 Should care be available free of charge to people who do not have the means to pay, or should we expect family members to provide this care?

4 What can be done to ensure that elderly and disabled people receive sufficient care?

2.5 Monohybrid inheritance – cystic fibrosis

Learning objectives:

- What is the difference between a gene and an allele?
- How is cystic fibrosis inherited?

Specification reference: 3.4.2

Link

You should read through Chapter 3 in *AS Human Biology* to remind yourself about cystic fibrosis.

You will recall from AS level that cystic fibrosis (CF) is an inherited disorder that affects plasma membranes. It occurs because people with CF have a faulty version of a protein called CFTR that transports chloride ions across cell membranes. CFTR is coded for by a **gene** on chromosome 7. A gene is a length of DNA that occupies a specific position on a chromosome, called a **locus**. Each gene codes for a specific protein or polypeptide. A particular gene may have more than one form that is slightly different from the others in its sequence of nucleotide bases. These different forms of genes are called **alleles**. You can see this in Figure 1.

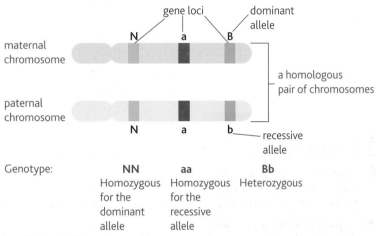

Figure 1 *A pair of homologous chromosomes*

The allele that codes for cystic fibrosis has three nucleotides missing. Otherwise, it is the same as the normal allele. The three missing nucleotides code for one amino acid, so this means that the faulty CFTR protein has one amino acid missing. However, this missing amino acid is very significant. Without it, the CFTR protein folds up into a different tertiary structure, resulting in a protein that will not function.

You also learned at AS level that human DNA is arranged in the form of chromosomes. Human body cells contain 23 pairs of chromosomes, making 46 chromosomes in all. A gamete contains one set of 23 chromosomes. This means that every individual receives one set of chromosomes from their mother, and one from their father. In other words, every individual has two genes that code for CFTR. However, the two alleles may be the same or different.

AQA Examiner's tip

Make sure you understand the meaning of the words homozygous, heterozygous, genotype and phenotype and that you use them correctly. Also make sure that you know the difference between a gene and an allele, and that you always use these terms correctly in examination questions.

Hint

Notice that we use the upper case letter, **F**, for the dominant allele and the lower case letter **f** for the recessive allele. Using the same letter for both alleles tells you that they are alleles of the same gene.

If an individual has two identical alleles of a gene, they are said to be **homozygous** for that gene. However, if the alleles are different, they are said to be **heterozygous**. An allele is said to be **dominant** if it is always expressed in the **phenotype**, or outward appearance and characteristics, of the organism. The allele for the normal CFTR protein can be identified with the symbol **F**. This allele is dominant. The allele for the faulty version of CFTR is **recessive**. It can be identified with the symbol **f**. This means that it is only expressed in the phenotype of an individual if two copies of the allele are present. The **genotype** of an individual is their genetic make-up. It tells you the alleles they carry. A person with the genotype **Ff** is heterozygous for the cystic fibrosis gene. A person who is **FF** or **ff** is homozygous.

Hint

- When doing a genetic cross, always set out your working in a clear, logical way, like the one used in Figure 2.

- It is helpful to draw a ring around the gametes so that you can identify them easily. Remember that gametes are haploid so they only contain one from a pair of alleles.

- You should set out the cross in a grid like the Punnett square shown here, so that you don't make mistakes. When you draw lines linking gametes to the offspring it is very easy to get confused – especially when you are solving harder problems than these.

One person in every 22 people of European descent is heterozygous for the CFTR gene. This means they have one normal CFTR allele and one faulty CFTR allele. They do not have CF because the normal allele is dominant. However, two people who are heterozygous for the CFTR gene may have a child with CF. This is because a person who is heterozygous passes on only one of their alleles into each gamete. If the child inherits one faulty CFTR allele from each parent, then they will develop CF. You can see how this occurs in Figure 2.

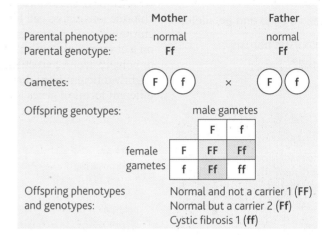

Figure 2 *The inheritance of cystic fibrosis*

You will notice from Figure 2 that two parents who are heterozygous for CF have a 1 in 4 or 25% chance of having a child with CF in any pregnancy. However, this is only a chance or probability. So two parents who are heterozygous might have four children, none of whom has CF, while another heterozygous couple might have two children, both with CF.

Notice that some of their children may be heterozygous, like their parents. They will not have CF but they could pass on CF to their own children. Sometimes we say that these heterozygous individuals are **carriers** of cystic fibrosis.

Monohybrid inheritance involves a cross between individuals in which only a single gene is being considered. Cystic fibrosis is an example of a monohybrid cross.

Summary questions

1. Copy and complete the table, adding the words that match the definitions:

	the position on a chromosome at which a gene is found
	a length of DNA that codes for a polypeptide
	the genetic make-up of an individual
	an individual who has two identical alleles of a gene
	an alternative form of a gene
	the outward appearance and characteristics of an organism
	an individual who has two different alleles of a gene

2. What are the chances of a woman with CF and a man who is homozygous and does not have CF having a child with CF? Set out your answer using the same layout as shown in Figure 2.

3. Cleidocranial dysostosis is a rare genetic condition caused by a dominant allele, **D**. A person with this condition has skeletal abnormalities and does not have a collar bone. A man is heterozygous for cleidocranial dysostosis. Does he have the condition? Explain your answer.

4. What is the chance of a man who is heterozygous for cleidocranial dysostosis and a woman who is homozygous for the recessive allele having a child with cleidocranial dysostosis? Set out your answer using the same layout as in Figure 2.

2.6 Sickle-cell anaemia

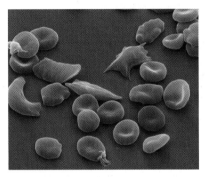

Figure 1 *Sickle-cell anaemia*

Sickle-cell anaemia is an inherited condition in which affected individuals produce an abnormal type of haemoglobin called haemoglobin S. You will remember from AS that haemoglobin is a protein found in red blood cells that carries oxygen. Sickle-cell haemoglobin has a very small difference in the order of amino acids in two of the polypeptide chains that make up each molecule. As a result, haemoglobin S is less soluble than normal haemoglobin, and crystallises in low oxygen concentrations. This means that the red blood cells of an affected individual become sickle-shaped when oxygen levels are low. You can see this in Figure 1.

These sickle-shaped cells have difficulty in passing through small blood vessels, so they can block capillaries. This means that less oxygenated blood reaches the affected parts of the body and as a result, cells die. This causes damage to tissues, and also causes severe pain in the affected individual. Sickle cells may block the flow of blood through capillaries in the lungs, causing lung tissue damage (acute chest syndrome) and severe pain in the arms, legs, chest and abdomen. It can also cause strokes, when parts of the brain tissue die because the cells do not receive enough oxygen. It also causes damage to most organs including the kidneys and liver. Children with sickle-cell anaemia often suffer damage to the spleen early in life, which leaves them vulnerable to certain bacterial infections.

The sickle-shaped cells are easily damaged and are rapidly broken down, causing anaemia. This means that the affected person has fewer red blood cells and a lower concentration of oxygen in their blood. People with sickle-cell anaemia are treated with blood transfusions, intravenous fluids, antibiotics and painkillers. However, without treatment, the condition is usually fatal in childhood.

■ Inheritance of sickle-cell anaemia

Sickle-cell anaemia is determined by a single gene with two alleles. However, these alleles are not dominant and recessive, like the alleles of the CFTR gene. The sickle-cell anaemia alleles are **codominant**. This means that both alleles contribute to the phenotype in a heterozygous individual. Because these alleles are codominant, we use a different convention when choosing symbols for the alleles. **Hb^A** is the allele for normal haemoglobin, and **Hb^S** is the allele for sickle-cell haemoglobin. A person with the genotype **$Hb^A Hb^S$** is heterozygous, and produces both kinds of haemoglobin in their red blood cells. However, they do not produce enough sickle-cell haemoglobin to show symptoms of sickle-cell anaemia, unless they are in conditions where oxygen is very limited. These people are said to have sickle-cell trait.

■ Sickle-cell haemoglobin and malaria

Sickle-cell anaemia is very rare in northern Europeans, but much more common among people in Africa or people with an African origin.

You will see from Figure 2 that the sickle-cell allele is particularly common in parts of Africa and India. These are also the areas where the disease malaria is very common. In these areas, people with normal haemoglobin and the genotype **$Hb^A Hb^A$** do not suffer from sickle-cell anaemia but are more likely to die of malaria. People with sickle-cell trait, **$Hb^A Hb^S$**, have some protection from malaria and are therefore more likely to survive.

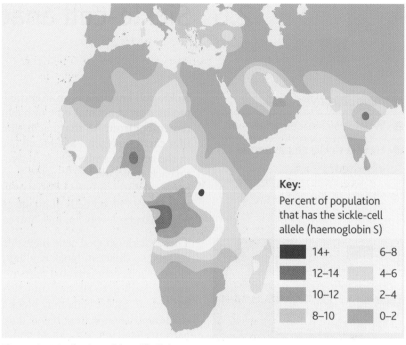

Figure 2 *Distribution of the Hbˢ allele*

■ Passing on sickle-cell anaemia

Because natural selection favours heterozygotes in areas where malaria is common, it is inevitable that children with sickle-cell anaemia will be born. You can see how sickle-cell anaemia is inherited in Figure 3.

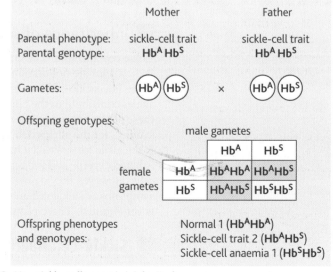

Figure 3 *How sickle-cell anaemia is inherited*

You will notice that two parents with sickle-cell trait have a 25% chance of having a child with sickle-cell anaemia. They also have a 25% chance of having a child with normal haemoglobin. However, half their children are likely to have sickle-cell trait, which gives them no symptoms of sickle-cell anaemia but gives them resistance to malaria.

Application and How science works

Cystic fibrosis gene protects against tuberculosis

Scientists wondered why cystic fibrosis is so common. The lethal disease strikes people who inherit two copies of a mutant allele. Historically, such people died before reproductive age – so the mutant allele would be expected to have gradually died out. Instead the allele has persisted for thousands of years, especially in people of European descent. In the UK, the disease strikes one in 2500, and 4% of people carry one mutant allele. Other racial groups have a far lower incidence. The scientists decided that Europeans must have derived some advantage from one copy of the allele that made up for losing the people with two copies – possibly protection from disease. Sickle cell anaemia is an example of such a disease that persists with high incidence in Africans, since having one copy of the allele protects against malaria.

Previous studies into the potential benefit of a single cystic fibrosis (CF) mutation focused on cholera or typhoid, because these diseases involve the protein that is mutated in CF, but little evidence had been found.

In the new study, scientists analysed data from historical death rates for cholera, typhoid and tuberculosis (TB). Cholera and typhoid simply did not kill enough people to explain the CF gene's persistence, they found. Even if one CF allele gave total protection against either disease, this would not be enough selective advantage to push the allele to modern European levels.

However, between 1600 and 1900, TB caused 20% of all deaths in Europe, an epidemic unparalleled elsewhere. The model showed that even if it gave only partial protection against TB, the CF allele would easily have reached its current levels in Europeans. TB has only been a problem in other continents for about 200 years, explaining why CF is less common in other racial groups.

CF patients and carriers of the CF allele have some resistance to TB. This could be because TB bacteria need a nutrient that CF patients do not make, the scientists suggest. The scientists believe that, although two copies of the CF allele kill, people with just one copy of the allele were more likely to survive when TB was a common disease.

© *New Scientist Magazine*

1 Explain why scientists would expect an allele that causes a disease that kills people during childhood to die out.

Summary questions

1 Use your knowledge of protein structure to explain why sickle-cell haemoglobin has a different tertiary structure from normal haemoglobin.

2 A man has sickle-cell trait and a woman has all normal haemoglobin. Is it possible for them to have a child with sickle-cell anaemia? Use a genetic diagram to explain your answer.

3 Use your knowledge of natural selection to explain why sickle-cell anaemia is very rare in people with northern European ancestry.

2.7 ABO and rhesus blood groups

Learning objectives:

- What are the different blood groups?

- How are they inherited?

Specification reference: 3.4.2

In Topic 2.6 you learned that people with sickle cell anaemia may need to be given a blood transfusion. In fact, there are many medical conditions and emergencies when blood transfusions can be used to save lives. However, when blood is transfused, it is important that people are given blood of the correct blood group. Blood groups are determined by the antigens present on the red blood cells. These antigens are inherited.

Rhesus blood groups

There are several antigens present in this blood group, but the main one is known as antigen D. People who have antigen D on their red blood cells are rhesus positive. This is coded for by the dominant allele, **D**. People who do not have antigen D on their red blood cells are rhesus negative. People who are rhesus positive may have the genotype **DD** or **Dd**. People who are rhesus negative have the genotype **dd**.

It is possible for a rhesus negative mother to be pregnant with a rhesus positive baby. As you learned in Topic 1.5, the mother's blood and the baby's blood do not mix in the placenta. However, it is possible for the baby's blood to come into contact with the mother's blood around the time of childbirth. If this happens, the rhesus antigen on the baby's red blood cells cause an immune response in the mother's blood. The mother's immune system will make antibodies against the rhesus antigen. This is unlikely to harm the baby, as it only happens at the very end of pregnancy.

If the mother becomes pregnant with a second rhesus positive baby, problems may be more severe. You will be aware that antibodies pass from the mother's blood to the baby's blood across the placenta. The mother's anti-rhesus antibodies pass across the placenta and attack the baby's red blood cells. This destroys them and causes severe anaemia. The condition is called **haemolytic disease of the newborn**. It varies in severity. In mild forms, the baby is born with anaemia and jaundice, but in severe cases the baby may develop respiratory arrest or heart failure, or may even be stillborn. This condition can be prevented by giving pregnant women who are rhesus negative an injection of anti-rhesus globulin, and repeating the injection after the birth of a rhesus positive baby. If this is done, the mother should be able to have further rhesus positive babies without any risk.

ABO blood groups

The best known blood group system is the ABO system. There are four blood groups in this system, A, B, AB and O. People with group A have A antigens on their red blood cells. People with group B have B antigens on their red blood cells, while people of group AB have both kinds of antigen. People with group O do not have either of these antigens on their red blood cells. People with different blood groups also have different antibodies in their blood plasma. You can see this in Table 1. When a person is given a blood transfusion, it is important that the antigens of the donor should be compatible with the antibodies of the recipient. If a person is given blood of the wrong blood group, their antibodies agglutinate or clump the donor red blood cells. This can block blood vessels and can be fatal.

ABO blood groups are inherited. They are determined by a single gene with three different alleles. This is an example of inheritance involving **multiple alleles**. Two alleles, I^A and I^B are **codominant**, while I^O is recessive. This is shown in Table 1.

Table 1 *ABO blood groups*

Blood group	Antigens on red blood cells	Antibodies in blood plasma	Possible genotypes
A	A	Anti-B	$I^A I^A$ or $I^A I^O$
B	B	Anti-A	$I^B I^B$ or $I^B I^O$
AB	A and B	Neither	$I^A I^B$
O	Neither	Anti-A and anti-B	$I^O I^O$

How science works

The discovery of ABO blood groups

In the 19th century it was known that when a human was given a transfusion of blood from another animal, the foreign red blood cells become clumped. The red blood cells broke open, releasing haemoglobin. This usually led to the death of the person who received the transfusion. In 1901–1903 Karl Landsteiner pointed out that a similar reaction often occurred when one human was given blood from another human. This work was largely ignored until he classified human blood groups into the A, B and O groups that we now know very well. (The AB blood group was discovered a little later by two other scientists.) Landsteiner showed that transfusions between individuals of groups A or B do not result in the destruction of new blood cells and that this occurs only when a person is transfused with the blood of a person belonging to a different group. Landsteiner realised, early in his career, that blood groups could be used to solve cases of disputed paternity, because ABO blood groups are genetically inherited. Later in Landsteiner's career, he worked with other scientists who discovered more about the antigens involved in blood grouping, and discovered the rhesus blood group system. He received the Nobel Prize in 1930.

Figure 1 *Karl Landsteiner*

1   Explain why red blood cells from an animal become clumped together when they are transfused into a human.

2   How could ABO blood groups be used to solve cases of disputed paternity?

Summary questions

1   A rhesus negative mother and a rhesus positive father have two daughters. The first is rhesus negative and the second is rhesus positive. Use a genetic diagram to show how each of the daughters inherited their blood group from their parents.

2   Use your knowledge of the immune system to explain how a rhesus negative mother produces antibodies against antigen D from her unborn baby.

3   Blood group AB is called the 'universal recipient'. Explain why.

4   In an emergency, blood of group O, rhesus negative, can be given safely to anybody. Explain why.

5   A mother who is blood group A and a father who is blood group B have two children. One is group O and the other is group AB. Use a genetic diagram to explain how these parents can have children with blood groups that are different from either of their parents.

6   A woman of blood group A gave birth to a baby with blood group O. She was not sure which of her boyfriends was the father. One boyfriend had group B and the other had group AB. Which of the boyfriends was the father? Use a genetic diagram to explain your answer.

2.8 Sex determination and sex-linked inheritance

The sex chromosomes

As you know, human cells contain 46 chromosomes. There are 23 different chromosomes, but we have one set of chromosomes from our mother and one from our father. These are arranged in homologous pairs. The chromosomes of a homologous pair look identical to each other, except for one pair – the sex chromosomes. Look at Figure 1. You can see that the X chromosome is much larger than the Y chromosome. Females have two X chromosomes, while males have one X chromosome and one Y chromosome.

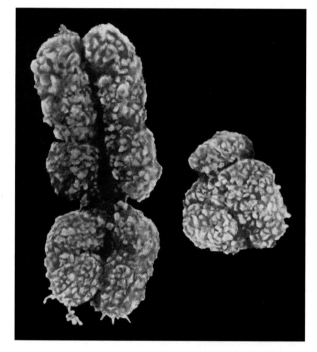

Figure 1 *The sex chromosomes. The X chromosome is the larger chromosome on the left, and the Y chromosome is on the right.*

The DNA in the human Y chromosome is 58 million base pairs long, but represents only about 0.4% of the total DNA in a human cell. It contains very few genes. One of the genes it does contain is known as SRY. This is the gene that triggers an embryo to develop as a male. Other genes on the Y chromosome are needed for normal sperm production.

Sex determination

Because females have two X chromosomes, every gamete they produce will contain a single X chromosome. However, the gametes produced by a male will differ from each other. Half will contain an X chromosome and half will contain a Y chromosome. The sex of the offspring will be determined by whether the sperm that fertilised the oocyte contains an X or a Y chromosome. You can see this in Figure 2.

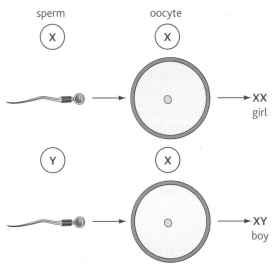

Figure 2 *How sex is determined in humans*

How science works

Y chromosomes give the name away

Leaving DNA evidence at the scene of a crime could help police to trace your name, even if your DNA has never been recorded on the national DNA database, a new study suggests.

Scientists at the University of Leicester examined 150 pairs of men. Each pair shared the same British surname but were not knowingly related. The researchers analysed their Y chromosomes, which are passed directly from father to son. The scientists found that a quarter of them were genetically connected, meaning they must have shared an ancestor more recently than 20 generations ago, or about AD 1300, when surnames were first used in the UK. The more uncommon the name, the more likely they were to have near-identical genetic profiles. People with names that are very common in the UK such as Smith, Jones and Taylor did not necessarily show genetic links. However, sharers of names such as Major (5500 occurrences in the UK) and Swindlehurst (650 occurrences) had almost identical patterns.

About 39 000 surnames are shared by 42% of British people. Males with these surnames have a 20% chance of being directly related through common ancestry. Forensic scientists could set up a database containing the Y chromosome profiles for all 39 000 names. If this was done, DNA from a crime scene would have a one in five chance of pointing to a particular surname.

© *New Scientist Magazine*

■ Sex linkage

Genes found on the sex chromosomes are said to be **sex linked**. There are very few genes on the human Y chromosome, so sex-linked genes are usually found on the X chromosome. An example of a sex-linked characteristic is Duchenne muscular dystrophy (DMD). This is a genetic disorder caused by an error in the dystrophin gene. Dystrophin is an important protein in muscle tissue. The condition causes progressive muscle weakness as the muscle cells break down and die. Symptoms usually appear in boys under the age of 6. The first signs are a loss of muscle tissue and weakness of the arms and pelvis. Children with the condition have progressive difficulties with mobility, and usually need a wheelchair from the age of about 11. By the time the child has reached his late teens or early twenties, the breathing muscles become affected and the condition becomes life-threatening. With very rare exceptions, this condition only affects boys. It occurs in one in every 3500 births in the UK.

DMD is caused by a recessive allele on the X chromosome. A woman who carries the allele for DMD does not usually show any symptoms of the condition, since she has a second X chromosome with a normal allele on it. We can use the symbol X^D for the normal (dominant) allele and X^d for the recessive DMD allele. Figure 2 shows how a normal woman who carries the DMD allele and a normal man can have a child with DMD.

	Mother	Father
Parental phenotype:	normal (carrier of DMD)	normal
Parental genotype:	$X^D X^d$	$X^D Y$

Gametes: (X^D) (X^d) × (X^D) (Y)

Offspring genotypes:

		male gametes	
		X^D	Y
female gametes	X^D	$X^D X^D$	$X^D Y$
	X^d	$X^D X^d$	$X^d Y$

Offspring phenotypes and genotypes:

Normal son 1 ($X^D Y$)
Son with DMD 1 ($X^d Y$)
Normal daughter who is a carrier of DMD 1 ($X^D X^d$)
Normal daughter who is not a carrier of DMD 1 ($X^D X^D$)

Figure 3 *The inheritance of Duchenne muscular dystrophy*

Key:
- ○ normal female
- □ normal male
- ▨ red-green colour blind male

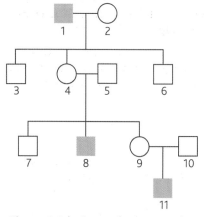

Figure 4 *Inheritance of red-green colour blindness*

Summary questions

1 A man with Duchenne muscular dystrophy cannot pass the condition on to his sons. Explain why.

2 Use a genetic diagram to show how it is possible for a girl to inherit DMD.

3 Figure 4 shows the inheritance of red-green colour blindness in a human family. What is the evidence that red-green colour blindness is: a recessive, b sex-linked?

4 Give the genotypes of individuals 3, 4 and 9. Use the letter 'B/b' to represent the normal and colour blind alleles.

5 Individual 7 from Figure 4 marries a woman with no family history of red-green colour blindness. What is the chance that they will have a child with red-green colour blindness?

2.9 Genetic counselling

Learning objectives:

- What is the role of a genetic counsellor?

- What are the ethical issues involved in embryo screening?

Specification reference: 3.4.2

People may be referred for genetic counselling if they have been diagnosed with a genetic condition. Alternatively, couples may have had a baby with a genetic condition, or they may be planning a pregnancy but are concerned that they may carry a genetic disorder. These people are usually referred for genetic counselling by their GP or by a hospital consultant.

Genetic counsellors are specially trained professionals who mainly come from a nursing or medical background. They have a thorough knowledge of genetic conditions and the impact they can have on a family. They try to help people understand the nature of a genetic condition, and how it affects an individual. Also, they can offer advice on whether the disorder can be prevented, or whether it is possible to diagnose the condition prenatally.

The first meeting between a genetic counsellor and a client involves collecting information about the family history. It is possible that genetic tests need to be carried out on the client(s) or other family members. These tests would be carried out by a clinical geneticist. After this, genetic counsellors can explain the information available to the client(s), and help them through associated emotional issues. Genetic counsellors never tell a client what to do, but support them by discussing issues and making information available.

Genetic screening

Link

Amniocentesis is discussed in Topic 1.5.

Some genetic conditions can be diagnosed by taking a sample of some cells from the person (maybe from some blood, or by washing out the mouth and collecting some cheek cells) and extracting their DNA. If the DNA base sequence of the gene causing the genetic disorder is known, the DNA can be examined to see if the individual has the genetic disorder. It may also be possible to tell whether the person is a carrier of a disorder (i.e. if they are heterozygous for a recessive disorder). You will learn more about how this is done in Topic 4.3. Some genetic conditions may be detected by testing for levels of specific enzymes, for example, in blood samples.

One problem with genetic screening is that we cannot screen for every genetic disorder at the moment. Also, many genetic conditions are caused by several different mutations and we are not able to screen for every single mutation.

Testing unborn babies

It is possible to examine DNA from a fetus by carrying out amniocentesis or chorionic villus sampling. Amniocentesis can be carried out between 15–20 weeks of pregnancy. You can see amniocentesis being carried out in Figure 1.

Ultrasound is used to locate the position of the fetus. Then a hypodermic needle is inserted through the mother's abdomen and into the uterus. A small sample of the amniotic fluid, in which the fetus is growing, is removed into a syringe. This fluid will have fetal cells in it. The cells may be cultured so that the fetus's chromosomes can be examined. Alternatively, the fetus's DNA can be extracted from the cells and tested to see if contains a gene for a genetic disorder. Amniocentesis is not usually performed before 15 weeks of pregnancy because of an increased risk of causing miscarriage.

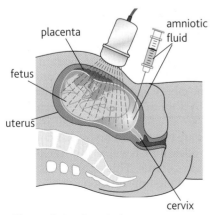

Figure 1 *Amniocentesis*

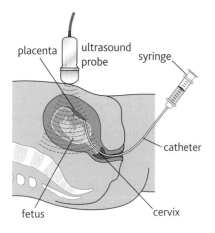

Figure 2 *Chorionic villus sampling*

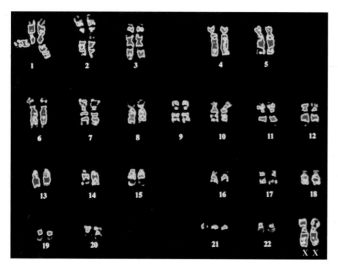

Figure 3 *A karyotype*

Another diagnostic test, called chorionic villus sampling (CVS), can be carried out from about 10 weeks of pregnancy. You can see this being done in Figure 2.

Ultrasound is used to guide a flexible tube, called a catheter, through the woman's vagina, through the cervix and close to the placenta. A syringe on the other end of the catheter is used to remove a small amount of tissue from the placenta. However, this test is sometimes done using a needle through the abdomen as in amniocentesis. This test gives a 1–2% chance of miscarriage, which is slightly higher than the risk of miscarriage when amniocentesis is carried out.

If, as a result of amniocentesis or CVS, a fetus is found to have a genetic disorder, the parents will be offered advice from a genetic counsellor. As a result of this advice the parents may choose to terminate the pregnancy.

■ Karyotyping

Cells from a fetus can be cultured in a special medium. This contains a chemical that stimulates the cells to divide by mitosis. However, a spindle inhibitor is also used, so that many of the cells will start to divide but will not complete mitosis. A photograph of the chromosomes is taken and the images manipulated on a computer screen. The chromosomes are arranged into their homologous pairs, starting with the largest pair and ending with the smallest. This arrangement is called a **karyotype**. You can see this in Figure 3.

A karyotype like this can detect abnormalities of the chromosomes, but not a disorder involving a single gene. Disorders involving individual genes can be detected using DNA analysis.

■ Embryo screening

If a couple is at risk of having a child with a genetic disorder, such as cystic fibrosis, they may choose to have a baby by embryo screening. This involves producing several embryos using IVF treatment, as you learned in Topic 1.8. The embryos are allowed to grow to the 4- or 8-cell stage. At this stage, the embryo is held steadily in a suction pipette, while a smaller, sharper pipette breaks open the protein covering of the embryo and sucks out a single cell. The DNA from this cell can be extracted and tested to see whether the alleles causing cystic fibrosis are present. Embryos that do not carry the allele for CF will be implanted into the woman's uterus. This technique is sometimes called pre-implantation genetic diagnosis (PGD).

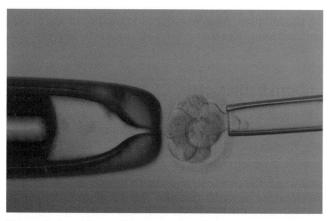

Figure 4 *Extracting a cell from an early embryo to analyse its DNA*

The DNA from one embryo may be tested for a large number of different genetic sequences. The embryo is not harmed, as it has only lost a single cell. However, it is a very new technique so nobody born using this technique is yet old enough to have had their own children. Therefore we cannot be certain that there is no long-term harm. The embryo is also unable to give consent to the treatment. Some people are unhappy about PGD because, like IVF, it involves creating a large number of embryos, many of which are not used. They feel that selecting embryos allows couples to have a 'designer baby'. However, embryo testing is regulated by the Human Fertilisation and Embryology Authority (HFEA) and is only allowed for serious genetic conditions.

How science works

Saviour siblings

Parents of children born with certain congenital diseases sometimes choose to have another baby using IVF and embryo screening. This baby, known as a 'saviour sibling', will have stem cells in its umbilical cord that can be extracted at birth. These stem cells are transplanted into the sick sibling. There is a small chance that the sick child's immune system could mount a response, but it is unlikely as the embryo would have been screened for the same key immune system genes. If the transplant is successful, the implanted stem cells will produce healthy white bloods cells capable of battling the disease.

It used to be illegal to use IVF and embryo screening in this way in the UK. Parents who wanted this procedure would have to go abroad. In May 2008 the procedure controversially became legal in the UK.

1 If another sibling was produced naturally, without using genetic screening, it is very unlikely that the stem cells from this baby could be used to treat the sick sibling. Use your knowledge of the immune system to explain why.

2 This kind of treatment is now allowed in the UK. Do you think this is a good idea? Give reasons for your answer.

Summary questions

1 You are a genetic counsellor. A couple have come to see you because they are planning a family. The woman has a brother with cystic fibrosis, but the man has no family history of cystic fibrosis. What information will you give them?

2 Genetic counsellors give people information about genetic conditions but they never tell their clients what they think they should do. Suggest why.

3 In the UK, pregnant women over the age of 35 are offered amniocentesis or CVS to screen for Down's syndrome in the fetus. However, some women decide not to take up this offer. Suggest why.

4 Look at the karyotype in Figure 3.

a What does this tell you about this individual?

b In what stage of mitosis was the photograph of the chromosomes taken?

5 Explain why, when culturing cells to produce a karyotype:

a the cells are stimulated to divide

b a spindle inhibitor is added.

2.10 Chi-squared test

- Why don't experimental results always fit the results we predict from genetic crosses?

- How can we test whether our experimental results are significantly different from predicted ratios?

Specification reference: 3.4.2

In Topic 2.5 you learned that cystic fibrosis is an example of monohybrid inheritance. You will remember that, when two heterozygous individuals interbreed, we expect a 3:1 ratio in the offspring. In other words, 75% of the offspring should show the phenotype determined by the dominant allele and 25% of offspring should show the phenotype determined by the recessive allele. We call this the expected **Mendelian ratio**. This term comes from the great scientist, Gregor Mendel, who carried out pioneering work in genetics. He spent years carrying out genetic crosses and, as a result, he worked out many of the rules of genetic inheritance.

We don't always obtain results that fit the Mendelian ratio exactly. This is because the ratio only tells you the chance of an individual with a certain phenotype being produced from a particular cross. So two parents who are heterozygous for CF have a 25% chance of having a child with CF at any pregnancy. If their first child has CF, there is still a 25% chance that their next child will have CF as well. In human families, because the numbers of children are fairly small, we do not usually find an exact 3:1 ratio in the offspring. There must be many couples in the UK who are both heterozygous for CF, but do not know it. They may have had two healthy children, and then no more.

Testing results

Sometimes large numbers of offspring are produced in genetic crosses. This is more likely to happen in genetic crosses using animals other than humans and plants. They may find that the ratio of phenotypes in the offspring does not fit the predicted ratio exactly. However, this could simply be the result of chance.

One way to test whether this is the result of chance is to use a statistical test, called the χ^2 test (pronounced Kie-squared). Here is an example:

Scientists were studying rats that had developed a resistance to warfarin, a rat poison. They bred two rats that were resistant to warfarin together and produced several litters of offspring. The numbers of resistant and non-resistant rats in these litters are shown in Table 1.

They thought that resistance could be caused by a dominant allele, and that the offspring were produced in a 3:1 ratio. We can check this by using a χ^2 test.

1 To carry out a χ^2 test, we first need a **null hypothesis**. The null hypothesis is always that 'there is no difference between observed and expected numbers'.

2 The expected numbers must be worked out. There were (27 + 13 =) 40 offspring in all, so if these showed a 3:1 ratio we would expect 30 resistant and 10 non-resistant rats. We put the observed and expected results in a table (Table 2).

3 We work out the value of χ^2 using the formula:

$$\chi^2 = \sum \frac{(O - E)^2}{E}$$

Table 1

Phenotype	Number of rats
resistant to warfarin	27
not resistant to warfarin	13

Table 2

Phenotype	Observed number of rats	Expected number of rats
resistant to warfarin	27	30
not resistant to warfarin	13	10

Where O = observed number

E = expected number

Σ = sum of

The best way to work out χ^2 is using a table (Table 3):

Table 3

Phenotype	O	E	$O-E$	$(O-E)^2$	$\dfrac{(O-E)^2}{E}$
resistant to warfarin	27	30	3	9	0.3
not resistant to warfarin	13	10	3	9	0.9
				Σ	1.2

For each genotype, we find the difference between the observed and expected numbers ($O - E$) and write them in the correct column. Note that you can ignore minus signs.

In the next column, we square the figure we obtained in the previous column, giving $(O - E)^2$. The reason we can ignore minus signs is because squaring the number in this column gets rid of any minus signs.

In the next column, we divide $(O - E)^2$ by E.

By adding all the numbers in the last column, we get the value for χ^2 which, in this case, is 1.2.

4 Now we have to look up this value of χ^2 on a table. You can see part of a χ^2 table in Figure 1.

You will notice that the table has a column labelled 'degrees of freedom'.

Probability (p)	0.90	0.80	0.70	0.50	0.30	0.20	0.10	0.05	0.02	0.01
	(90%)	(80%)	(70%)	(50%)	(30%)	(20%)	(10%)	(5%)	(2%)	(1%)
Degrees of freedom					χ^2					
1	0.016	0.064	0.15	0.46	1.07	1.64	2.71	3.84	5.41	6.64
2	0.21	0.45	0.71	1.39	2.41	3.22	4.61	5.99	7.82	9.21
3	0.58	1.01	1.42	2.37	3.67	4.64	6.25	7.82	9.84	11.34
4	1.61	2.34	3.00	4.35	6.06	7.29	9.24	11.07	13.39	15.09

Figure 1 A χ^2 table

This is always one less than the number of categories you have in your data. We had two categories, resistant and non-resistant, so in this case we have one degree of freedom.

We look along the χ^2 values in the row for one degree of freedom to find the value of χ^2 closest to 1.2. You will see that the closest value is 1.07. The rows of χ^2 values have numbers at the top, called 'probability (p)'. This number tells you the probability of getting this value of χ^2 by chance alone. In this case, 1.07 is in the column headed 0.3. This tells you that these observed numbers would be expected 3 times out of 10, or 30% of the time simply by chance. In other words, it is very likely that there is no significant difference between our observed and expected numbers, and our null hypothesis is very likely to be true. So we can **accept** our null hypothesis.

It is usual in biology to accept our null hypothesis if our χ^2 value gives a probability (p) of 0.05 or higher. If p is less than 0.05, then the observed and expected results are **significantly different**, and we must **reject** our null hypothesis.

Hint

The χ^2 formula looks scary, but it really isn't. Working it out just involves simple arithmetic. You don't need to know how the test works – you just need to know how to use it and when to use it. It's a bit like driving a car – you need to know how to use the pedals, gear stick and steering wheel, but you don't need to understand how they work.

Summary questions

1 In cats, short hair is dominant over long hair. A breeder crossed a long-haired and a short-haired cat many times, and in all they had 46 kittens. Twenty-five had short hair and 21 had long hair. Use the chi-squared test to determine whether these results are consistent with a 1:1 ratio.

2 Use suitable genetic symbols to show how a long-haired cat and a short-haired cat could produce long- and short-haired kittens in approximately equal numbers.

2.11 Gene mutation

Learning objectives:

■ How do new alleles arise?

■ What causes changes in DNA?

Specification reference: 3.4.2

■ Link

You may want to look back at Chapter 7 in *AS Human Biology* to remind yourself of the properties of DNA.

You already know that a gene is a length of DNA that codes for one polypeptide, and that alleles are different forms of a gene. This means that changes in the DNA that makes up one gene will result in new alleles.

You will remember from AS Level that DNA is made of a double helix. It is like a twisted rope-ladder with pairs of organic bases forming the 'rungs' of the ladder. The sequence of these four bases, A, C, G and T, carries the coded information that cells use to make polypeptides and proteins. You will learn more about this in Topic 3.1.

Whenever DNA replicates, there is a risk that a mistake may be made. The risk that a new nucleotide may be added is very low, but mistakes can happen. However, almost all mistakes in copying DNA are corrected by the DNA polymerase enzymes that copy the DNA. Uncorrected errors in the DNA are called **mutations**.

Some mutations will have no effect, because they occur in so-called 'junk' DNA. We know that a large proportion of our DNA does not appear to code for polypeptides, and so mutations in this DNA would not have any effect. However, some parts of 'junk' DNA seem to stay remarkably the same from generation to generation, which means that some of this 'junk' DNA probably does have a function that we do not yet understand.

■ Types of mutation

Look at the upper part of Figure 1. You can see that a piece of DNA has replicated, but a mistake has occurred. One base pair has been deleted. We call this type of mutation a **deletion**. You will learn in Topic 3.1 that DNA codes for the structure of polypeptides in the form of a triplet code. Three bases in DNA codes for one amino acid. So, if one base-pair is missed out, this changes the way the code is read. We say that a deletion mutation is a **frame-shift mutation**. This is because all the triplets are changed after the point of mutation. This kind of mutation produces a very big change in the primary structure of the protein that is coded for by the mutated allele. The resulting protein will have a very different tertiary structure, and is likely to be non-functional.

There are many different mutations of the CFTR gene that can lead to cystic fibrosis. The most common mutation involves the deletion of just three nucleotides that code for the amino acid phenylalanine. As a result, the CFTR protein coded for has an altered tertiary structure.

Another kind of mutation is a **substitution**. Look at the lower part of Figure 1. You can see that the original DNA has replicated, and a mistake has occurred. However this time, one base has been changed for a different one, so one base-pair has changed to another. This mutation will only affect a single triplet. Therefore, just one amino acid in the whole protein will be altered.

You might think that changing just one amino acid in a protein will not make much difference, since most proteins have a very large number of amino acids in their structure. However, this is not the case. You will recall from Topic 2.6 that sickle cell anaemia results from haemoglobin S, which differs from normal haemoglobin by a change in just one amino acid in two of its four polypeptide chains. The triplet code CTT in the

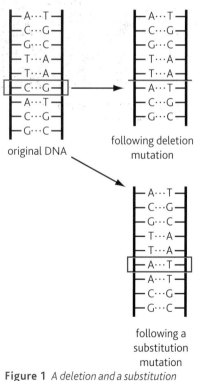

original DNA

following deletion mutation

following a substitution mutation

Figure 1 *A deletion and a substitution mutation*

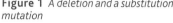

DNA has changed to CAT. CTT is the code for glutamate, while CAT codes for valine. These amino acids have very different R-groups, making haemoglobin S much less soluble than normal haemoglobin.

It is possible for a substitution mutation to occur, but without causing any change in the protein coded for. There are more triplet codes than amino acids, so some amino acids have more than one DNA triplet code. In the example given for sickle cell anaemia, had the triplet CTT changed to CTC, there would have been no change in the amino acid coded for. This is because both CTT and CTC code for glutamate.

Changes in the base-sequence of genes are usually referred to as **point mutations**. Deletion and substitution are just two examples of this. Chromosome mutations may also occur. These are changes in the number and structure of chromosomes. Non-disjunction, which you studied at AS level, is an example of a chromosome mutation.

Mutagens

DNA mutations occur naturally at a very low rate. However, the rate at which they occur may be increased by environmental factors known as **mutagens**. These include certain types of radiation such as ultraviolet, X-rays, alpha and beta radiation. Chemicals such as nitrous acid, and chemicals in diesel exhaust and cigarette smoke may also act as mutagens.

Figure 3 *Cigarettes also contain chemical mutagens*

Figure 2 *Tanning beds or sunbeds use harmful ultraviolet radiation*

Summary questions

1. In Topic 2.9 you learned that genetic screening may be carried out on individuals or on embryos to detect alleles that cause genetic disorders. Use the information in this topic to explain why it is not possible to detect all alleles causing genetic disorders.

2. Is the deletion mutation causing sickle cell anaemia a frame-shift mutation? Explain your answer.

3. Explain why changing just one amino acid in a whole protein can result in a protein which does not function.

4. The examples of mutagens given in this topic are also examples of carcinogens. Explain why carcinogens cause cancer.

2.12 Meiosis and variation

Learning objectives:

- How does meiosis contribute to variation?
- How else does variation occur?

Specification reference: 3.4.2

You will remember from AS Topic 8.4, that gametes, or sex cells, are made by meiosis. This is a two-stage division in which a diploid cell divides to produce four haploid gametes. You can see the stages of meiosis in Figure 1.

Interphase
Immediately before meiosis, DNA replicates so that the cell now contains four, rather than the original two, copies of each chromosome.

Prophase I
The chromosomes shorten and fatten and come together in their homologous pairs to form a bivalent. The chromatids wrap around one another and attach at points called chiasmata. The chromosomes may break at these points and swap similar sections of chromatids with one another in a process called crossing over. Finally the nucleolus disappears and the nuclear envelope disintegrates.

Metaphase I
Centromeres attach to the spindle and the bivalents arrange themselves randomly on the equator of the cell with each of a pair of homologous chromosomes facing opposite poles.

Anaphase I
One of each pair of homologous chromosomes is pulled by spindle fibres to opposite poles.

Telophase I
In most animal cells a nuclear envelope re-forms around the chromosomes at each pole, but in most plant cells there is no telophase I and the cell goes directly into metaphase II.

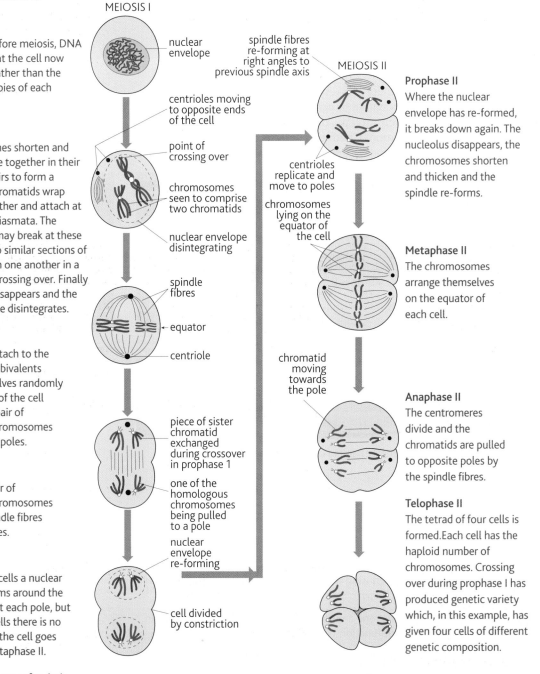

Prophase II
Where the nuclear envelope has re-formed, it breaks down again. The nucleolus disappears, the chromosomes shorten and thicken and the spindle re-forms.

Metaphase II
The chromosomes arrange themselves on the equator of each cell.

Anaphase II
The centromeres divide and the chromatids are pulled to opposite poles by the spindle fibres.

Telophase II
The tetrad of four cells is formed. Each cell has the haploid number of chromosomes. Crossing over during prophase I has produced genetic variety which, in this example, has given four cells of different genetic composition.

Figure 1 *The stages of meiosis*

The first division of meiosis involves the homologous chromosomes pairing together to form a **bivalent**. Chromatids of homologous chromosomes wrap around each other at places called **chiasmata**. Here, the chromatids break and equivalent pieces of chromatid are exchanged. This process is called **crossing over** and is shown in Figure 2. The second division of meiosis is like mitosis. In this stage the chromatids separate from each other.

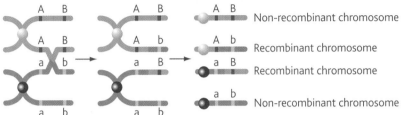

Figure 2 *Crossing over*

How meiosis brings about variation

- Crossing over during prophase 1 produces new combinations of alleles.
- Independent assortment of homologous chromosomes – in metaphase 1 pairs of homologous chromosomes line up on the spindle randomly. Each pair of chromosomes separates independently of any other pair. The assortment occurs when the chromosomes move apart in anaphase 1.
- Independent assortment of chromatids in anaphase 2 again occurs randomly, so each pair of chromatids separates independently of any other pair.

> ### Link
>
> Look back at Chapter 8 of *AS Human Biology* and the beginning of this book. Check that you understand the meaning of the terms chromatid, centromere, homologous pair, spindle fibres, centriole and equator.

> ### AQA Examiner's tip
>
> You do not need to memorise the details of the stages of meiosis, just understand them in sufficient detail to appreciate how crossing over, independent assortment and random fertilisation lead to new combinations of alleles.

In **arrangement 1**, the two pairs of homologous chromosomes orientate themselves on the equator in such a way that the chromosome carrying the allele for brown eyes and the one carrying the allele for blood group A migrate to the same pole. The alleles for blue eyes and blood group B migrate to the opposite pole. Cell ❶ therefore carries the alleles for brown eyes and blood group A while cell ❷ carries the ones for blue eyes and blood group B.

In **arrangement 2**, the left-hand homologous pair of chromosomes is shown orientated the opposite way around. As this orientation is random, this arrangement is equally as likely as the first one. The result of this different arrangement is that cell ❸ carries the alleles for blue eyes and blood group A, whereas cell ❹ carries ones for brown eyes and blood group B.

All four resultant cells are different from one another. With more homologous pairs the number of possible combinations becomes enormous. A human, with 23 such pairs, has the potential for $2^{23} = 8\,388\,608$ combinations.

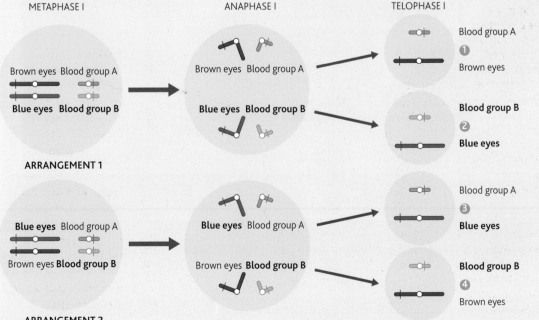

Figure 3 *How independent assortment of chromosomes contributes to variety in gametes and hence the offspring*

■ Fertilisation and variation

A further source of variation occurs at fertilisation. Each gamete produced by an individual is genetically different from any other. There are usually many different male gametes, and it is purely random which one of these will fertilise the oocyte. As a result, the genetic make-up of every new individual is unique.

■ Summary questions

1 Copy and complete the table to show the differences between mitosis and meiosis.

Mitosis	Meiosis
a one-stage division	
	produces daughter cells with half the number of chromosomes as the parent cell
	homologous chromosomes pair together forming bivalents in prophase 1
chiasmata are never formed and crossing over does not occur	
daughter cells are genetically identical to the parent cell	
two daughter cells are formed	

2 A student wrote in an exam, 'Homologous chromosomes are the same as each other because they have the same genes'. Explain what is wrong with this statement.

2.13 Continuous and discontinuous variation

You are already aware that living things show variation in their phenotype. You have also learned that variation is important, as natural selection favours phenotypes that are well adapted to their environment. Evolution occurs as a result of variation and natural selection.

There are two main kinds of variation: discontinuous variation and continuous variation.

Discontinuous variation

Discontinuous variation occurs when you can put organisms into definite categories. A good example of this is the ABO blood group system. Look at Figure 1.

You will notice that people can be categorised into one of these groups depending on which ABO blood group they are. There are no intermediate categories. Discontinuous variation results from one gene with two or three different alleles. Environmental factors have little effect on these characteristics.

Continuous variation

Some kinds of variation between individuals do not fall into distinct categories. An example is human height. It would not be easy to classify the students in your class into 'tall' 'medium' or 'short', for example. You would have a few people who are much shorter than the others, and a few very tall people. However, most people would be intermediate in height. If you plotted a graph showing the height of a large number of people, you would obtain a graph something like the one shown in Figure 2.

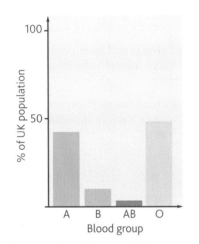

Figure 1 *Discontinuous variation illustrated by the percentage of the UK population with blood groups A, B, AB and O*

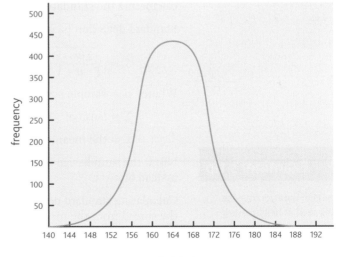

Figure 2 *Graph of frequency against height for a sample of humans*

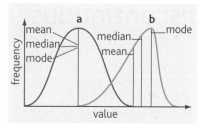

Figure 3 *Mean, median and mode for*
a *a normal distribution,*
b *a skewed distribution*

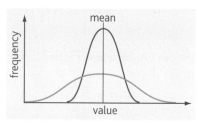

Figure 4 *Two different normal distribution curves with the same mean, median and mode*

Characteristics like height are determined by many genes, not just one. Each of these genes may have several alleles. We call this **polygenic inheritance**. Your height will depend on the mixture of dominant and recessive alleles that you inherit for each of the genes involved in height.

Environmental factors may also play a part in characteristics that show continuous variation. For example, if children have a very poor diet in childhood, or suffer from a series of illnesses, they may not grow to be as tall as they might have done if they were healthy or had a good diet.

■ Mean, median and mode

The **mean** is the 'average' of a set of values. It is found by adding together all the values and dividing by the number of individual values.

The **median** is the middle value of a set of values. For example, if there are 25 values in order in a list, the median would be the thirteenth value on the list, as this comes exactly half-way. If there is an even number of values, the median is the average of the two middle values.

The **mode** is the value that occurs most frequently in a set of values.

The graph in Figure 2 shows a **normal distribution**. The curve is symmetrical about a central value and the mean, median and mode are all the same.

Look at Figure 3. Graph (a) shows a normal distribution. However, graph (b) shows a **skewed distribution**. You will notice that, in this case, the mean, median and mode all have different values.

■ Standard deviation

Suppose that you measured the heights of a large number of students in one college, and then measured the heights of another large group of students in a different college. Then you analyse the data, and find that both populations have exactly the same mean, median and mode. Does this mean that both the populations are the same? Look at Figure 4.

You will notice that both populations here have the same mean, but population (a) shows much less variation from the mean than population (b). We usually measure the spread about the mean of a population by calculating the **standard deviation**.

Standard deviation (s) is calculated using the formula:

$$S = \sqrt{\frac{\Sigma(x - \bar{x})^2}{n - 1}}$$

Where n = sample size

Σ = 'sum of'

$\bar{x}$ = the mean.

NB: If the sample size is greater than 30, n is used in this formula instead of $n - 1$.

Calculating standard deviation is useful because it gives you a measure of the spread from the mean. A small standard deviation tells you that there is little variation from the mean, while a large value tells you there is a great deal of variation in the population.

If the population is normally distributed, then about 68% of the population will have a value that lies within one standard deviation of the mean. About 95% of the population will have a value that lies within two standard deviations of the mean, and almost 100% of the population will have a value that lies within three standard deviations of the mean.

Calculating the standard deviation

Using the standard deviation formula is not as complicated as it looks. Let us look at an example.

A group of eight students were surveyed to determine their heights which were as follows: 165 cm, 172 cm, 171 cm, 168 cm, 190 cm, 181 cm, 184 cm, 177 cm. Let the height be x.

1   Calculate the mean height ($\bar{x}$). This is the sum of all the values divided by the number of values.
    $165 + 172 + 171 + 168 + 190 + 181 + 184 + 177 = 1408$.
    $1408/8 = 176$ so $\bar{x} = 176$

2   Subtract the mean height from each of the measured heights $(x - \bar{x})$. This gives: $-11, -4, -5, -8, 14, 5, 8, 1$.

3   Square all of these values $(x - \bar{x})^2$. This gives: $121, 16, 25, 64, 196, 25, 64, 1$.

4   Add all of the squared values together:
    $\Sigma(x - x)^2 = 512$

5   Divide the sum of all the squares by the sample size $n - 1$.
    $$\sqrt{\frac{\Sigma(x - \bar{x})^2}{n - 1}} = \frac{512}{7} = 73.1$$

6   Take the square root:
    $$s = \sqrt{\frac{\Sigma(x - \bar{x})^2}{n - 1}} = 8.56$$

So the standard deviation is 8.56 for this set of data.

For larger samples, you might find it useful to put your values into a table, like Table 1.

Table 1 *The heights of 14 male students*

Height/cm	Number of male students
191	1
188	1
185	2
183	1
175	5
173	1
168	2
165	1

Summary questions

1   Use Figure 5 to name the kind of variation shown by human birth mass.

2   What does Figure 5 show about the genetic control of birth mass in humans?

3   Use the information in Figure 5 to explain how natural selection favours intermediate birth mass in humans.

4   Find the mean, median and mode of the data in Table 1.

5   Calculate the standard deviation of the sample in Table 1.

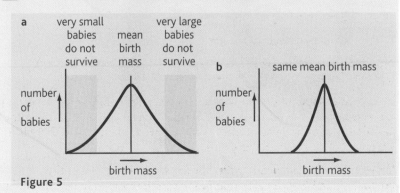

Figure 5

2.14   Genes and the environment

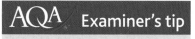
- How does the environment affect our phenotype?

- How can we determine the effect that environment has on our phenotype?

Specification reference: 3.4.2

You learned in Topic 2.5 that our phenotype does not always depend only on our genotype. Sometimes it is affected by the environment. You will remember that human height is not only determined by the different combinations of alleles that we inherit, but also environmental factors, like the diet we eat, or exposure to disease.

Twin studies

When scientists are trying to find out the relative effects of genetics and the environment on a particular factor, it is very useful to study twins. There are two different kinds of twins:

- Monozygotic (MZ) twins develop from a single fertilised oocyte (or zygote) that splits early in development to form two genetically identical individuals. Any differences between monozygotic twins are likely to be the result of the environment.

- Dizygotic (DZ) twins develop from two different fertilised oocytes, or zygotes. Although they share about 50% of their genes, they are no more alike than any other brothers and sisters. Differences between dizygotic twins might be the result of genetic or environmental factors.

Twins are said to be **concordant** for a characteristic if they are both similar for that characteristic, and **discordant** if they are different for that characteristic. Look at the data shown in Table 1.

Table 1 *Concordance for various characteristics between MZ and DZ twins*

Characteristic	Concordance/%	
	MZ	DZ
blood group	100	66
measles	95	87
breast cancer	6	3

Figure 1 *These monozygotic twins show differences due to their environments*

You will notice that MZ twins have 100% concordance for blood group. This will not surprise you, because we know that blood group is determined by genes and that the environment cannot influence this. DZ twins have a fairly high concordance, because they are related and share many of their genes, but they do not always have the same blood group as their twin. Breast cancer, on the other hand, has a low concordance for both types of twin. Genetic factors can play a part, as is shown by the concordance for MZ twins being 6% but only 3% for DZ twins. However, this low concordance tells us that environmental factors are very important in determining whether a person will develop breast cancer.

Table 1 also shows that measles has a high concordance between MZ twins. You might think at first that this indicates that genetic factors cause measles. However, this would be wrong. We know that measles is caused by a virus. This is also shown because DZ twins also have a very high concordance for measles. Twins are living in the same environment, so if one twin develops measles, it is very likely that the other one will also become infected. Therefore, when looking at concordance between MZ twins, who have identical genes, it is helpful to also study the concordance between DZ twins, who are genetically different.

There have also been cases where MZ twins have been separated early in life and raised in different families. Scientists sometimes study the concordance between MZ twins raised together (when they share both their genes and their environment) with the concordance between MZ twins raised apart (when they share the same genes but have a different environment). Look at the data in Table 2.

Table 2 *Results of one study to determine concordance between twins for IQ score*

Concordance for IQ score/%		
MZ reared together	MZ reared apart	DZ reared together
86	72	60

This shows that genetic factors must be important in IQ, since the concordance for MZ twins reared together is high, and is higher than for DZ twins reared together. Also, MZ twins reared in different environments have a higher concordance than DZ twins reared together. However, there are some problems in reaching conclusions here. Firstly, MZ twins reared apart do not necessarily have a completely different environment. They might experience environments which are essentially similar, in terms of such factors as mental stimulation, diet and housing quality. Secondly, the IQ test used might not give reliable measurements, so differences between twins' scores might be the result of a poorly constructed test.

Summary questions

1 Explain why monozygotic twins are always the same sex.

2 When comparing concordance between MZ and DZ twins, scientists usually only include DZ twins that are of the same sex. Suggest why.

2.15 Prader-Willi syndrome

Learning objectives:

■ Can environmental factors cause inherited changes in the phenotype?

■ What is Prader-Willi syndrome?

Specification reference: 3.4.2

Recently, scientists have found that the same allele can have a different effect, depending on which parent it was inherited from. For example, Prader-Willi syndrome is a genetic condition that affects one in every 12 000–15 000 people. It affects both sexes and all ethnic groups. It is caused by deletion of some of the genetic material on chromosome 15. However, a child only develops Prader-Willi syndrome if the defective chromosome 15 is inherited from the father.

The main features of a child born with Prader-Willi syndrome are:

■ characteristic facial features (may include almond-shaped eyes, down-turned mouth, eyes close together, eyes pointing in different directions, thin upper lip)

■ developmental delay including mild to moderate learning difficulties

■ feeding problems in infancy leading to low weight gain

■ underdevelopment of the sex organs and delayed puberty

■ poor muscle tone

■ rapid weight gain between the ages of 1 and 6 years. This can lead to serious obesity if no steps are taken to reduce this

■ extreme hunger and obsession with food

■ infants and children are usually happy and loving, with few behaviour problems. However, older children and adults may show behaviour problems, including temper tantrums and violent outbursts

■ obsessive/compulsive behaviour, stealing and lying

■ usually short in stature, with small hands and feet

■ often fair skin and light-coloured hair

■ may have speech problems.

There is no cure for PWS, but many of the symptoms can be treated. A key priority is to manage the person's weight. Parents of children with PWS need to give the child a balanced, low-energy diet with plenty of vitamins and minerals. They may need to keep food cupboards and the refrigerator locked. Growth hormone helps the child to grow taller and gain more muscle mass. Sex hormones may be used to encourage more normal sexual development at puberty, and advice is given about behavioural management. Speech and language therapy will help if there are speech disorders, and special educational programmes can assist the child to develop to her/his full potential.

■ Epigenetic imprinting

As already mentioned, the chromosome 15 with its deleted region behaves differently depending on whether it comes from the father or the mother. During gametogenesis, certain genes in both sperm and oocytes are modified by the addition of methyl ($-CH_3$) groups. This process is called **epigenetic imprinting**. The term 'epigenetics' is used to describe inherited changes in the DNA that do not involve a change in the DNA base sequence. As a result, the chromosome is 'labelled' as maternal or paternal, and this means they have different effects on the developing embryo. Scientists do not yet understand this process fully. The process must be reversible. If a man passes a chromosome to his daughter it will be epigenetically imprinted as 'paternal'. However, if this daughter passes the same chromosome onto her child, it will become imprinted as 'maternal'.

You can see a pedigree showing the inheritance of PWS in one family in Figure 1. You will notice that individuals who inherit the defective chromosome from their mother do not develop PWS.

Key:
☐ ○ unaffected (male and female)
⊡ ⊙ healthy carriers (male and female)
⬛ ⬤ affected (male and female)

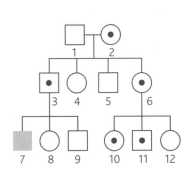

Figure 1 *Inheritance of PWS in one family*

Epigenetic changes may be brought about by the environment. For example, some heavy metals and pesticides can cause changes in DNA methylation. DNA methylation also decreases as cells age. This is an area of current research.

Application and How science works

Prader-Willi syndrome and obesity

Prader-Willi syndrome can cause many problems. If diet is not strictly regulated then there is a good chance the sufferer could become morbidly obese. One man from Hastings is blaming the combination of his disorder and his local council for his morbid obesity. Chris Leppard weighs 40 stone at the age of 25. He claims he has been unable to move from his armchair for over 6 months because of his size. Chris is unable to walk up the stairs or fit through his front door at the council house he shares with his mother. He wants the local council to either move him to a bungalow or increase the size of his front door. If this were to happen he feels as though he would have a better chance of losing weight as he would be able to go outside and exercise. Hastings council have not moved Chris and his mother or made the doorway bigger, but they are working with the family to find a solution. Chris used to eat over 5000 calories a day.

1 Should Hastings council do more to help Chris Leppard? Give reasons for your answer.

2 Do you think Chris Leppard is to blame for his obesity problem?

Summary questions

1 Look at the pedigree in Figure 1. Individual 7 has PWS but his brother and sister do not have PWS. Explain how this has occurred.

2 Individual 11 marries a woman with no family history of PWS. The couple approach a genetic counsellor because they would like to have a family, but they are concerned that they might have a child with PWS. What information will the genetic counsellor give them?

1 The inheritance of ABO blood groups is controlled by three alleles of the same gene, I^A I^B and I^O. The alleles I^A and I^B are codominant. Both I^A and I^B are dominant to the allele I^O.

(a) Explain what is meant by an allele. *(1 mark)*

(b) (i) Copy and complete the table to show the missing genotypes.

Blood group phenotype	Possible genotypes
A	$I^A I^A$,
B	$I^B I^B$,
AB	
O	

(ii) Children of blood groups A and O were born to parents of blood groups A and B. Draw a genetic diagram to show the possible ABO blood group phenotypes of the children which could be produced from these parents. *(5 marks)*

AQA, 2007

2 Tongue-rolling and red-green colour blindness are two genetically controlled conditions which occur in humans.

Tongue-rolling is controlled by the dominant allele, **T,** while non-rolling is controlled by the recessive allele, **t.**

Red-green colour blindness is controlled by a sex-linked gene on the X chromosome. Normal colour vision is controlled by the dominant allele, **B,** while red-green colour blindness is controlled by the recessive allele, **b.**

(a) Copy and complete the genetic diagram to show the possible genotypes and phenotypes which could be produced from the following parents. *(4 marks)*

	Female Colour blind and heterozygous for tongue-rolling	Male Normal colour vision and non-roller
Parental genotypes		
Genotypes of gametes		
Genotypes of children		
Sex and phenotypes of children		

(b) Explain why a higher percentage of males than females in a population is red-green colour blind. *(1 mark)*

(c) Sex-linked genes on the Y chromosome have been found in humans and other animal species. Suggest and explain **one** piece of evidence which would support the presence of such a gene. *(1 mark)*

AQA, 2007

3 In humans, cystic fibrosis is caused by a recessive allele, **f.**

(a) What is an *allele*? *(1 mark)*

(b) A man and woman are both heterozygous for the cystic fibrosis allele. They have one healthy son but would like to have another child. What is the probability that they will produce a girl who has cystic fibrosis? Show your working. *(2 marks)*

(c) Sperms are produced by meiosis. Give **two** ways in which differences in sperms are a result of meiosis. *(2 marks)*

AQA, 2003

4 The allele for Rhesus positive, **R**, is dominant to that for Rhesus negative, **r**. Haemophilia is a sex-linked condition. The allele for haemophilia, **h**, is recessive to the allele for normal blood clotting, **H**, and is carried on the X–chromosome. **Figure 1** shows the Rhesus blood group phenotypes in a family tree where some individuals have haemophilia.

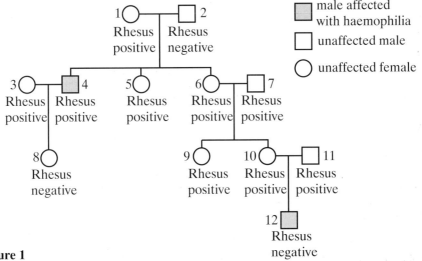

Figure 1

(a) (i) Use the information in **Figure 1** to give **one** piece of evidence that the allele for the Rhesus negative condition is recessive.

   (ii) Explain the evidence from the cross between individuals 3 and 4 that the gene controlling Rhesus blood group is **not** sex-linked. *(3 marks)*

(b) Give the full genotype of
   (i) individual 6;
   (ii) individual 12. *(2 marks)*

(c) What is the probability that the next child of couple 10 and 11 will have the same genotype as the first child? Show your working. *(3 marks)*

AQA, 2000

5 Familial polyposis is a hereditary disease. In this disease small wart-like growths appear on the inner surface of the large intestine during childhood. The risk that one of these warts will develop into a malignant tumour in teenage years is so high that affected children are often treated by having that part of the large intestine surgically removed.

(a) Explain what is meant by a *malignant tumour.* *(2 marks)*

(b) Familial polyposis is caused by a dominant allele, **A**. The allele is **not** located on a sex chromosome.

   In a couple, the male is heterozygous for familial polyposis and the female is unaffected. What is the probability that a child born to this couple would have familial polyposis? Use a genetic diagram to explain your answer. *(3 marks)*

(c) Serious diseases caused by dominant alleles are relatively uncommon compared with diseases caused by recessive alleles.
   Suggest an explanation for this. *(2 marks)*

AQA, 2000

6 Many body processes are affected by ageing (senescence). For example, cardiac output gradually reduces, the basal metabolic rate falls and the performance of the nervous system gradually declines.

Explain:

(a) the reduction in cardiac output; *(1 mark)*

(b) the fall in basal metabolic rate; *(1 mark)*

(c) the decline in the performance of the nervous system. *(2 marks)*

AQA, 2000

3 The management structure of cells

3.1 Protein synthesis

You should already know that the order of bases in DNA forms the genetic code. Three bases form a triplet code that codes for one amino acid. So the order of bases in the DNA codes for the sequence of amino acids in a protein.

You should also know that proteins are made by the ribosomes, in the cell cytoplasm and on the rough endoplasmic reticulum. Yet the DNA in a cell is found inside the nucleus. This means that the information in the DNA has to be transferred to a 'messenger molecule' in the form of RNA, so that proteins can be made.

Transcription

This takes place in the cell nucleus. You can see the process of transcription in Figure 1.

1 A section of DNA in the nucleus of the cell 'unzips'. The hydrogen bonds between the DNA bases are broken by an enzyme.

2 The two DNA strands separate and one strand becomes a template. Nucleotides of RNA, which are present in the nucleus, attach to the exposed bases of the template strand by complementary base-pairing.

3 The enzyme **RNA polymerase** joins the RNA nucleotides together to form a single strand of **messenger RNA (mRNA)**.

4 Once the mRNA has been formed, an enzyme 'zips up' the DNA molecule again.

5 Pieces of the mRNA, called **introns**, are spliced out of the mRNA. These sections are not required to make the final protein.

6 The mRNA leaves the nucleus via a pore in the nuclear envelope.

Translation

This stage of protein synthesis happens in the cytoplasm. Another kind of RNA, called **transfer RNA (tRNA)** is involved. Look at Figure 2.

You will see that tRNA has three unpaired bases at the bottom of the molecule, called an **anticodon**. The anticodon determines the kind of amino acid that the tRNA will attach to. There are at least 60 different tRNA molecules, each with a different anticodon. The amino acid attaches at the top of the tRNA molecule, where there are three unpaired bases for the amino acid to attach to.

Figure 3 shows the stages involved in making a polypeptide.

1 A ribosome attaches to the start of the mRNA molecule. Three bases on the mRNA form a **codon** that codes for one amino acid. Two codons fit into the ribosome at any one time.

2 The tRNA molecules with the complementary anticodons to the mRNA codons enter the ribosome, and bind by complementary base-pairing. They bring their specific amino acids.

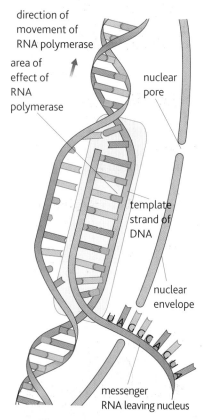

direction of movement of RNA polymerase

area of effect of RNA polymerase

nuclear pore

template strand of DNA

nuclear envelope

messenger RNA leaving nucleus

Figure 1 *Transcription*

3   The two amino acids carried by the tRNA molecules come very close to each other. They join to each other by a **peptide bond**.

4   The first tRNA is now released from the amino acid. It leaves the ribosome and can now bring another amino acid of the same kind to the ribosome.

5   The ribosome moves along the mRNA and a new tRNA enters the ribosome, bringing its specific amino acid.

6   This continues until the ribosome reaches a **stop** or **non-sense codon**. This does not code for any amino acid. It acts like a 'full stop'. The ribosome, the last tRNA and the polypeptide chain all separate from each other.

The genetic code

Look at Table 1. This shows all the codons in mRNA and the amino acids they code for.

Table 1 *The genetic code. The base sequences shown are those on mRNA*

First position	Second position				Third position
	U	C	A	G	
U	Phe	Ser	Tyr	Cys	U
	Phe	Ser	Tyr	Cys	C
	Leu	Ser	Stop	Stop	A
	Leu	Ser	Stop	Trp	G
C	Leu	Pro	His	Arg	U
	Leu	Pro	His	Arg	C
	Leu	Pro	Gln	Arg	A
	Leu	Pro	Gln	Arg	G
A	Ile	Thr	Asn	Ser	U
	Ile	Thr	Asn	Ser	C
	Ile	Thr	Lys	Arg	A
	Met	Thr	Lys	Arg	G
G	Val	Ala	Asp	Gly	U
	Val	Ala	Asp	Gly	C
	Val	Ala	Glu	Gly	A
	Val	Ala	Glu	Gly	G

The genetic code is said to be **degenerate**. This means that there is more than one codon for most amino acids. For example, leucine has six codons while glycine has four.

The genetic code is also **non-overlapping**. This means that each base is part of only one codon. So the sequence ACGCUA is read as ACG–CUA, not ACG–CGC–GCU, etc.

It is also a **universal** code. The code is the same in all organisms, whether humans or bacteria. This is useful in genetic engineering, which you will learn more about in Chapter 4.

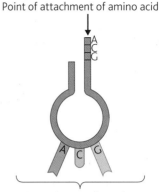

Point of attachment of amino acid

Anticodon – this sequence of ACG means that the amino acid cysteine will attach to the other end of this tRNA molecule. This anticodon will combine with the codon UGC on a mRNA molecule during the formation of a polypeptide. The mRNA codon UGC therefore translates into the amino acid cysteine.

Figure 2 *The structure of tRNA*

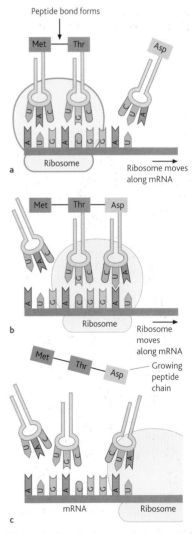

Figure 3 *Translation*

Hint

Remember the base-pairing rules that you learned at AS level, and remember the differences between DNA and RNA. Cytosine pairs with guanine, and thymine in DNA pairs with adenine in RNA. However, there is no thymine in RNA, so adenine in DNA pairs with uracil in RNA.

Hint

Remember that it is new RNA *nucleotides* that pair with the exposed bases on DNA – not RNA bases.

Applications and How science works

Genetic code discoveries

In 1955 the British scientist, Francis Crick, who was the co-discoverer of the structure of DNA, proposed his 'Adapter Hypothesis'. He believed that some structure, not yet discovered, carried amino acids and put them together in the right order, according to the base sequence in the nucleic acid strand.

George Gamov, a Russian scientist, proposed that the genetic code must be based on three letters. He concluded this because there are 20 different amino acids used to make proteins. There are four bases in DNA. If the code was based on two bases, there would only be 4^2 combinations, which is 16 – not enough to code for 20 amino acids. However, if the code was based on triplets, there would be 4^3 combinations, which is 64 – more than enough to code for all the amino acids. The fact that codons did consist of three DNA bases was first demonstrated in an experiment in 1961 by Francis Crick and Sidney Brenner, a South African scientist.

Two scientists working in the USA, Marshall W. Nirenberg and Johann H. Matthaei, tried to work out the genetic code. They ground up bacterial cells so that there were no cells present, but the mixture contained everything needed for protein synthesis, such as ribosomes. They put this 'cell-free' mixture in a test tube, then added specific sequences of mRNA and amino acids. In the first experiment they used messenger RNA consisting only of uracil nucleotides. They found that this produced a protein made only of the amino acid phenylalanine. Next, they worked out that CCC was the code for the amino acid proline.

Another US scientist, Har Gobind Khorana, made more synthetic RNA molecules. The first one he made was a strand repeating the two nucleotides UCUCUC. This produced a strand of amino acids, reading serine-leucine-serine-leucine alternately. Eventually, by making a series of synthetic mRNA molecules, the whole genetic code was worked out.

Robert Holley, also a scientist in the US, discovered transfer RNA. In 1965 he worked out its structure. He realised that tRNA was the unknown molecule that Crick had suggested in his 'Adapter Hypothesis' 10 years earlier.

In 1968, Nirenberg, Khorana and Holley were awarded the Nobel Prize in Physiology or Medicine.

1 What would be in the 'cell-free' extract, apart from ribosomes, that is necessary for protein synthesis?

Summary questions

1 The DNA base sequence on the template strand of a piece of DNA reads: CGGTAACTG. Write down the base sequence of the mRNA formed from it.

2 What kind of reaction occurs when two amino acids join together?

3 Use Table 1 to give the tRNA anticodon for tryptophan (Trp).

4 Give **two** differences between DNA replication and transcription.

3.2 Gene regulation

Learning objectives:

- How is the production of proteins controlled?
- Are genes active all the time?

Specification reference: 3.4.3

The proteins made in protein synthesis are very important for cell functions.

- They may be **structural proteins**, such as keratin in hair, or collagen in bone, the walls of blood vessels and connective tissue.
- They may be proteins in cell membranes, such as the antigens of the ABO blood groups, the CFTR protein that transports chloride ions out of the cell, or **receptors** in cell membranes that respond to specific molecules, such as hormones. You will learn more about this later.
- They may be proteins with specific biochemical functions, such as haemoglobin that carries oxygen in red blood cells.
- They may be **enzymes** that control specific reactions, such as the reactions in respiration, or RNA polymerase in protein synthesis.

It is important that genes are regulated. This is because some genes are only needed in certain cells. For example, haemoglobin does not need to be made in a bone cell. Also, some genes are only required at a certain stage in life. For example, there are genes required in embryonic development that are not required after a certain stage of embryonic development is reached. Furthermore, proteins do not need to be made all the time. It is a waste of energy and resources for a gene to be active all the time. Therefore, genes can be 'switched' on and off.

Testosterone

You learned in Topic 2.1 that testosterone is the main male sex hormone. It is responsible for most of the male secondary sex characteristics. Testosterone circulates in the blood, so it is in contact with every cell in the body. However, it has its effects only on certain **target cells**.

Testosterone is a steroid hormone so it is related to lipids. This means that it can move relatively easily across cell membranes. You can see this in Figure 1.

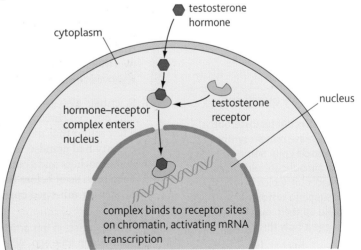

Figure 1 *How testosterone regulates transcription*

- Inside its target cells it combines with a receptor molecule in the cytoplasm. This receptor molecule is only found inside these cells.

Link

Remember from *AS Human Biology*, Topic 8.1, that DNA in our chromosomes is wound around proteins called histones. This complex of DNA and histone proteins is called **chromatin**.

■ The hormone–receptor complex travels into the cell nucleus and binds to another receptor on the chromatin.

■ This stimulates specific genes to be transcribed into mRNA.

■ Acetylation and methylation

Acetyl groups may be added to histone proteins to remove positive charges. This reduces the affinity between the histone proteins and the DNA. As a result, the enzyme RNA polymerase and other factors needed for transcription can bind to the DNA more easily. Therefore, in most cases:

■ adding acetyl groups to the histone proteins stimulates transcription

■ removing acetyl groups from the histone proteins represses transcription.

DNA methylation occurs when methyl groups ($-CH_3$) are added to cytosine bases on the DNA. This represses transcription of the DNA. The pattern of methylation can be passed on to daughter cells after cell division. This is an important factor in epigenetics – which you will remember is the term used for inherited changes in the DNA that do not involve a change in the DNA base sequence. Methylation is an important mechanism for 'switching' genes on and off during embryonic development, when cells are differentiating.

You will remember from Topic 2.15 that Prader-Willi syndrome is inherited when a defective chromosome 15 is inherited from the father, but not when the same chromosome is inherited from the mother. This happens because methyl groups are added to some genes depending on whether they come from the mother or the father. Methylated genes are effectively 'switched off'.

How science works

Men's bad habits could affect the health of their sons

Scientists have found the smoking and nutritional habits of men early in life may have an affect on the health of their sons and grandsons. This could be due to epigenetic modifications.

A study was conducted using British couples who had babies in the early 1990s. Some of the fathers in the study had started smoking before reaching puberty. Prior to puberty the body would have been more sensitive to environmental stress. The scientists found that men who started smoking before puberty had sons who were overweight at aged 9. This was not the case for daughters. Data was removed for fathers who smoked at the time the baby was conceived.

Scientists also looked at data from a remote area in Europe. The data covered three generations and assessed the diets of the people. They found that the grandfathers who were short of food prior to puberty had grandsons who lived longer. This was also the case for grandmothers and granddaughters. Trans-generational effects like this have been seen before in women, but never before in men. Scientists think that inherited epigenetic changes could be the cause of disease.

1   The scientists in this article were looking at historical data and surveys done some time ago. What are the advantages and disadvantages of this approach?

2   Do you think it is likely that the recent increases in obesity and diabetes are the result of inherited epigenetic changes? Give reasons for your answer.

Summary questions

1   One effect of testosterone is to increase protein synthesis in muscle cells. Some athletes use anabolic steroids, such as stanozolol, which is similar in shape to testosterone, to build up their muscle tissue. Suggest how this works.

2   It is possible for the DNA of two monozygotic twins to be identical genetically but not epigenetically. Explain how.

3.3   Genes and cancer

Learning objectives:

■ What is the difference
  between a benign and a
  malignant tumour?

■ How can genes cause cancer?

■ How can environmental
  factors increase the chances
  of developing cancer?

Specification reference: 3.4.3

You will recall from AS level and from Topic 2.3 that cancer may develop when mutations occur in proto-oncogenes and in tumour suppressor genes. These genes normally regulate cell division, but the mutated genes lead to uncontrolled cell division.

■ Tumour suppressor genes

Tumour suppressor genes cause a tumour to form when they are inactivated, or 'turned off'.

■ Some tumour suppressor genes help to control cell division. Because our cells are diploid, we have two copies of each gene. A person may inherit a mutated tumour suppressor gene, but have a normal allele as well. In other words, they may be heterozygous for this tumour suppressor gene. As long as they have a normal allele no cancer develops. However, when the normal allele mutates, cancer can develop.

■ Some tumour suppressor genes code for proteins that 'proofread' DNA and repair any errors that may happen when the DNA replicates. If these tumour suppressor genes mutate the proteins are not produced. This means that mutations in the DNA, including mutations in tumour suppressor genes and oncogenes, can be left uncorrected. This can allow cancer to develop.

■ Some tumour suppressor genes cause cells with badly damaged DNA to 'commit suicide'. The term for this is **apoptosis**, or programmed cell death. An example of this kind of tumour suppressor gene is p53. If the p53 gene is not working properly, cells with damaged DNA may continue to grow and cancer may develop as a result.

■ Benign and malignant tumours

There are two different types of tumour that can form. Benign tumours will not spread to surrounding cell types. They can, however, grow very large. As a benign tumour grows, the pressure it puts on surrounding cells can cause damage.

Malignant tumours cause a lot more damage than benign tumours. A malignant tumour will invade the surrounding cells and can spread very quickly. Only malignant tumours are considered to be cancerous.

■ Proto-oncogenes

Proto-oncogenes normally control cell division, stimulating them to divide only when necessary. If a proto-oncogene mutates into an oncogene, it can lead to uncontrolled cell division, or cancer.

■ Some proto-oncogenes code for proteins, called growth factors, which stimulate cells to grow. If they mutate into oncogenes, excessive amounts of these growth factors may be produced.

■ Some proto-oncogenes code for receptors on cell membranes into which these growth factors fit. If they mutate, they may become oncogenes that produce cell membrane receptors that are always 'on' and stimulate cell division, even when the growth factor is not present.

- Some proto-oncogenes act as signal transducers, which are intermediate between the growth factor receptor on the plasma membrane and the DNA in the cell nucleus. These are normally only active when a growth factor has stimulated the cell. If the proto-oncogene mutates, the signal transducer protein is permanently 'switched on', stimulating cell division. A well-known oncogene of this type is known as *ras*. This is found in many different kinds of cancer.

- Some proto-oncogenes code for transcription factors that act on DNA and control which genes are transcribed into RNA to synthesise proteins. If the proto-oncogene mutates, these transcription factors are produced in large amounts, stimulating cell division.

- Other proto-oncogenes code for proteins that cause a cell to commit apoptosis when it becomes abnormal. If the proto-oncogene mutates, the protein produced does not have this effect. Instead, abnormal cells may be allowed to grow and cancer may develop.

Methylation and cancer

Gene mutations cause changes in the DNA sequences of proto-oncogenes and tumour supressor genes. However, more gradual changes may take place. You learned in Topic 3.2 that increasing methylation of DNA represses transcription of genes.

Genes have a **promoter region** which is the part that RNA polymerase attaches to. This also determines which strand of the DNA becomes the template for mRNA transcription. If the promoter region has too many methyl groups added, the gene cannot be transcribed. This kind of change happens gradually, and happens at a different rate in different cells. Scientists believe that methylation may be increased by certain environmental factors, such as smoking.

Oestrogen and cancer

About 60% of breast cancers have oestrogen receptor proteins. Oestrogen is a steroid hormone, so it stimulates gene transcription in the same way as testosterone. Oestrogen passes through the plasma membrane and binds to a specific oestrogen receptor protein inside the cell cytoplasm. This binds to another molecule called a **coregulator** and the whole complex binds to the chromatin. This stimulates gene transcription.

Breast cancer that is 'oestrogen positive' and contains oestrogen receptor proteins can be treated by **tamoxifen**. You can see how tamoxifen works in Figure 1.

Tamoxifen is similar in shape to oestrogen, so it binds to the oestrogen receptor protein. However, the coregulator does not bind to it. Although the tamoxifen/oestrogen receptor complex binds to the chromatin, it cannot cause transcription without the coregulator being present.

Tamoxifen is very effective in treating oestrogen-positive breast cancer in the short term. However, after it has been used for 2 or 3 years the cancer often develops resistance to it. After that, the woman may be prescribed a different drug that reduces the amount of oestrogen circulating in her body.

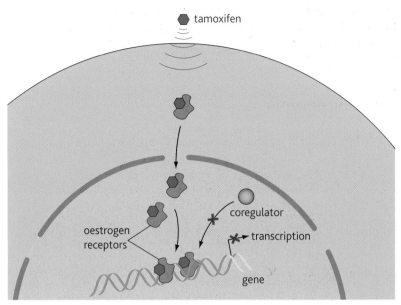

Figure 1 *How tamoxifen works*

How science works

Alcohol and breast cancer

Look at Figure 2. This shows the incidence of breast cancer per 100 women in developed countries, according to the number of alcoholic drinks consumed each day.

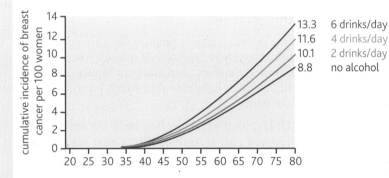

13.3	6 drinks/day
11.6	4 drinks/day
10.1	2 drinks/day
8.8	no alcohol

Figure 2 *The incidence of breast cancer according to the number of alcoholic drinks consumed each day. Data from Cancer Research UK*

1. Describe the pattern shown by this graph.

2. Calculate the percentage increase in risk of developing breast cancer by the age of 80 for a woman consuming six alcoholic drinks a day compared to a woman who does not drink alcohol. Show your working.

3. Does this data show that drinking alcohol causes breast cancer? Give reasons for your answer.

Summary questions

1. Some women have breast cancer cells which have a receptor protein called HER2 on their cell membranes. A growth factor naturally present in the body (called human epidermal growth factor) fits into the HER2 receptor and stimulates the cells to divide. The cancer drug, herceptin, attaches to the HER2 protein.

   a Explain how herceptin is effective against breast cancer in these women.

   b Explain why herceptin cannot be used to treat all cases of breast cancer.

3.4 Enzymes in cells

Learning objectives:

- How do enzymes control cellular reactions?

- How can a reaction product inhibit an enzyme?

- What is an uncompetitive inhibitor?

Specification reference: 3.4.3

Link

It would be helpful to look back at Topic 7.4 of *AS Human Biology* to remind yourself of how genes control characteristics.

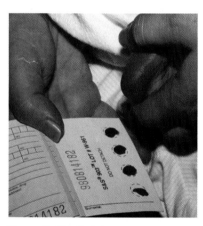

Figure 2 *The Guthrie test*

AQA Examiner's tip

You do not need to learn the specific examples on this page. You simply need to understand that enzymes control metabolic pathways, and that the absence of an enzyme may result in a specific condition.

Missing enzymes

You will recall from AS level that many genes code for proteins that act as enzymes. These enzymes have specifically shaped active sites and control metabolic pathways within cells. Figure 1 shows one of these metabolic pathways. If certain enzymes are faulty as the result of a mutation, then different conditions can result.

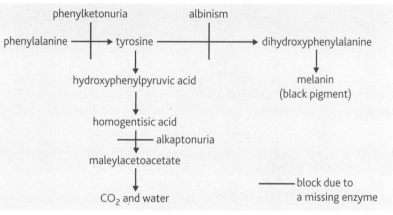

Figure 1 *A metabolic pathway that leads to phenylketonuria, alkaptonuria and albinism*

Alkaptonuria results when a person is unable to break down homogentisic acid. Homogentisic acid builds up in the tissues and urine. It is black, making the person's urine dark in colour. The black colour can also be seen in the sclera of the eye.

Albinism results when a person does not produce the enzyme that converts tyrosine to dihydroxyphenylalanine. As a result, there is insufficient or no melanin in the eyes, skin and hair. Melanin is a dark pigment so people with albinism have pale skin which burns easily in the sun, and virtually white hair.

Phenylketonuria (PKU) is caused when a person lacks the enzyme that converts the amino acid phenylalanine into tyrosine. As a result, phenylalanine build ups in the blood and other body tissues, particularly the brain, where it can cause severe and irreversible damage. If PKU is not detected early, progressive developmental delay and severe learning difficulties can occur. Other problems may include epilepsy and eczema. There is no cure for PKU but the problems associated with the disease can be prevented by giving the person a diet that is very low in phenylalanine. In the UK, all newborn babies are screened for PKU using a heel-prick blood sample.

How science works

The Guthrie test

The Guthrie test was developed in 1962 by an American doctor, Robert Guthrie, to screen for PKU. The test detects whether a blood sample has an unusually high concentration of phenylalanine in it. A sample of blood is collected on a piece of filter paper. You can see this in Figure 2.

An agar gel plate is covered in a suspension of the bacteria, *Bacillus subtilis*. A compound called B-2-thienylalanine is also placed on the agar plate. This substance normally inhibits the growth of *Bacillus subtilis*. However, if extra phenylalanine is present, the bacteria are able to grow. Discs containing blood samples are placed on the agar plate containing the bacteria. The plates are incubated for a day. If a disc contains blood from a person without PKU there will be a region around the filter paper disc where bacteria have not grown. However, if the disc contains blood from a person with PKU the bacteria will be able to grow close to the disc.

1 Give two examples of aseptic technique that would need to be observed while carrying out the laboratory test.

2 Explain why bacteria will grow close to a disc containing blood from a person with PKU, but will not grow close to a disc containing blood from a healthy person.

End-product inhibition

As already mentioned, metabolic pathways have many steps. Each step is controlled by a single enzyme. These pathways are regulated by **end-product inhibition** (Figure 3). This means that when the final product in the pathway has been produced in sufficient quantity, it inhibits an enzyme early in the pathway. In this way, a product can control the rate at which it is synthesised.

An example of this is the pathway by which bacteria synthesise the amino acid isoleucine from another amino acid, threonine. This is shown in Figure 4.

How a product can inhibit an enzyme

Figure 5 shows how a product may inhibit an enzyme. It acts as a **non-competitive inhibitor**. It fits into the enzyme's tertiary structure, but not the active site. However, in doing this, it changes the shape of the active site so that the enzyme can no longer function. When there is a shortage of the product again, it is released from the site on the enzyme. The enzyme returns to its normal shape, and can continue catalysing as normal.

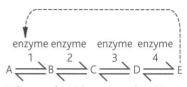

When product E has accumulated in sufficient quantity, it inhibits enzyme 1. This prevents more product E being made

Figure 3 *End-product inhibition*

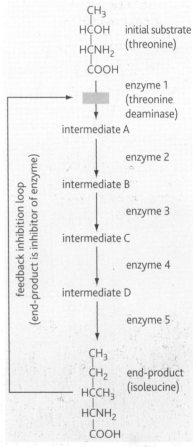

Figure 4 *The pathway that produces isoleucine from threonine*

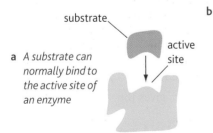

a *A substrate can normally bind to the active site of an enzyme*

substrate

active site

b *A non-competitive inhibitor binds to the enzyme at a location away from the active site, but alters the tertiary structure of the enzyme so that the active site is no longer fully functional*

non-competitive inhibitor

Figure 5 *Non-competitive inhibition*

Summary questions

1 PKU is caused by a recessive allele. Two parents, neither of whom has PKU, have a baby with PKU. What is the chance that their next baby will have PKU?

2 Explain the advantage to the cell of end-product inhibition.

1   (a)   Copy and complete **Table 1** to show **three** differences between the structure of DNA and the structure of RNA.   *(3 marks)*

Table 1

Structure of DNA	Structure of RNA
1	1
2	2
3	3

(b)   **Table 2** shows the genetic code as triplets of bases found in mRNA;

Table 2

	U	C	A	G	
U	UUU ⎫ Phe UUC ⎭ UUA ⎫ Leu UUG ⎭	UCU ⎫ UCC ⎬ Ser UCA ⎪ UCG ⎭	UAU ⎫ Phe UAC ⎭ UAA Stop UAG Stop	UGU ⎫ Cys UGC ⎭ UGA Stop UGG Trp	U C A G
C	CUU ⎫ CUC ⎬ Ile CUA ⎪ CUG ⎭	CCU ⎫ CCC ⎬ Pro CCA ⎪ CCG ⎭	CAU ⎫ His CAC ⎭ CAA ⎫ Glu CAG ⎭	CGU ⎫ CGC ⎬ Arg CGA ⎪ CGG ⎭	U C A G
A	AUU ⎫ AUC ⎬ Ile AUA ⎪ AUG Met	ACU ⎫ ACC ⎬ Thr ACA ⎪ ACG ⎭	AAU ⎫ Ser AAC ⎭ AAA ⎫ Glu AAG ⎭	AGU ⎫ Ser AGC ⎭ AGA ⎫ Arg AGG ⎭	U C A G
G	GUU ⎫ GUC ⎬ Val GUA ⎪ GUG ⎭	GCU ⎫ GCC ⎬ Ala GCA ⎪ GCG ⎭	GAU ⎫ Asp GAC ⎭ GAA ⎫ Glu GAG ⎭	GGU ⎫ GGC ⎬ Gly GGA ⎪ GGG ⎭	U C A G

(i)   What is the sequence of amino acids in the peptide coded for by the following length of mRNA?

     A   G   A   C   C   G   G   C   U   G   G   A

(ii)   What is the sequence of bases in DNA which, when transcribed, gives the above length of mRNA?

(iii)   A mutation occurred in this DNA strand so that adenine in the DNA was replaced by guanine. Using **Table 2**, explain why this mutation had no harmful effect on the organism.   *(3 marks)*

AQA, 2000

2   **Figure 1** shows part of a molecule of mRNA bound to a ribosome.

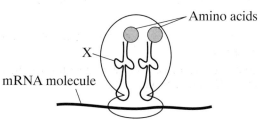

Amino acids

X

mRNA molecule

Figure 1

(a) (i) Molecule **X** carries an amino acid molecule to the ribosome. Name molecule **X**.

   (ii) The mRNA codon below molecule **X** is AUC. Give the sequence of bases in molecule **X** which would bind to the MRNA at this site. *(2 marks)*

(b) Explain the roles of mRNA and molecule **X** in producing a protein. *(3 marks)*

AQA, 2000

3   (a) **Figure 2** shows a molecule of transfer RNA (tRNA).

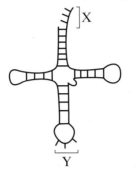

Figure 2

Describe the function of

(i)   part **X**;

(ii) part **Y**. *(2 marks)*

(b) Explain why the genetic code is described as

(i)   non-overlapping;

(ii) degenerate. *(3 marks)*

AQA, 2003

4   **Figure 3** shows the mean daily fat consumption for a number of different countries and the death rate from breast cancer in those countries.

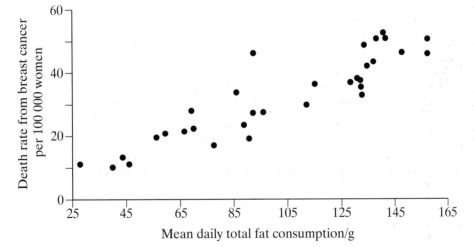

Figure 3

(a) (i) Describe the relationship shown by **Figure 3**.

   (ii) Explain why the death rate from breast cancer is given per 100 000 women. *(3 marks)*

(b) Do these data show that high fat consumption causes breast cancer? Give reasons for your answer. *(3 marks)*

(c) Tumour suppressor genes and oncogenes may be involved in the development of tumours. Describe how. *(6 marks)*

New genes for old

4.1 PCR

Learning objectives:

■ How can copies of DNA be made?

Specification reference: 3.4.4

Link

You may find it helpful to look back at Topic 7.5 in *AS Human Biology* to remind yourself about DNA replication.

At AS level you learned about the structure of DNA and the complementary base-pairing that holds the two helices together. You also learned how DNA replication takes place. You will remember that DNA copies itself with great accuracy, mainly because of the base-pairing rules, but also because of the 'proofreading' carried out by the enzyme DNA polymerase.

PCR stands for the **polymerase chain reaction**. It is a method of replicating many copies of DNA in a test tube. This is very useful in research when a sample of DNA may be extremely tiny. It is usually carried out in very small tubes in a machine called a **thermocycler**. You can see a thermocycler in Figure 1.

Figure 1 *A thermocycler*

PCR is very similar to semi-conservative DNA replication inside cells, except that heat is used to break open the hydrogen bonds between the bases in the DNA double helix. It uses **primers** to mark the beginning and end of the section of DNA to be copied. Primers are short, single-stranded nucleic acid sequences, about 20 nucleotides long. It also uses DNA polymerase. However, because heat is needed to split the DNA into single strands, a special heat-stable DNA polymerase is used. This is obtained from a bacterium that lives in hot springs.

To carry out PCR, the sample of DNA to be copied is placed in a test tube, together with:

- DNA polymerase enzyme
- primers
- DNA nucleotides.

The process of PCR is shown in Figure 2.

1 First, the mixture is heated to 93 °C. This breaks the hydrogen bonds between the two strands of DNA and makes the DNA single-stranded.

2 The DNA is cooled to 55 °C. This allows the primers to attach to the ends of the DNA sequence that needs to be copied.

3 The DNA is then heated to 72 °C (the optimum temperature for the DNA polymerase enzyme used). The enzyme joins new nucleotides on to the DNA strands, producing two identical molecules from the original DNA.

- The whole cycle takes about 2 minutes. The cycle is repeated as many times as necessary.

How science works

Discovering PCR

One Friday night in April 1983, Kary Mullis was driving to his weekend cabin. He was working for the Cetus Corporation in California. While he was driving along the road he had a brainwave. The thought struck him that a pair of primers and DNA polymerase could be used to make unlimited copies of a sequence of DNA in a test tube. Before this, DNA could only be copied using cells. This method gave a limited number of copies and took a great deal of effort.

Kary Mullis named his new method the 'polymerase chain reaction'.

At first, he had to add DNA polymerase every thermal cycle because the polymerase enzyme was denatured during the heating stage. In 1986, he started to use DNA polymerase from *Thermus aquaticus* (called Taq polymerase). This made the process much cheaper and faster.

Kary Mullis was awarded the Nobel Prize for chemistry in 1993.

How science works

Bio-prospecting

Kary Mullis was working for the Cetus Corporation when he had the brainwave to use Taq polymerase. The Taq polymerase came from a bacterium called *Thermus aquaticus* that lived in a hot spring in the Yellowstone National Park. This bacterium was chosen because the polymerase within it would need to be heat resistant due to the temperatures of the spring. Taq was first identified in 1966 and since then has been very useful in scientific research. Despite this, no royalties have been paid to Yellowstone National Park.

Officials working for the National Park Service feel very strongly that they should not have been exploited in this way. They think there should be a 'benefit sharing' policy put in place that means National Parks would benefit from any commercially valuable

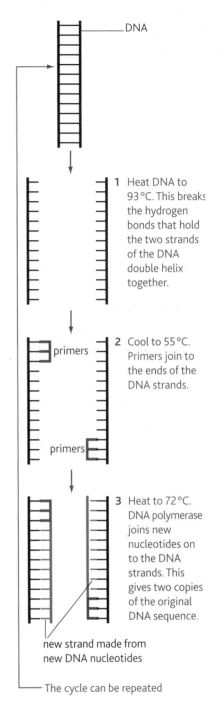

1 Heat DNA to 93 °C. This breaks the hydrogen bonds that hold the two strands of the DNA double helix together.

2 Cool to 55 °C. Primers join to the ends of the DNA strands.

3 Heat to 72 °C. DNA polymerase joins new nucleotides on to the DNA strands. This gives two copies of the original DNA sequence.

new strand made from new DNA nucleotides

The cycle can be repeated

Figure 2 *PCR. Note that the DNA sequence to be copied, and the primers, are both shown much shorter than real size*

Figure 3 *US microbiologist Tom Brock taking a sample from near a hot spring, hopefully containing* Thermus aquaticus. *Brock and a colleague discovered this bacterium in 1969*

discoveries on their land. Any living organisms found in the parks cannot be patented, but results of research can be. The commercial use of specimens from the park is prohibited.

The most active area for commercialisation is 'bio-prospecting'. This is the study of microbes for enzymes that have industrial applications. Most of this research has focused on 'extremophiles' – microbes that have evolved in extreme environments like the hot springs of Yellowstone.

1 Should the Cetus corporation pay a proportion of the profits they have made from Taq polymerase to the Yellowstone National Park? Give reasons for your answer.

2 Suggest how organisms such as *Thermus aquaticus* have evolved to live in such a hot environment.

3 Taq polymerase, like other enzymes, is a protein. Suggest how the bonds in its tertiary structure might differ from DNA polymerase found in a human cell.

4 Enzymes found in 'extremophiles' are very useful in many industrial processes. Suggest why.

■ Uses of PCR

PCR has many uses, including:

- making many copies of the DNA extracted from a single embryo cell, so that enough DNA can be obtained for genetic analysis
- making many copies of the tiny quantity of DNA that can be extracted from fossils and mummified remains
- making many copies of DNA found in small samples at a crime scene, e.g. vaginal swabs from rape victims or spots of blood found at a murder scene. The DNA can then be analysed using DNA fingerprinting.

Summary questions

1 Suggest how the two DNA strands are separated in a human cell when DNA replication takes place.

2 Explain why two different primers are needed.

3 Starting from just one DNA molecule, how many copies will be present after eight cycles of PCR? Show your working.

4 Explain why the DNA polymerase used in PCR needs to be heat-stable.

5 When PCR is carried out on a forensic sample, human DNA may be mixed with bacterial DNA. Explain how primers ensure that only the human DNA is multiplied.

6 The DNA code is universal. Explain how using DNA polymerase from a bacterium is evidence for this statement.

4.2 Making recombinant DNA

Learning objectives:

- What is genetic engineering?
- How are genetically modified organisms produced?

Specification reference: 3.4.4

Genetic engineering is a collection of techniques by which genes are altered or transferred from one organism to another. The modified DNA is called **recombinant DNA**. Genetic engineering involves:

- obtaining or isolating the gene to be transferred
- transferring the gene into the new cells
- identifying the modified cells.

Isolation of a gene

One useful application of genetic engineering is in the production of human insulin to treat Type 1 diabetes. Before genetic engineering was available insulin from pigs was used. This differs from human insulin by only one amino acid. However, this kind of insulin sometimes causes an immune response in diabetics. Now, the gene for human insulin can be inserted into bacterial cells. The bacteria are grown in large numbers and the insulin is extracted. Because this is human insulin, which is identical in every person, diabetics are unlikely to show an immune response.

First, the gene for human insulin needs to be obtained. There are three different ways of doing this:

The gene may be cut out of human DNA. First of all, the gene for human insulin can be located in human DNA using a gene probe. You will learn more about this in Topic 4.3. Then the DNA is cut using a **restriction enzyme**. These enzymes are naturally found in bacteria. They cut DNA whenever they find a specific base sequence. Many restriction endonucleases cut DNA, leaving a staggered cut or **sticky ends**. You can see this in Figure 1.

AQA Examiner's tip

Reverse transcriptase is used to make a DNA copy of RNA. Be very careful with how you express this – many students incorrectly say that reverse transcriptase 'turns RNA into DNA'.

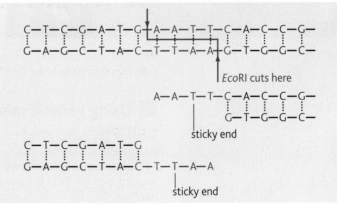

Figure 1 *EcoR1 is a restriction enzyme that recognises the base sequence shown in red. It cuts the DNA as shown, leaving overhanging ends called sticky ends*

Hint

Reverse transcriptase is so-called because it carries out the reverse of transcription. You will recall that in transcription, a DNA template is used to make a complementary RNA copy. Reverse transcriptase does the opposite. It uses an RNA template to synthesise a complementary DNA copy.

The gene may also be made from mRNA. For example, the cells in the pancreas that produce insulin have a large amount of mRNA coding for insulin. This mRNA can be isolated. You may remember from AS that HIV, the virus that causes AIDS, contains an enzyme called **reverse transcriptase**. It uses this enzyme to make a DNA copy from its RNA. Genetic engineers use reverse transcriptase to make a single-stranded DNA copy of the mRNA from the pancreas cells. This single-stranded DNA is called cDNA (which stands for 'copy DNA'). The cDNA is then made into double-stranded DNA.

Figure 2 *DNA synthesising machine*

■ You will remember from Topic 3.1 that three bases in DNA code for one amino acid. We know the triplets that code for each amino acid. We also know the sequence of amino acids in human insulin. Therefore, we can work out the DNA base sequence that codes for human insulin. This DNA sequence can then be made using a machine that synthesises DNA, or 'gene machine'. You can see a 'gene machine' in Figure 2.

Sometimes the polymerase chain reaction is used to make more copies of the gene.

■ Inserting DNA into new cells

A **vector** is used to transfer the isolated gene into a new host cell. There are several different kinds of vector, including viruses, but a common type of vector is a **plasmid**. You may remember from AS level that a plasmid is a circular piece of DNA found in many bacteria. The DNA in the plasmid is cut open using a restriction enzyme. Sometimes this is the same one that was used to isolate the gene. If not, nucleotides are added to the cut plasmid and to the isolated gene so that they have matching sticky ends.

The gene is inserted into the cut plasmid using another enzyme, **DNA ligase**. You can see this in Figure 3.

Bacterial cells and plasmids are mixed together. The bacteria are treated so that many of them take up a modified plasmid, or **recombinant plasmid**.

■ Using genetic markers

Some of the bacterial cells do not take up a plasmid, or take up a non-recombinant plasmid. Scientists hoping to produce human insulin want to make sure they only have bacterial cells that contain the gene for human insulin. It is not possible to tell by looking at the cells which ones contain the insulin gene. However, you may notice in Figure 3 that the plasmid used contains an antibiotic resistance gene. This gene gives resistance to a specific antibiotic, such as tetracycline. This is called a **genetic marker**. The scientists can mix the bacteria with the antibiotic that the marker gene gives resistance to. This will kill all the bacteria that do not contain a plasmid. Marker genes may also code for other characteristics that are easy to identify, such as production of a fluorescent protein or a specific enzyme.

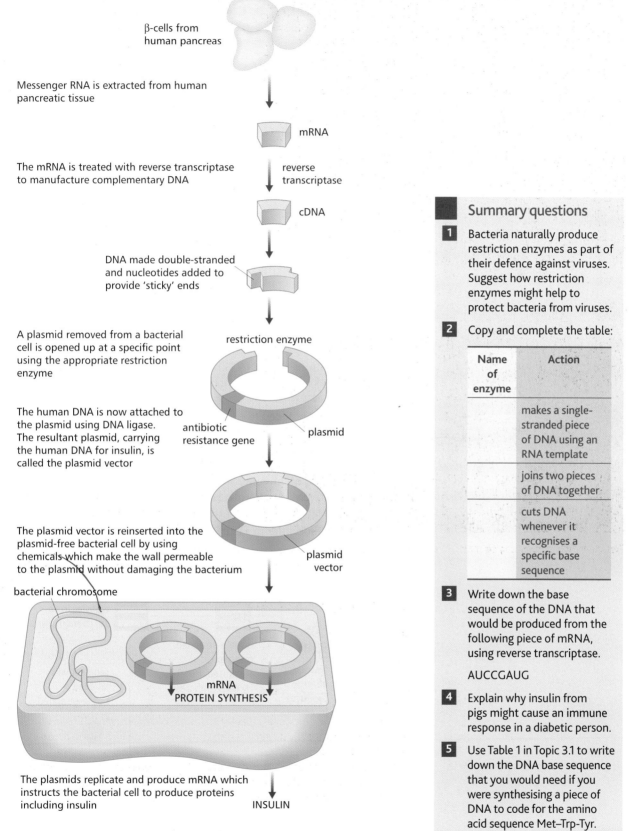

β-cells from
human pancreas

Messenger RNA is extracted from human
pancreatic tissue

mRNA

The mRNA is treated with reverse transcriptase
to manufacture complementary DNA

reverse
transcriptase

cDNA

DNA made double-stranded
and nucleotides added to
provide 'sticky' ends

A plasmid removed from a bacterial
cell is opened up at a specific point
using the appropriate restriction
enzyme

restriction enzyme

The human DNA is now attached to
the plasmid using DNA ligase.
The resultant plasmid, carrying
the human DNA for insulin, is
called the plasmid vector

antibiotic
resistance gene

plasmid

The plasmid vector is reinserted into the
plasmid-free bacterial cell by using
chemicals which make the wall permeable
to the plasmid without damaging the bacterium

plasmid
vector

bacterial chromosome

mRNA
PROTEIN SYNTHESIS

The plasmids replicate and produce mRNA which
instructs the bacterial cell to produce proteins
including insulin

INSULIN

Figure 3 *Summary showing one way of using genetic engineering to make human insulin*

Summary questions

1. Bacteria naturally produce restriction enzymes as part of their defence against viruses. Suggest how restriction enzymes might help to protect bacteria from viruses.

2. Copy and complete the table:

Name of enzyme	Action
	makes a single-stranded piece of DNA using an RNA template
	joins two pieces of DNA together
	cuts DNA whenever it recognises a specific base sequence

3. Write down the base sequence of the DNA that would be produced from the following piece of mRNA, using reverse transcriptase.

   AUCCGAUG

4. Explain why insulin from pigs might cause an immune response in a diabetic person.

5. Use Table 1 in Topic 3.1 to write down the DNA base sequence that you would need if you were synthesising a piece of DNA to code for the amino acid sequence Met–Trp–Tyr.

4.3 Gene probes and gene libraries

Learning objectives:

■ What is a gene probe?

■ What is a gene library?

Specification reference: 3.4.4

■ Gene probes

A gene probe is a single-stranded sequence of DNA or RNA that is complementary to the nucleotide sequence of a specific gene. It is specifically labelled, i.e. made radioactive or fluorescent. This means that it will bind to a complementary DNA sequence, and allow this sequence to be identified. For example, it may be used to detect the cystic fibrosis allele in DNA from an embryo. One way of using a DNA probe is as follows:

1 The DNA to be tested is extracted and if a bigger sample is required, PCR is used.

2 The DNA is digested into pieces using a restriction endonuclease.

3 The DNA fragments are separated using a process called **gel electrophoresis**. The DNA is placed in a well in a gel, and an electric current is applied. DNA fragments move towards the positive electrode. The smaller pieces move more quickly through the gel, so the pieces of DNA are separated according to size.

4 The gel containing the DNA is treated with an alkali, so that the DNA becomes single-stranded. The single-stranded DNA is then transferred to a membrane.

5 The gene probe is added to the membrane and time is allowed for it to bind to its complementary sequence.

6 The membrane is washed. If the probe is radioactive, the membrane is placed against a piece of photographic film. If the probe has bound to the DNA on the membrane, 'fogging' will occur on the photographic film. Alternatively, if the probe is fluorescent, it will show up when the membrane is viewed under ultraviolet light.

You can see this procedure in Figure 1.

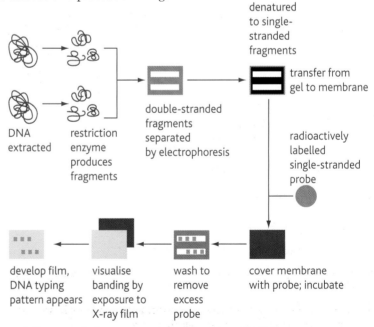

Figure 1 *One way of using a gene probe to detect a specific DNA sequence*

■ Gene libraries

A gene library is a set of thousands of DNA fragments from one organism. The DNA of the organism is cut using a restriction endonuclease. Each fragment is inserted into a plasmid to make a recombinant plasmid. The recombinant plasmids are then inserted into bacterial cells. Alternatively, mRNA is isolated from cells. Reverse transcriptase is used to make a DNA copy of the mRNA. Each piece of DNA is added to a separate plasmid and inserted into a bacterial cell.

Each clone of bacteria is stored in a separate well in a multi-well dish. They may be stored, frozen, for a long time. Later, if a particular gene is required, the DNA of the bacteria can be tested using a gene probe. You can see this in Figure 2.

Another way of locating a desired gene is to test each clone of bacteria using an antibody specific to the protein produced by the gene of interest.

Gene libraries are being used to store genetic material from many different organisms. If the gene library contains all the DNA from the organism, it is said to be a **representative** library. This means that it is highly probable that a specific gene from the organism can be found in at least one of the fragments.

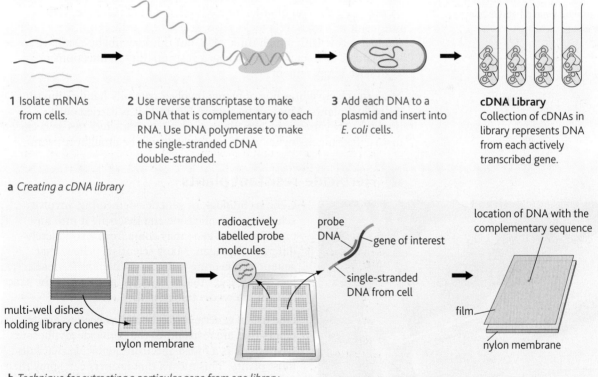

1 Isolate mRNAs from cells.

2 Use reverse transcriptase to make a DNA that is complementary to each RNA. Use DNA polymerase to make the single-stranded cDNA double-stranded.

3 Add each DNA to a plasmid and insert into *E. coli* cells.

cDNA Library
Collection of cDNAs in library represents DNA from each actively transcribed gene.

a *Creating a cDNA library*

multi-well dishes holding library clones

nylon membrane

radioactively labelled probe molecules

probe DNA — gene of interest

single-stranded DNA from cell

location of DNA with the complementary sequence

film

nylon membrane

b *Technique for extracting a particular gene from one library*

Figure 2 *Making a gene library*

Summary questions

1 Why does a gene probe need to be fluorescent or radioactive?

2 Suggest why pieces of DNA move towards the positive electrode in gel electrophoresis.

3 How would an antibody enable a scientist to identify which clone of bacteria contains the gene for a specific protein?

4.4 Selective breeding and gene technology

■ Selective breeding

You will recall from AS level that humans have used selective breeding for thousands of years to domesticate plants and animals. For example, bread wheat was selectively bred from wild grasses, beef and dairy cattle were selectively bred from the wild aurochs and domestic dogs were selectively bred from wolves. In each case, humans chose the organisms with desirable characteristics and used these to breed from. Over hundreds and thousands of years, they produced plants and animals that were very different from their ancestors.

However, selective breeding is a slow process. It is also not reliable, because animals and many crop plants reproduce sexually. For example, a farmer may breed a cow that produces a high yield of milk with a bull whose mother and sisters also produced a high yield of milk. However, he cannot be sure that all the female offspring will produce high milk yields. This is because an organism only passes on half its alleles in each gamete, and the combination of alleles that an individual inherits may not result in a high milk yield. Also, selective breeding in this way may produce a cow with a high milk yield, but this cow might also have some undesirable characteristics, such as a susceptibility to infections or an aggressive temperament.

Modern selective breeding of plants involves carefully hand-pollinating individual plants. This can be seen in Figure 1. It is carried out to produce new varieties of crop plants with such characteristics as better disease resistance, higher yields, or the ability to be drought resistant.

■ Herbicide-resistant plants

Gene technology, or genetic engineering, involves selecting a specific gene and inserting it into an organism. The gene may come from a completely different organism. Many scientists believe that genetic engineering is not very different from selective breeding, except that it is much quicker and the exact gene of interest is transferred.

One use of genetic engineering has been to create soya plants that are resistant to the herbicide glyphosate. Glyphosate is a broad-spectrum herbicide that kills most plants. This means that farmers can spray their field with glyphosate, knowing that it will kill any weeds but leave the crop unharmed. Some scientists think that this is a useful development. They argue that it reduces the use of herbicides as just one kind of herbicide is needed to kill all the weeds. Also, glyphosate is a herbicide that breaks down very quickly in the soil. However, other scientists are

Figure 1 *Modern selective breeding. A biologist is using tweezers to remove part of the female reproductive organ in an organic soya plant*

concerned that pollen from the crop might pollinate weed species. If this happened, the gene for herbicide resistance might transfer into weed species, creating 'superweeds' that are difficult to control. They are also concerned that, when the glyphosate resistance gene is added, it may disrupt another gene that is important for the plant to function properly. There are also concerns that encouraging the use of a broad-spectrum herbicide reduces the plant foods available for non-pest species, thereby reducing biodiversity. Furthermore, the same company manufactures both the genetically modified seeds and the herbicide glyphosate.

Increasing milk yields

Bovine somatotrophin (BST), also called bovine growth hormone, is a protein hormone produced in the pituitary glands of cattle. The hormone can also be produced synthetically. The process of making synthetic BST is rather like the process described in Topic 4.3 for making human insulin. The gene that codes for BST is inserted into a bacterial cell, and then the genetically modified bacteria are grown in large numbers. BST is extracted from the bacterial cells. When this purified hormone is injected into cattle it increases milk production. It is thought to do this by increasing the number of cells in the udder, which is where milk is produced.

The hormone is not approved for use in the EU. One concern is that it increases the incidence of mastitis, an udder infection, and lameness. There are also concerns for animal welfare. Some people are concerned that milk from BST-treated cows might contain hormones that cause harm to humans consuming the milk, but there is no firm evidence for this.

Application and How science works

Milk yields

The table shows the results of three different investigations into the use of BST. The researchers carrying out the studies measured the % increase in milk yield when BST was used, and the % increase in feed efficiency. Feed efficiency is the milk produced, in kg, per kg of feed eaten.

Location	Increase in milk yield/%	Increase in feed efficiency/%
A	21.8	8.2
B	8.3	2.7
C	17.8	9.3

1 Suggest how the increase in milk yield was measured.

2 Explain why it is important to measure feed efficiency.

3 Suggest **two** explanations why the results for the three locations are very different.

Summary questions

1 Give **two** differences between selective breeding and gene technology.

2 Suggest why BST must be given to cows by injection, rather than with their food.

4.5 Ethical issues

Learning objectives:

- What are the ethical issues associated with gene technology?

Specification reference: 3.4.4

Benefit or curse?

Many scientists think that genetic engineering is a useful set of techniques that can be used for both good or bad purposes. In other words, they think that the techniques themselves are potentially useful, but we need to consider carefully what they should be used for.

One major concern at the moment is growing enough food to feed everybody in the world. Climate change means that farmers are experiencing natural disasters such as flooding or drought. Genetic engineering can be used to produce varieties of crops that are resistant to these changes. Intensive farming requires the application of considerable amounts of nitrogen fertiliser, which contributes strongly towards global warming. Genetic engineers are researching the production of crops that do not need fertilisers, by transferring genes for nitrogen fixation into various crop species. Genes for disease resistance can also be added to crops, to reduce the need for herbicides and pesticides. Many people think that genetic engineering is little different from selective breeding, except that it is a more predictable and rapid process. Although it involves transferring genes from one species to another, they point out that different species share a large proportion of their genes anyway.

People who oppose genetic engineering argue that we cannot be sure that there are no long-term health risks. When a new gene is inserted, it may disrupt a regulatory gene, and may lead to genes that are not normally active being expressed. This could have unknown effects, including the production of proteins to which the consumer may be allergic. There are concerns that genes may spread from one species to another, for example, in the pollen of genetically modified (GM) crops. This pollen could transfer the gene to non-GM varieties or even to weed species. Biotechnology companies have tried to overcome this concern by inserting so-called 'terminator' genes into GM crop plants. This means that GM crops would produce sterile seeds. Opponents of genetic engineering also argue that the production of genetically modified organisms mainly increases profits for the large biotechnology companies, without significant benefits for the consumer. They argue that the 'terminator' technology just described increases profits, as farmers have to buy new seeds every year. They believe that genetic engineering is very different in ethical terms from selective breeding, as it transfers genes from one species to another, rather than from one member of a species to another. It is also irreversible: once a gene is transferred, it is very difficult to remove.

Sanctity of the species

Species are the product of millions of years of evolution. Some people argue that it is morally wrong to produce GM organisms that contain genes from other organisms. This is why genetic engineering is tightly regulated, and currently there is international agreement that human cells should not be modified. The reason for this is that we do not wish to introduce new genes into the human genetic line. Many people feel that if it is important to keep the human genetic line unaltered, then we should treat other living organisms in the same way. However, other people argue that we have already altered the genome of many species through selective breeding. In this case, we have selected genes that are already present in the species, rather than introducing new ones. Also, GM crops are made to be sterile, so that they cannot interbreed with wild relatives and transfer their genes. Similarly, GM bacteria are developed so that they could not survive if they 'escaped' from the laboratory.

How science works

Seed patents

A large international agricultural company has recently won a legal battle against a farmer in Canada. The farmer was accused of growing a patented GM rapeseed without paying for it. The farmer claimed that he had not planted the seeds and that they must have grown on his land through natural cross pollination. The court, however, found him guilty of breaching the agricultural company's patent. The farmer was ordered to hand over any of the patented crops or seeds he had left in his possession.

People who oppose GM crops believe that this court action means that all farmers whose crops become accidentally pollinated by GM plants could now have legal action taken against them. They also believe that farmers wishing to market their grain as non-GM would be less likely to complain if their crops became contaminated by GM plants.

Figure 1 *A researcher studying genetically modified rapeseed plants*

1. Is it right for large agricultural companies to be able to patent a variety of crop plant?

2. Was it fair for the farmer to be prosecuted?

AQA Examiner's tip

When you are asked in an exam to evaluate the use of genetic engineering, stick to factual, scientific arguments. Do not write vaguely, such as 'We must not play God' or 'It is unnatural'.

Big business?

Developing genetically modified organisms costs a great deal of money in research and development. Most of these costs are paid by private biotechnology companies, who therefore expect to make a profit out of their investment. When they develop a new GM crop variety, for example, they patent it. This means that nobody can use this variety unless they pay a fee to the company who first developed it. The biotechnology companies say that patents encourage them to invest in research and development, as it guarantees them a fair financial return. In the UK, patents last for 20 years. In return, the holder of the patent has to publish details of their invention so that other people can learn from it.

People who oppose genetic engineering argue that we should not patent genes or crop varieties. They argue that major biotechnology companies have too much control over farming and the food supply of other countries. If farmers in poorer countries become dependent on GM crops, which could be withdrawn later, this could lead to mass starvation and major political problems.

How science works

Media campaigns against GM crops

In December 2007 the government's chief scientific adviser, Sir David King, criticised the way in which some of the British media reported issues concerned with GM food. He said Britain's failure to adopt GM crops had cost the economy a huge amount of money. In particular he criticised BBC's *Today* programme and the *Daily Mail* and accused them of campaigning against GM crops. The media can have a huge impact on public reactions to controversial issues such as genetic engineering.

1. What is the role of the media in informing people about new scientific developments?

2. Do you agree that media coverage has influenced the views of people in the UK towards genetic engineering?

Figure 2 *The Super Lamb Banana on display in Liverpool. A light-hearted comment on the possible future of genetically modified food*

Application and How science works

What do you think?

The following are examples of genetic engineering. For each one, think about the ethical issues that are raised. Evaluate whether you think it is right to use genetic engineering in each case.

- Genes may be introduced into fruit, such as a banana, to produce an oral vaccine. The GM banana can be given to children to eat instead of giving them a vaccination by injection. This could be used in developing countries, as it does not require refrigeration like traditional vaccines, and does not need to be administered by trained medical professionals.

- The gene for rennet has been inserted into yeast. This means that large supplies of rennet can be obtained from yeast. Rennet is an enzyme required for making cheese. Before this GM yeast became available, all the rennet used for cheese-making was extracted from the stomachs of calves that were slaughtered for meat production.

- Human genes may be inserted into pigs so that the cells of these pigs have human antigens on their surface. These GM pigs could be killed and their organs could be used for human transplants. There are not enough human organs available for transplant, but organs from non-GM pigs would be rejected by the human immune system.

- The Enviropig is a transgenic pig that produces low-phosphorus manure. It is genetically engineered to produce the enzyme phytase (which normal pigs do not produce). As a result its manure is much less harmful to the environment. Normal pig manure contains high levels of phosphates that can cause eutrophication in streams and rivers.

- Golden rice is a GM variety of rice that produces high levels of vitamin A. It was developed for use in third world countries where rice forms a major component of the diet. A lack of vitamin A can lead to night-blindness.

- GM bacteria are used to produce factor VIII – a blood-clotting factor. People with classical haemophilia need regular injections of factor VIII. Before GM bacteria were available, factor VIII was obtained from blood transfusions. This meant that people with haemophilia were at high risk of developing diseases such as hepatitis and AIDS from infected blood transfusions.

- The Oncomouse is a GM mouse that has an active oncogene. This means that it is highly likely to develop cancer, and is therefore very useful in cancer research. It can be used to test drugs that might prevent cancer developing in humans.

Summary questions

1 Many consumers in the UK do not want to buy food containing GM ingredients. Are they right to be concerned? Give reasons for your answer.

2 Describe how a gene coding for a specific antigen could be inserted into a banana.

4.6 The human genome project

Learning objectives:

- What is the human genome project?

- What information does it give us?

Specification reference: 3.4.4

Figure 1 *DNA sequencers used for the Human Genome Project*

The Human Genome Project (HGP) is the largest international collaborative project that has ever taken place in biology. It started in 1990. Thousands of scientists all over the world took part in this project to find the base sequence of the DNA that makes up a human being. The project was completed in April 2003. There are three billion bases in the human DNA sequence. If it was written down on paper it would fill 200 telephone directories.

The UK contribution to this project was led by Sir John Sulston. He founded the Wellcome Trust Sanger Institute in Cambridge. It was here that one-third of the human genome was sequenced. It was particularly important that this work was funded by governments and by charities, such as the Wellcome Trust. As a result, the information is freely available to scientists and the public all over the world.

At the same time, another draft sequence of the human genome was produced. This one was the work of a private organisation run by Dr J. Craig Venter in the USA. However, this information was also placed in the public domain.

What does the HGP tell us?

The HGP has given us a list of the nucleotide sequence in the DNA of one human being. However, we still do not know very much about what this means. You will remember that some of our DNA consists of 'junk' or non-coding DNA. Other sequences code for regulatory genes, concerned with 'switching' genes on and off.

The collection of all proteins in the body of an organism is called the **proteome**. The next task for scientists, now that they have the sequence of nucleotides in human DNA, is to work out the structure and functions of the proteome that the genome codes for. This is much more complex. First, they will have to work out the amino acid sequences that the DNA codes for. However, you will remember from AS level that simply knowing the sequence of amino acids in a protein does not tell you what its shape is, or what function it has in a cell.

Furthermore, there are still a few parts of the DNA sequence that have not been decoded. The part of the chromosome close to the centromeres consists of millions of base pairs that are very repetitive. These are difficult to sequence with current technology. At the ends of the chromosomes are more very repetitive sequences known as telomeres, and these are also very difficult to sequence with present technology. Apart from these, there are a few gaps in the human DNA sequence that has been published, but scientists hope to sequence these difficult regions in the next few years.

The base sequence published by the HGP is the combined result of analysing DNA from a small number of anonymous donors. This means that every individual human will have a slightly different base sequence from this.

What are the benefits of the HGP?

It is hoped that this detailed knowledge of the human genome will lead to medical and biotechnological advances. For example, it should be possible to develop tests that will inform individuals whether they have a predisposition to certain health conditions, such as cancer or coronary heart disease. It is hoped that this could also lead to new treatments for disease. It is also possible to compare human DNA sequences with those of other organisms, and find evolutionary relationships.

However, there are some concerns about the information we can obtain from sequencing the human genome. Some people are concerned that insurance companies or mortgage lenders may, at a future date, demand that their clients undergo genetic testing. The results of these tests could be used to assess the future health of the client, and the cost of obtaining insurance, or obtaining a mortgage, could be adjusted to take account of this. These tests could also be used by employers to check that they do not employ people who are at high risk of developing serious health conditions. Some people think it would be useful to know whether they have a predisposition to a particular disease in later life, as this would allow them to make careful lifestyle choices so that they could reduce the risk of developing the disease. However, other people think that there is no point in telling a person they have an increased risk of developing a disease, as this can worry them. They point out that not everybody with a particular genetic predisposition to a disease or condition goes on to develop that disease.

◼ How science works

Low-cost personal DNA readings are on the way

Earlier this week, genomics pioneer Craig Venter revealed an almost complete sequence of his genome, while that of James Watson, co-discoverer of DNA's double-helix structure, has been available on the web since late June.

It costs almost $1 million to read Watson's genome. However, many scientists believe that the most pertinent information could be gleaned by sequencing the 1% of the genome that codes for proteins. This might be done for as little as $1000 per person.

To push the field forward, a geneticist called George Church has launched the Personal Genome Project (PGP), in which he and nine other volunteers have signed up to have the protein-coding regions of their genome made available on the web, along with their medical records, photographs of facial features and the results of a questionnaire about their health and personal habits. Within a few months, he aims to start scaling up to 100 000 volunteers. As well as raising questions about the protection of personal genetic data, there are concerns about how people will use information that, at this stage, even trained medical geneticists find extremely difficult to interpret.

Church's PGP is intended in part as a resource to help researchers investigate the biological consequences of individual DNA sequence variation. To facilitate this, the relevant genetic, medical and lifestyle information will be placed in a public database. Church does not intend to include the names of the 100 000 volunteers, but he will warn them that it may be possible for them to be identified from the information that they provide.

For this reason, the PGP is drawing up procedures for obtaining informed consent, detailing ways in which participants' personal information might, in theory, be misused. These range from discrimination in employment or insurance to more fanciful scenarios such as someone synthesising DNA that matches part of your genome, then planting it at a crime scene. The project will even have an entrance examination to ensure that people understand the potential risks of participating.

© *New Scientist Magazine*

Summary questions

1 Many scientists believe it is important that the results of the HGP are available on the internet for everybody to use. Suggest why.

2 Finding the nucleotide sequence of the DNA of a human is just the first stage of the HGP. Explain why.

1 Evaluate the benefits of the PGP to scientists.

2 Would you be willing to volunteer your DNA for the PGP? Give reasons for your answer.

4.7   Uses of sequenced DNA

Malaria vaccines

The Sanger Institute at Hinxton in Cambridgeshire is funded by the Wellcome Trust. This is where a large part of the human genome was sequenced. Scientists working at the Sanger Institute are now working on several different projects involving genome sequencing.

One of the organisms whose DNA has been sequenced is *Plasmodium falciparum*. This is one of four different *Plasmodium* parasites that cause malaria. Every year there are about 515 million cases of malaria. Of these, between 1 and 3 million people die and most of these deaths are due to *P. falciparum* infection. There is currently no vaccine.

Scientists are confident that it should be possible to develop a vaccine against malaria. It is clear that people can develop immunity to the disease. Most of the deaths from malaria are among children, because adults who have been exposed to *P. falciparum* infection several times during their childhood can develop resistance that makes them less likely to die from the disease when they get infected again. However, although a vaccine should be possible, there are more than 5000 genes in the genome of *P. falciparum*, so it is difficult for scientists to decide which of these genes might be a target for a vaccine.

The search for a *P. falciparum* vaccine may be aided by high throughput sequencing. High throughput uses the combination of robotics, sensitive detectors and high-tech data-processing equipment so that scientists can sequence genomes much faster than ever before. It took many years to sequence the first *P. falciparum* genome. This was from a specific strain that had been kept in culture in the lab for many years. Scientists at the Sanger Institute are now sequencing the genomes of hundreds of different *P. falciparum* parasites, including parasites taken directly from patients in Africa and never cultured in the lab. Comparing all these genomes will allow scientists to understand for the first time how much genomic variation there is between *P. falciparum* parasites taken from different countries, different villages in the same country, or even from different patients in the same village. It will also allow scientists to identify the genes that show the most variation between parasites. This indicates the genes that are under the greatest selection pressure which is a common response to pressure from the human immune system, so these genes can be investigated further as targets for a vaccine.

Identifying genes that might be a suitable target for a vaccine is just the first stage. Scientists then need to express the proteome and test whether it is targeted by the immune system in patients naturally infected with malaria. Scientists at the Sanger Institute are working with centres in Africa and Asia to carry out these tests. Once a vaccine passes these early tests there are then several stages of clinical field trials that test for vaccine safety and effectiveness. It may be some years before an effective anti-malaria vaccine is readily available, but sequencing DNA enables scientists to focus their efforts on the most likely targets.

GenBank

GenBank is an international, open access database of all publicly available DNA sequences and the protein sequences that they code for. Sequences are deposited from all over the world. Some of these are sequences for a single gene, but other sequences may be for entire genomes. The DNA sequencing technology has improved enormously since the first draft of the human genome was published in 2003. It took 13 years to produce this draft sequence. However, with faster sequencing becoming available, the database at GenBank is growing rapidly. More sequences were deposited at GenBank in the first 3 months of 2008 than in the whole of the past 20 years.

Figure 1 *Chimpanzee DNA differs from human DNA by only a few per cent*

Uses of sequenced DNA

Sequenced DNA allows scientists to study the similarities and differences between organisms of different but closely related species. For example, human DNA differs from chimpanzee DNA by only a few per cent. Scientists at the Sanger Institute are able to compare the DNA of modern humans with the DNA of earlier species, such as Neanderthals. This can help us to understand evolutionary relationships.

In the near future, it is likely that scientists will be able to sequence the DNA of a particular person, and use this information to identify genes that predispose that individual to specific diseases. That person could then make an informed lifestyle choice to reduce the risk of developing the disease. For example, a person could discover they have a predisposition to skin cancer and take better care to stay out of the sun.

Ethical issues

Some people are concerned about who can access information about their sequenced DNA. For example, if a young person has their DNA tested to see whether they have a predisposition to a type of cancer, they might wish to ensure that their employer does not obtain this information. An unscrupulous employer might decide that s/he would not employ people who are at risk of a serious disease. Life insurance companies might also refuse insurance to people with a predisposition to certain diseases.

Another concern is that people may find it hard to cope with the results of genetic tests. A young person may become severely depressed to know that they have an increased chance of developing serious disease in middle age.

Some genes that have been sequenced have been patented. Pharmaceutical companies have done this so that they can be sure they will benefit from the sale of any medical treatments developed as a result of their work. These companies say that researching diseases and sequencing DNA is expensive, so it will not be carried out unless they can be sure of gaining financial benefits. Other people believe that nobody should patent a gene, and that information should be shared freely via organisations like GenBank.

Medical treatments in the future are likely to be targeted at specific mutations. For example, a cancer drug may be targeted at a particular mutated oncogene. This will be beneficial for people who have this mutation, as such a drug is likely to be more effective. However, it may mean that people whose cancer is the result of an unusual mutation have much poorer treatment options.

Summary questions

1. Suggest why genes in *Plasmodium* that show the most variation between different parasites are the best genes to investigate as a target for a vaccine.

2. Use your knowledge of the immune system to explain why adults who have been exposed to several different strains of malaria are less likely to die from malaria than young children.

3. Explain how sequencing DNA can help scientists to understand evolutionary relationships.

4. In the future, it might be possible to sequence the genome of a young child as soon as the child is born. What would be the advantages and disadvantages of doing this?

AQA Examination-style questions

1   The polymerase chain reaction is a process used in a laboratory to make large quantities of identical DNA from a very small sample. The process is summarised in the flowchart.

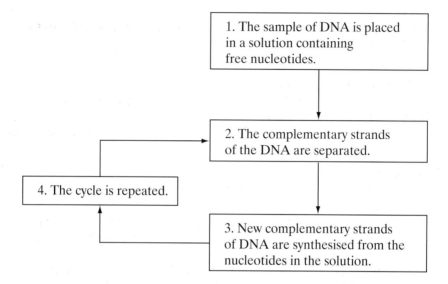

1. The sample of DNA is placed in a solution containing free nucleotides.

2. The complementary strands of the DNA are separated.

4. The cycle is repeated.

3. New complementary strands of DNA are synthesised from the nucleotides in the solution.

(a) (i) At the end of one cycle, two molecules of DNA have been produced from each original molecule.

How many DNA molecules will have been produced from one molecule of DNA after 5 complete cycles?

(ii) Suggest **one** practical use of this technique.   *(2 marks)*

(b) Give **two** ways in which the polymerase chain reaction differs from transcription.   *(2 marks)*

AQA, 2000

2   A restriction endonuclease cuts DNA at a particular base sequence. The restriction endonuclease, *Bam* H1, recognises the sequence of six bases as shown in the diagram and cuts the DNA to form sticky ends. The arrows show where *Bam* H1 cuts the DNA.

G   G   A   T   C   C
C   C   T   A   G   G

(a) Draw the sticky ends which are produced when *Bam* H1 has cut the DNA.   *(1 mark)*

(b) Describe how the two polynucleotide chains of DNA are normally held together.   *(2 marks)*

(c) The enzyme DNA ligase is used to join together pieces of DNA from different sources. Explain why the DNA to be joined together must be cut with the **same** restriction endonuclease before DNA ligase is used.   *(2 marks)*

AQA, 2000

3 (a) A herbicide, glyphosate, works by blocking the activity of an enzyme needed for amino acid synthesis. A mutant petunia plant was found which contained a gene giving resistance to glyphosate. This gene was transferred into tomato plants.

    (i) Describe how the gene may have been transferred from the mutant petunia into tomato plants.

    (ii) Suggest how the gene from the mutant petunia may give resistance to glyphosate. *(4 marks)*

  (b) (i) Explain how growing herbicide-resistant plants may lead to a higher yield of tomatoes.

    (ii) Describe **one** environmental risk associated with growing herbicide-resistant tomato plants. *(3 marks)*

AQA, 2000

4 (a) Describe how DNA is replicated in a cell. *(4 marks)*

The gene for alpha-1-antitrypsin was isolated from human white blood cells and multiplied by the polymerase chain reaction.

  (b) Explain the reason for each of the following stages in the polymerase chain reaction.

    (i) heating the sample of DNA above 90 °C

    (ii) adding primers

    (iii) using polymerase enzymes obtained from bacteria that live at high temperatures. *(5 marks)*

  (c) One cycle of the polymerase chain reaction takes 5 minutes to complete. Starting with a single gene, calculate the number of copies of the gene produced in 40 minutes. *(1 mark)*

  (d) Genetic engineering has been used to introduce the human gene for alpha-1-antitrypsin into sheep. The sheep produce alpha-1-antitrypsin in their milk.

    (i) Give **one** advantage of producing alpha-1-antitrypsin in this way.

    (ii) Suggest **three** reasons why people may be concerned about using genetic engineering in this way. *(3 marks)*

AQA, 2001

5 A new variety of tomato has been produced by genetic engineering. This variety contains a synthetic gene that blocks the action of a natural gene that would make the fruit soften rapidly once ripe. It also contains a marker gene.

The marker gene added by the scientists makes this variety of tomato resistant to the antibiotic, kanamycin. It is possible that this gene could be taken up by disease-producing bacteria in the human gut. In humans, kanamycin is used to treat certain types of gut infections.

Using information from the passage, explain the advantages and disadvantages of putting this new variety of tomato on the market. *(5 marks)*

AQA, 2001

Drugs can affect how we perceive the world around us

5.1 Sensory and motor neurones

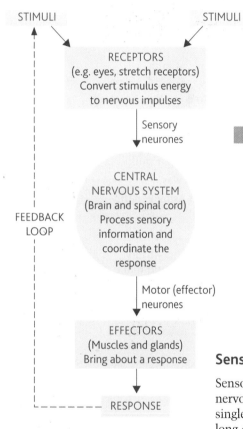

Figure 1 *Interrelationships of components of the nervous system*

The central nervous system

The **central nervous system** is made up of the brain and spinal cord. The **peripheral nervous system** consists of all the other parts of the nervous system: the nerves and the receptors. Figure 1 shows how the components of the nervous system fit together.

Stimuli are changes in the environment, either inside the body or outside. You will see that stimuli are detected by **receptors**. These send impulses along **sensory neurones** to the central nervous system. Here, the information is processed and the response is coordinated. Impulses are sent along **motor neurones** to **effectors**, such as muscles and glands, to bring about a response.

Neurones

The cells in the nervous system that carry nerve impulses are called **neurones**. Neurones have a **cell body** containing a nucleus, and many other organelles such as mitochondria, ribosomes and endoplasmic reticulum. **Dendrites** are extensions of the cell body. They have a large surface area to receive nerve impulses which they pass on to the cell body. The **axon** is a single, long projection from the cell body that carries nerve impulses away from the cell body. The axon of many neurones is surrounded by **Schwann cells**. These grow in circles around the axon. As they do this, the cytoplasm of the Schwann cell is squeezed into a very thin layer. The cell forms a fatty protection, rather like a bandage, around the axon. This is called the **myelin sheath**. You can see this in Figure 2.

Sensory and motor neurones

Sensory neurones carry nerve impulses from receptors to the central nervous system. They have many dendrites that join together to form a single long dendron that brings the impulse towards the cell body, and a long axon that carries impulses away from the cell body.

Motor neurones carry nerve impulses away from the central nervous system towards an effector. They have many short dendrites and a single axon that may be very long. You can compare a sensory and a motor neurone in Figure 3.

The different parts of the nervous system all work together. For example, receptor cells in the eye detect light. This causes impulses to pass along sensory neurones to the brain, which is part of the central nervous system. Here, the impulses are interpreted and as a response, if necessary, impulses are sent along motor neurones to effectors such as muscles. The process of interpreting images from the receptor cells in the eye is called **visual perception**. You will learn more about this in Topic 5.9.

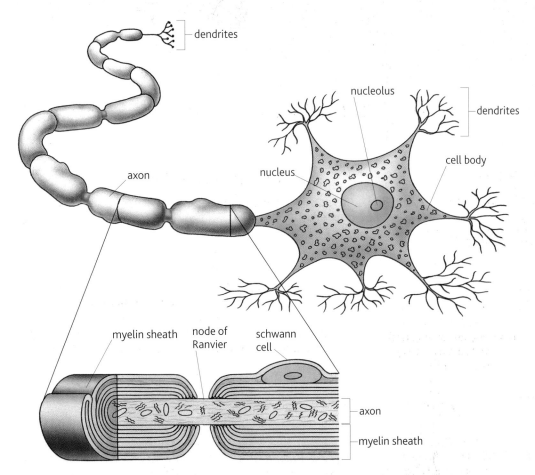

Figure 2 *A motor neurone with its myelin sheath*

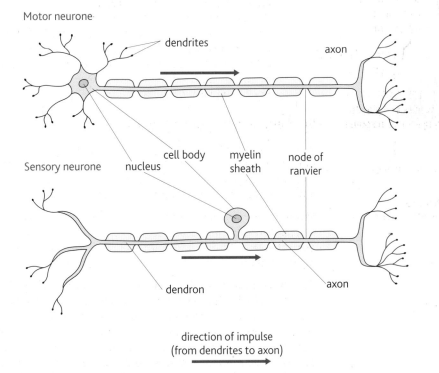

Figure 3 *Sensory and motor neurones*

The discovery of neurones

In 1838 Theodore Schwann and Matthias Schleiden proposed the Cell Theory. This stated that the cell was the basic functional unit of all living things. However, at this time they did not think that the cell theory applied to the brain and nervous system.

In the 1820s, Johannes Evangelista Purkyne (1787–1869) made many discoveries about the nervous system, including calculating the size of a neurone. His most famous discovery was the cerebellar cells in the brain. These are large cells, and were the first neurones to be identified. He made clear drawings of these cells and gave a detailed description.

In 1836, Gabriel Gustav Valentin (1810–1863) identified the cell body and nucleus of neurones.

Otto Friedrich Carl Dieters (1834–1863) had better microscopes available to him and produced the most accurate description yet of a neurone, complete with axon and dendrites.

A much better staining method was discovered by Camillo Golgi (1843–1926). The Golgi staining methods allowed the structure of a neurone to be seen much more clearly. He wrote descriptions of nerve tissue, but did not appreciate that it was made up of many separate neurones. He believed that nervous tissue must be made of a continuous network, rather than separate cells, so that communication could take place.

Santiago Ramón y Cajal (1852–1934) was the first person to suggest that the neurone was the functional unit of the nervous system. In 1887, Cajal started using Golgi's method of staining nervous tissue and found a way to improve it. He made extremely detailed drawings and descriptions of nerve tissue.

The term 'neurone' was not introduced until 1891. The axon was given its name by Rudolph von Kollicker, and the dendrites were named by Wilhelm His. In 1897 Sir Charles Sherrington described the junction between nerve and muscle, and named it the 'synapse'.

1. You have met the names Purkyne, Golgi and His before. Can you remember where?

2. Explain why it was necessary to use a good staining method to observe nerve tissue.

Summary questions

1. Give **two** differences between the structure of a sensory and a motor neurone.

2. Which of the following are: a receptors b effectors?

   eye, pancreas, ear, leg muscle, diaphragm

5.2 Action potentials

When a nerve impulse passes along a neurone, changes occur in the electrical charge across the plasma membrane of the neurone. An influx of sodium ions into the sensory nerve ending causes its membrane to **depolarise** and create a **generator potential**. Once a generator potential exceeds a **threshold value** an **action potential** is generated.

Resting potential

When a neurone is not carrying a nerve impulse it is said to be **at rest**. However, this does not mean that the cell is not using any energy. When the neurone is at rest, the inside of the neurone is negatively charged with respect to the outside. This difference in charge is called the **resting potential**. It is about −70 mV. We say that the membrane is **polarised**. This potential difference across the membrane is achieved by the distribution of ions.

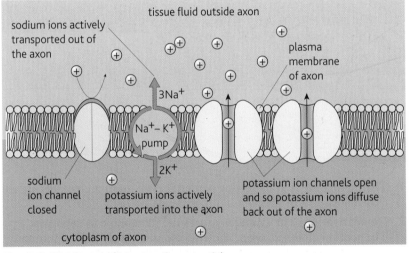

Figure 1 *Distribution of ions at resting potential*

Look at Figure 1. You can see there are several kinds of protein molecules in the neurone membrane. There is a sodium–potassium (Na^+–K^+) pump that uses ATP to pump three sodium ions (Na^+) out of the cell in return for two potassium ions (K^+) into the cell. There are also sodium and potassium channels. Inside the cell are negatively charged protein ions.

The inside of the neurone is negatively charged compared to the outside when the neurone is at rest for the following reasons:

- There are negatively charged protein ions inside the cell which cannot pass through the plasma membrane.

- The sodium–potassium pump moves three sodium ions out for every two potassium ions that are moved in, so that there are more positively charged ions outside the neurone than inside.

- The sodium channels are closed when the neurone is at rest, and only allow very limited diffusion of sodium ions back into the cell. However, the potassium channels are not all completely closed when the neurone is at rest. Therefore, potassium ions diffuse out of the neurone along their concentration gradient, adding to the concentration of potassium ions outside the neurone. Overall, the membrane is about 100 times more permeable to potassium ions than to sodium ions when the neurone is at rest.

Link

It would be helpful for you to look back at Topics 3.3, 3.4 and 3.5 in *AS Human Biology* to remind yourself of the structure of plasma membranes and the different kinds of protein in them.

What happens during an action potential?

The sodium and potassium channels in the neurone membrane are open or closed, depending on the potential difference, or voltage, across the membrane. Therefore they are said to be **voltage gated channels**. The events that occur during an action potential are shown in Figure 2.

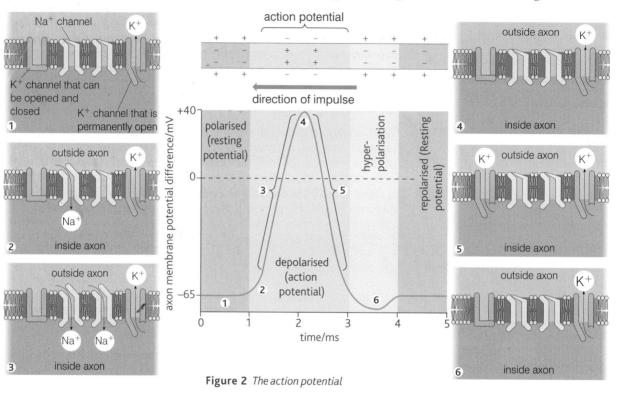

Figure 2 *The action potential*

1   The membrane is at its resting potential. The sodium channels are all closed but some of the potassium channels are open.

2   **Depolarisation**. The stimulus causes the membrane to depolarise. The sodium channels open, allowing sodium ions to diffuse into the neurone. These ions are positively charged so they cause the potential difference across the membrane to reverse. Inside the neurone now becomes positively charged compared to outside.

3   Once some sodium ions enter this causes more sodium channels to open. As a result, even more sodium ions enter the neurone.

4   **Repolarisation**. Now that the potential difference across the membrane has changed, the sodium channels close and the potassium channels start to open. Potassium ions diffuse out of the axon.

5   As potassium ions diffuse out of the axon, more potassium channels open. This causes even more potassium ions to diffuse out of the axon.

6   So many potassium ions leave the neurone that there is a temporary 'overshoot'. This is called **hyperpolarisation**. The potassium channels now close, and the sodium–potassium pump restores the resting potential.

For a very short time after the sodium and potassium channels close, they are unable to open again. This happens because of the change in shape of the proteins. This period lasts only for a few milliseconds, but during this time the neurone cannot conduct another action potential. Therefore, this period is called the **refractory period**. The membrane of the neurone is unable to respond to a stimulus that would normally cause an action potential.

Summary questions

1   Explain why sodium ions cannot pass through a potassium channel, and why potassium ions cannot pass through a sodium channel.

2   The sodium–potassium pump is an example of active transport. Explain why.

3   The cell body of a neurone contains many mitochondria. Explain why.

5.3 Propagation of nerve impulses

Learning objectives:

- How do nerve impulses pass along a neurone?

- What factors affect the speed of transmission of action potentials?

Specification reference: 3.4.5

Once an action potential has been set up, it spreads rapidly along the length of the neurone. As one part of the membrane becomes depolarised, sodium ions move into the neurone. These sodium ions then cause the depolarisation of the membrane immediately ahead. You can see this in Figure 1.

a *The resting state*

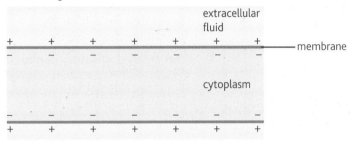

b *Depolarisation of the membrane causing the flow of sodium ions into the neurone*

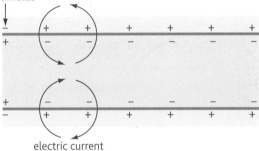

c *The inflow of sodium ions causes a wave of depolarisation to spread along the membrane*

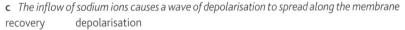

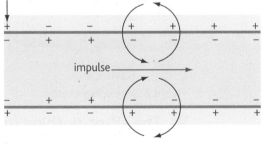

d *The process repeats itself along the axon*

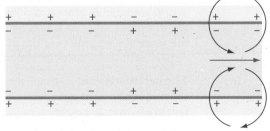

Figure 1 *Propagation of an action potential*

In an unmyelinated neurone, a nerve impulse passes along the neurone as a **wave of depolarisation**.

■ Propagation of a nerve impulse in a myelinated neurone

Myelinated neurones have Schwann cells wrapped around their axons, forming a myelin sheath. Myelin does not allow ions to pass through. However, there are small gaps between the Schwann cells. These gaps are called **nodes of Ranvier**. Voltage gated sodium and potassium channels are found in the neurone membrane in the nodes of Ranvier. Sodium and potassium ions can enter and leave the neurone through these channels, so action potentials can occur here.

In a myelinated neurone, sodium ions enter at one node. They pass along the neurone to the next node, where they cause further depolarisation. You can see this in Figure 2.

This is called **saltatory conduction**. The impulse 'jumps' from node to node. This means that an action potential passes much more quickly along a myelinated neurone than along an unmyelinated neurone.

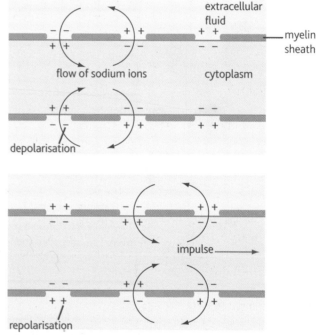

Figure 2 *Saltatory conduction*

■ The all-or-nothing rule

A stimulus has to be of a certain intensity, called a **threshold value**, to cause an action potential. If a stimulus is below the threshold value no action potential is generated. However, if the stimulus is greater than the threshold value an action potential occurs. The action potential is the same size regardless of the intensity of the stimulus. In other words, either a stimulus causes an action potential or it does not.

The nervous system distinguishes between larger and smaller stimuli by the **frequency** of action potentials. A larger stimulus causes more action potentials to be generated within a particular time period.

Factors affecting the speed of transmission of action potentials

Action potentials pass along a particular neurone at a constant speed. They do not speed up or slow down. However, the speed of transmission varies between neurones.

- Myelinated neurones transmit impulses more quickly than unmyelinated neurones
- Wider diameter axons transmit impulses more quickly than narrower diameter axons. This is because wider axons have a smaller surface area to volume ratio, so they lose fewer ions by leakage.

Importance of the refractory period

The **refractory period** is very important in keeping impulses flowing in one direction only. Impulses cannot pass both ways along a neurone because the membrane in the region behind the action potential is in its refractory period.

The refractory period also makes sure that action potentials are separated from each other (are discrete), and that there is a maximum number of action potentials that can pass along a membrane in a particular period of time.

Application and How science works

Multiple sclerosis

Multiple sclerosis (MS) is a condition of the central nervous system. It affects around 85 000 people in the UK. Almost twice as many women are affected as men, and it is most commonly diagnosed when a person is between 20 and 40 years old. There is no cure for MS, although there are treatments that help to manage the symptoms.

MS is thought to be an autoimmune disease. The body's own immune system attacks the myelin sheath on neurones in the central nervous system. Bare patches form on neurones, called scleroses or **plaques**. These areas cannot conduct nerve impulses. Plaques usually start developing in the optic nerve, parts of the brain and the upper part of the spinal cord.

The most common symptoms of MS are muscle weakness in the arms and legs, tingling sensations like 'pins and needles', abnormal muscle spasms, difficulty in moving, visual problems, problems speaking or swallowing, fatigue and pain, and bladder and bowel difficulties.

Other theories for the cause of MS include that it might be caused by a virus, dietary deficiencies in childhood, or by a metabolic problem.

Some people with MS experience progressive difficulties, with the disease gradually worsening until the person is severely disabled. Others have periods when disease symptoms are frequent, followed by spells of remission when the symptoms are better or even absent altogether. About 10% of people with MS end up in a wheelchair.

1 Explain why a person's own immune system does not normally attack the myelin sheath of their neurones.

2 Scientists have found that if one identical twin develops MS, there is a 30% chance that the second twin will also develop MS. For non-identical twins, there is only a 4% chance that the second twin will develop MS if the first twin develops the disease. Does this suggest that MS might have a genetic cause? Explain your answer.

Summary questions

1 Charcot-Marie-Tooth disease (CMT) is an inherited condition. In some kinds of CMT the myelin sheath surrounding sensory and motor neurones gradually breaks down, causing damage to the neurone so that it cannot conduct action potentials properly. Suggest the symptoms that a person with CMT will have if: a the motor neurones leading to the leg muscles are affected, b the sensory neurones going from the fingers to the brain are affected.

5.4 Synapses

Learning objectives:

- How are impulses passed from one neurone to another?

Specification reference: 3.4.5

When two neurones meet, there is a tiny gap between them called a **synaptic cleft**. The point where two neurones meet is called a **synapse**. You can see this in Figure 1.

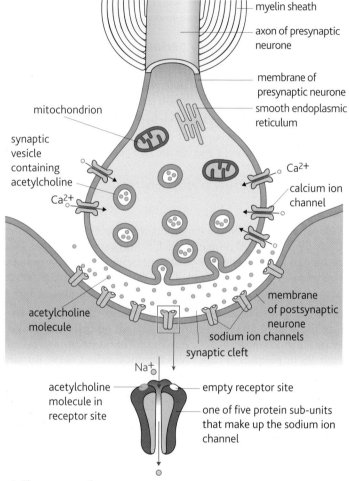

Figure 1 *The structure of a synapse*

Summary questions

1 Suggest what would happen if acetylcholinesterase was not present in the synaptic cleft.

2 A drug used to treat Alzheimer's disease works by inhibiting acetylcholinesterase. Suggest how this drug may reduce the symptoms of Alzheimer's disease.

The end of the axon of the presynaptic neurone is swollen to form the **synaptic knob**. This contains many vesicles containing a chemical called a **neurotransmitter**. In most synapses this neurotransmitter is **acetylcholine**. Synapses that use acetylcholine are called **cholinergic** synapses. There is a synaptic cleft between the synaptic knob and the membrane of the postsynaptic neurone. The membrane of the postsynaptic neurone has many sodium channels. Impulses are passed across the synaptic cleft by the neurotransmitter chemical.

■ How impulses pass across a synapse

You can see how impulses pass across a synapse in Figure 2.

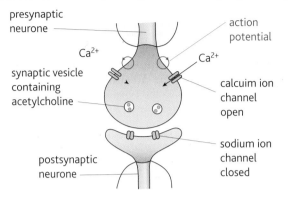

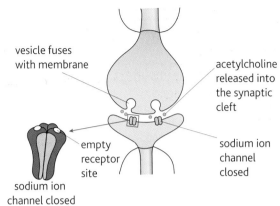

1 An action potential arrives in the presynaptic neurone. This causes calcium channels to open and calcium ions (Ca^{2+}) enter the synaptic knob.

2 The calcium ions cause the vesicles of acetylcholine to fuse with the presynaptic membrane. Acetylcholine is released into the synaptic cleft and diffuses across to the postsynaptic membrane.

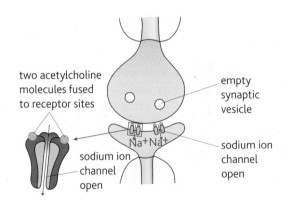

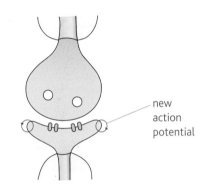

sodium ions (Na^+)

3 Acetylcholine binds to specific receptor sites on the sodium channels in the postsynaptic membrane. This causes the sodium channels to open, allowing sodium ions to enter the postsynaptic neurone.

4 If enough sodium ions enter to overcome the threshold value, an action potential is generated in the postsynaptic neurone.

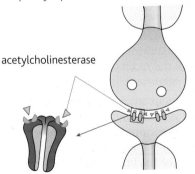

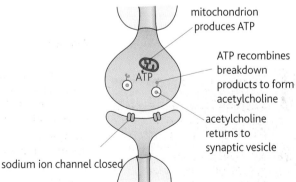

5 An enzyme in the cleft, called acetylcholinesterase, hydrolyses acetylcholine. The breakdown products diffuse back across the cleft and back into the synaptic knob. The sodium channels close once the acetylcholine is broken down.

6 ATP produced in the mitochondria is used to re-form acetylcholine from the breakdown products. This is stored in vesicles in the synaptic knob until it is needed again.

Figure 2 *How impulses pass across a cholinergic synapse*

5.5 Neuromuscular junctions

The place where a motor nerve meets a muscle is called a **neuromuscular junction**. This is a kind of synapse where the presynaptic neurone connects with a muscle. You can see this in Figure 1. You will notice that it looks very similar to the type of synapse you have already studied, where one neurone meets another.

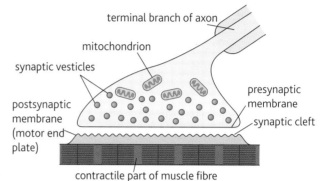

Figure 1 *A neuromuscular junction*

There is a gap between the motor neurone and the membrane around the muscle called the **sarcolemma**. The part of the sarcolemma that is in contact with the synapse is called the **motor end plate**. When an action potential reaches the neuromuscular junction, vesicles of acetylcholine fuse with the presynaptic membrane in the same way as in a normal synapse. The acetylcholine diffuses across the gap and fits into specific receptors in the postsynaptic membrane. This allows sodium ions to enter the muscle which brings about contraction.

■ Drugs and synapses

Many drugs affect synapses. There are several ways in which these drugs may have an effect:

■ Some drugs are similar in shape to the neurotransmitter. This means that they can fit into the receptor sites on the postsynaptic membrane. This is how nicotine (in cigarettes) works.

■ Other drugs are able to block the receptors in the postsynaptic membrane, preventing the neurotransmitter from binding. Curare is a drug that causes paralysis. It blocks the receptors for acetylcholine in the postsynaptic membrane at the neuromuscular junction.

■ Some drugs stimulate the release of a neurotransmitter so that more neurotransmitter than usual is released into the synaptic cleft. Caffeine, ecstasy (MDMA) and cocaine are all examples of this type of drug. For example, ecstasy stimulates the release of the neurotransmitter serotonin in parts of the brain.

■ Some drugs inhibit the action of the enzyme in the synaptic cleft that breaks down the neurotransmitter. The neurotransmitter is therefore present for longer. Myaesthenia gravis is a condition in which people suffer muscle weakness because they do not produce enough acetylcholine at the neuromuscular junction. It is treated using a drug called pyridostigmine bromide. This inhibits the enzyme acetylcholinesterase, so acetylcholine remains present in the cleft for longer.

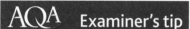

Botox

Clostridium botulinum is a bacterium that causes a severe kind of food poisoning called botulism. It produces several kinds of neurotoxin called botulinum. Botulinum stops vesicles of acetylcholine fusing with the presynaptic membrane. As a result, action potentials cannot pass across the synaptic cleft and the muscle becomes paralysed. Botulinum toxin is one of the most powerful toxins known. About 1 μg is lethal to humans by causing paralysis of muscles, including the muscles involved in ventilating the lungs.

One kind of botulinum toxin is used by medical practitioners as 'Botox' to treat conditions such as wrinkles.

Hyperhidrosis is a condition in which a person produces excessive amounts of sweat. This can be treated by using botulinum toxin to target the nerves that lead to the sweat glands.

Figure 2 *Botox*

1. Name the muscles involved in ventilating the lungs.
2. Explain how paralysis of these muscles would kill a person.
3. Is it right to use a powerful toxin like botulinum to treat conditions such as wrinkles? Give reasons for your answer.

Why do we have synapses?

Synapses are very important because they control the passage of nerve impulses in the nervous system. They do this in various ways:

- They ensure that a nerve impulse can pass in one direction only.
- They allow one neurone to stimulate several other neurones.
- The threshold value allows unimportant stimuli to be 'filtered out'. If a stimulus produces low-frequency impulses there will not be enough neurotransmitter released to set up an action potential in the postsynaptic membrane.
- It is thought that synapses in the brain are important in learning and memory.

Poison darts

Some Indian tribes in South America use poison from certain species of frog to make poison darts. The poison is usually obtained by roasting the frogs over a fire. However, one toxin called batrachotoxin is so powerful that they only need to dip the dart into the back of the frog without killing it. You can see the frog in Figure 3.

Figure 3 *Phyllobates terribilis, a frog found in South America that is the source of the powerful toxin batrachotoxin*

Just 100 μg of this poison is enough to kill an adult human. Batrachotoxin acts on sodium channels in the membranes of neurones. It increases the permeability of the resting cell membrane to sodium ions, but does not affect potassium or calcium ion concentrations.

1. Use the information above to suggest the effect that batrachotoxin has on neurones.
2. Suggest the advantage to the *Phyllobates* frog of producing batrachotoxin.

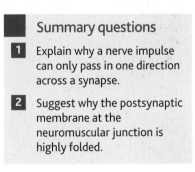

Summary questions

1. Explain why a nerve impulse can only pass in one direction across a synapse.
2. Suggest why the postsynaptic membrane at the neuromuscular junction is highly folded.

5.6 Drugs and the brain

Learning objectives:

■ What affect do recreational drugs have on the body?

■ What affect do recreational drugs have on the brain?

Specification reference: 3.5.5

You learned about synapses in Topic 5.4. Many drugs work by interacting with synapses. Some of these drugs are useful in medicine. However, some of these drugs are used recreationally and may have harmful effects.

■ Nicotine

Nicotine is similar in shape to acetylcholine. This means that it binds to acetylcholine receptors in the postsynaptic membrane.

Nicotine stimulates the synapses in the ganglia of the sympathetic nervous system. You will learn more about the sympathetic nervous system in Topic 6.1. As a result it causes the heart rate and breathing rate to increase, and blood pressure rises. In the brain, synapses that use acetylcholine are stimulated. The user's reaction time becomes faster and they feel more energised.

Nicotine also increases the release of dopamine in the brain. Dopamine is an example of a monoamine transmitter. This neurotransmitter is found in the limbic system, or 'reward pathway' in the brain. By stimulating these neurones, nicotine brings on pleasant feelings that encourage the person to use the drug again.

■ Cocaine

Cocaine affects synapses that use dopamine as a neurotransmitter. After dopamine has been released into the synaptic cleft it is transported back into the presynaptic neurone so that it can be used again. The dopamine is taken back into the presynaptic neurone via specific dopamine transporters. These are shown in Figure 1. Cocaine fits into these dopamine transporters and prevents dopamine from binding to them. As a result, dopamine remains in the synaptic cleft, causing repeated stimulation of the postsynaptic neurone. The user experiences pleasant feelings that encourage them to take the drug again.

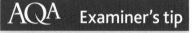

AQA Examiner's tip

The **limbic system** is the part of the brain associated with emotions, long-term memory and behaviour. You do not need to know this term for your exam, but you do need to be able to explain that drugs can affect the area of the brain associated with emotion, long-term memory and behaviour.

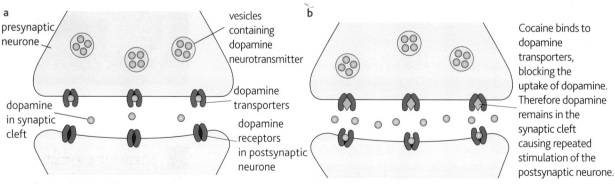

Figure 1 *How cocaine affects the brain*

■ LSD

LSD is a synthetic drug derived from a fungus that causes the disease ergot in cereal crops. It causes powerful hallucinations. LSD acts mainly on synapses that use another kind of monoamine neurotransmitter, serotonin. It is similar in shape to serotonin.

There are two different kinds of serotonin receptors on postsynaptic membranes: type 1 and type 2. When LSD fits into type 1 serotonin receptors, it opens sodium channels in the postsynaptic neurone and stimulates the postsynaptic neurone. However, when LSD fits into type 2 serotonin receptors on postsynaptic membranes, it blocks these receptors and prevents serotonin from binding. This therefore causes inhibition of the postsynaptic neurone.

Marijuana

Inhibitory neurotransmitters are normally present in dopamine synapses, inhibiting the release of dopamine. The body naturally produces a cannabinoid substance called anandamide. This binds to cannabinoid, or THC receptors on the neurones that produce this inhibitor. This stops these neurones releasing the inhibitor and allows dopamine to be released by the dopamine-releasing neurones. This is shown in Figure 2.

Marijuana contains a chemical called THC which is very similar in shape to anandamide. THC binds to the cannabinoid or THC receptors and suppresses the release of the inhibitory neurotransmitter. As a result, dopamine is released.

Anandamide breaks down very quickly in the body so it causes only a temporary release of dopamine. THC, however, remains for much longer, causing a greater release of dopamine and a lasting 'high'.

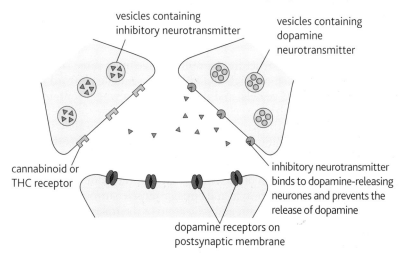

a *How an inhibitory neurotransmitter normally controls the release of dopamine*

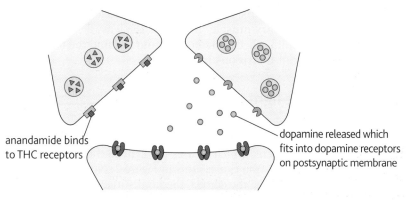

b *How anandamide causes the release of dopamine*

Figure 2 *How THC in marijuana affects synapses*

Summary questions

1. Use the information about nicotine on this page to explain why smoking can cause heart disease.

2. Suggest why anandamide breaks down very quickly in the body but THC does not.

5.7　The human eye

Learning objectives:

■ How is the human eye structured?

■ How does the human eye focus?

■ What does the term 'accommodation' mean?

Specification reference: 3.4.5

Figure 1 shows the structure of the human eye.

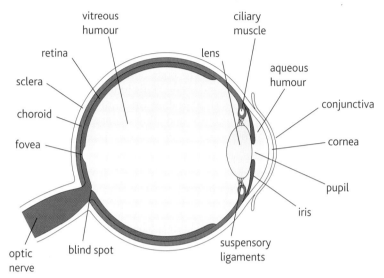

Figure 1 *Section through a human eye*

The eye is a spherical structure. Its wall is made up of three layers: the outer **sclera** which is tough and fibrous; the middle **choroid layer** which contains blood vessels; and the inner **retina** which contains the light-sensitive rod and cone cells. At the front of the eye the sclera is transparent, forming the part known as the **cornea**. This allows light to enter the eye. The conjunctiva is a membrane that surrounds the cornea. It is lubricated by watery tear fluid.

The **iris** is a modification of the choroid layer at the front of the eye. It contains smooth radial muscle and circular muscle surrounding a hole called the **pupil**. Contraction and relaxation of the iris muscles change the size of the pupil and allow variable amounts of light to enter the eye. The light-sensitive receptor cells in the retina – the rods and the cones – connect to sensory neurones that leave the retina at the blind spot and form the optic nerve.

Just behind the pupil is a **lens**, held in place by **suspensory ligaments** attached to a ring of **ciliary muscle**. In front of the lens the eye is filled with a clear, watery fluid called **aqueous humour**. Behind the lens is a clear jelly-like fluid called **vitreous humour**.

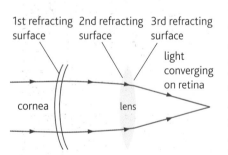

Figure 2 *The three refracting surfaces in the human eye*

Examiner's tip

The AQA specification says that you need to understand 'the transmissive properties of the eye'. This means that you need to know which parts of the eye are transparent, to allow light through, and which parts refract the light.

■ Focusing the light

When light rays enter the eye they are **refracted** or bent. First of all the light is refracted by the cornea at the front of the eye. The light is then further refracted by the lens, which focuses the light so that it forms a clear image on the retina. You can see this in Figure 2. Notice that there are three refracting surfaces: the cornea, the front surface of the lens and the rear surface of the lens.

Accommodation

Light is refracted to a greater or lesser extent depending on how far the object is from the eye. Light rays from a distant object are almost parallel and these need a thin lens to focus them on the retina. When you look at a distant object the ciliary muscles relax, pulling the suspensory ligaments taut. The lens, which is made of flexible proteins, is pulled into a thin shape. When you look at a near object the light rays enter the eye at a more acute angle. The ciliary muscles contract and the suspensory ligaments become looser, allowing the lens to return to its natural round, fat shape. This refracts the light rays more so that they are focused on the retina. The way in which the ciliary muscles cause the lens to change shape is called **accommodation**.

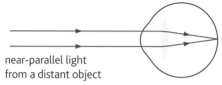

near-parallel light
from a distant object

a *Light from a distant object is focused on the retina by a flattened lens*

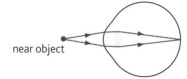

near object

b *Light from a near object is focused on the retina by a near-spherical lens*

Figure 3 *Accommodation*

Hint

The lens acts a little like something you may have played with as a child, called 'potty putty'. It will form a roughly spherical shape unless it is stretched. So, when you look at a distant object, the lens needs to be pulled into a thin shape. However, to view a near object, the lens is allowed to form a fat, round shape.

The ciliary muscles surround the lens in a ring. The suspensory ligaments surround the lens and join it to the muscle. When the ciliary muscle contracts it gets fatter, like any other muscle. This means the suspensory ligaments become looser and the lens can fall into a spherical shape. When the ciliary muscle relaxes, it gets thinner, pulling on the suspensory ligaments and making the lens thinner.

Remember it is only the ciliary muscles that contract or relax – *not* the suspensory ligaments.

Application

Cataracts

As people get older, the proteins in the lens of the eye start to denature. The lens becomes less elastic so it is more difficult to focus on objects at different distances from the eye. This means that many older people need glasses for both reading and distance vision.

As these proteins denature, they sometimes clump together and form a cloudy area that stops light passing through. This is called a cataract. You can see this in Figure 3.

A cataract may be treated by removing the lens, using a local anaesthetic. An artificial lens may be inserted instead. The best replacement lenses actually change shape and can undergo some accommodation.

1. What do we mean when we say that a protein has denatured?
2. When a person has a cataract removed but no new lens fitted, they can still see blurred images even when they are not wearing glasses. Explain why.

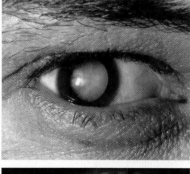

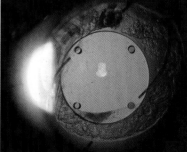

Figure 4 *A cataract (top) and a replacement lens*

Summary questions

1. Suggest the function of the aqueous humour and the vitreous humour.
2. Suggest the function of the muscles attached to the outside of the eye.

5.8 The retina

The structure of the retina

You learned in Topic 5.7 that the retina is the part of the eye that contains the light-sensitive cells – rods and cones. These are called **receptor cells**. A receptor cell is able to detect changes in the environment and converts sensory information into a form that can be transmitted to the central nervous system. You can see how the rod and cone cells are arranged in the retina in Figure 1.

You will notice that light has to travel through several layers of neurones to reach the light-sensitive rod and cone cells. Underneath the rods and cones is a layer of cells containing the pigment, melanin. This absorbs any light that is not absorbed by the rods and cones. Your pupil looks black because you are looking at this black layer at the back of the eye.

Figure 1 The structure of the retina

Rods

Figure 2 shows a rod cell. The outer segment contains many vesicles of the light-sensitive pigment, rhodopsin. When light hits a rod cell it causes the rhodopsin to split into opsin and retinal. This process is called **bleaching**. As a result of this, the permeability of the rod cell to sodium ions is altered. The rod cell sets up a **generator potential**.

If this is enough to overcome the threshold value, an action potential is generated in the sensory neurone which carries the impulse along the optic nerve to the brain. After this the rhodopsin is regenerated. The regeneration of rhodopsin requires energy from ATP.

Rod cells are extremely sensitive to low light intensities, so they are most useful for vision in dim light. However, they cannot distinguish between different wavelengths, meaning that they produce images in only black and white. This is called **monochromatic vision**. You will notice from Figure 1 that several rod cells synapse with just one sensory neurone. This means that rod cells have **low visual acuity**. They cannot discriminate very well. For example, it is difficult to read a newspaper in dim light.

The grouping of rod cells together also increases their sensitivity. This is called **summation**. If each of several rod cells results in a tiny amount of neurotransmitter being released into the synaptic cleft, these small amounts 'add together' so that there is enough to set up an action potential in the sensory neurone.

There are many more rods in the retina than cones. They are found evenly spread across the retina, although there are very few at the fovea.

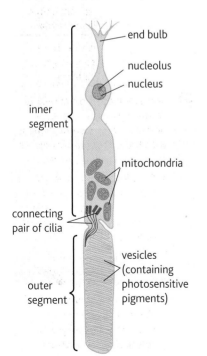

Figure 2 Structure of a single rod cell

Cones

Cone cells have a similar structure to rods. However, they contain a different pigment, called iodopsin. There are three different kinds of iodopsin, each sensitive to red, green or blue light. Different colours are perceived depending on the proportion of the different types of cone that are stimulated. For example, if red and green cones are stimulated equally, the brain interprets the colour as yellow. This is called the **trichromatic theory** of colour vision. You can see this in Figure 3.

Cone cells are not as sensitive to light as rods so they only work in high light intensity. There are also fewer cones than rods in the retina. Most of these are concentrated at the fovea which you can see in Topic 5.7, Figure 1. This is the part of the retina where the centre of the image is focused when you look straight at an object. However, most cones have an individual connection to a sensory neurone, meaning that cones have much greater visual acuity than rods. This is shown in Figure 4.

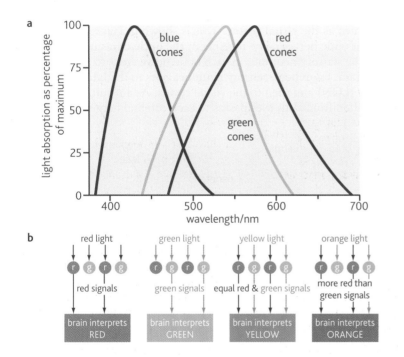

Figure 3 a The different wavelengths absorbed by the different kinds of cone. b How stimulation of different types of cone allows the brain to interpret colours

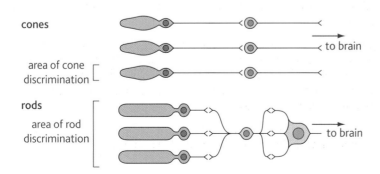

Figure 4 Comparing the visual acuity of rods and cones

5.9 Visual perception

Learning objectives:

- How does our brain interpret what we see?
- Which parts of the brain are involved in interpreting impulses from the eye?
- What are the bottom-up and top-down theories of visual perception?

Specification reference: 3.4.5

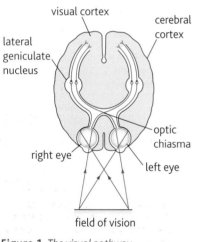

Figure 1 *The visual pathway*

The visual pathway

You learned in Topic 5.8 that light is detected by the rods and cones in the retina. As a result, these receptor cells send impulses along the optic nerve to the brain. Each eye has its own **visual field**. This is the region of the environment from which the eye collects light. The visual fields of our two eyes overlap, because both eyes are close together and forward-facing. This is called **binocular vision** and is shown in Figure 1. Light from the right-hand side of the visual field focuses on the left side of each retina, and light from the left-hand side of each visual field focuses on the right side of each retina.

Action potentials from the rods and cones in the retinas pass along the optic nerve to the occipital lobe at the back of the brain. They enter an area known as the **visual cortex**. This is shown in Figure 2. The left and right optic nerves come together at the **optic chiasma**. Here, about half of the sensory neurones in each optic nerve cross to the other side of the brain. From here, sensory neurones pass to the **lateral geniculate nucleus (LGN)** and then to the visual cortex. As a result, sensory information from both eyes passes to each cerebral hemisphere.

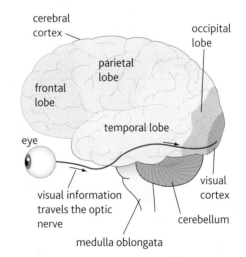

Figure 2 *The location of the visual cortex in the human brain*

Each eye is looking at the environment from a slightly different position. This means that the images received from each eye will be very slightly different. The visual cortex processes both images to produce a single perceived image.

Binocular vision is very useful for judging distances. The brain is able to judge depth and distance from a comparison of the two different images received from each eye.

The visual cortex is the part of the brain that interprets (or **perceives**) what we see. The images are actually recognised when impulses pass to the visual association area. Each part of the visual cortex is devoted to one part of the retina, so the brain can recognise which part of the retina the impulses have come from.

Bottom-up theories of visual perception

These theories propose that perception occurs from using all the information available to the eyes. There is no need to process the information, as the eyes themselves receive sufficient information. This information includes the following features:

- As mentioned above, each eye has a slightly different view. By combining these two different images, the brain is able to perceive depth and distance.
- The closer an object is to the eye, the more the eyes are converged (turned inwards).
- Closer objects are in front of more distant objects.
- Objects closer to the eye can be seen in more detail than those further from the eye.
- Parallel lines converge (come together) as they spread into the distance.
- As you move along, for example when you drive a car, the point you are moving towards remains stationary, while the rest of the view moves away from this point. This gives us information on our speed and direction of movement.
- Clues about distance and speed can be gained from the environment.

Figure 3 *Clues about distance and speed can be gained from the environment. For example, we know that the cars in the distance are further away because they look smaller*

Top-down theories of visual perception

These theories are also called constructivist theories. They state that perception does not just rely on the information from the senses, but also past knowledge, experience and expectations. Evidence for these theories comes from studying some visual illusions.

When we look at a person in the distance or a person near to us, we perceive them both as being the same size even though they cast different images on the retina. Top-down theorists explain that this is because of our past experiences. Similarly, objects seen from different angles cast different images on the retina, but the brain recognises them as the same object.

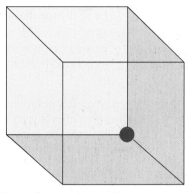

Figure 4 *An ambiguous figure. Most people can see this cube in two different ways, either with the red dot at the front or with the red dot at the back*

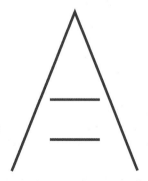

Figure 5 *A distortion. Most people see the top line as being wider than the lower line, because we use clues about depth band distance from our memory to interpret the figure. Actually, both lines are the same length*

Figure 6 *Hidden triangle. In this illusion, we can see a triangle that is not really there*

Summary questions

1. How can the visual cortex detect the strength of a visual stimulus?

2. Suggest how binocular vision would have been an advantage for ancestors of humans living in trees.

AQA Examination-style questions

1   (a) Explain how the eye is able to detect an object that is purple.   *(1 mark)*

    (b) When we look directly at an object, its image is focused on part of the retina known as the fovea. Explain how the fovea provides more information to the brain than do other parts of the retina.   *(2 marks)*

    (c) The eyes of short-sighted people focus light rays from a distant object in front of the retina instead of on it, producing a blurred image on the retina. Laser surgery can be used to correct this defect. **Figure 1** shows the results of one treatment in which part of the cornea is removed by laser surgery. After the surgery, the surface of the cornea heals to give a new, smooth surface, with a smaller curvature.

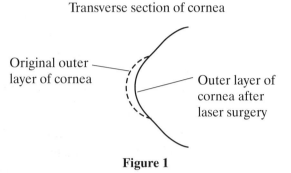

Figure 1

        Explain how the surgery corrects the patient's short-sightedness.   *(2 marks)*

    (d) Blindness can be caused by a cataract, a condition in which the lens of the eye becomes cloudy. If the lens is removed, the patient can see again, but vision is not as good as before the cataract developed. Explain why this is so.   *(2 marks)*

AQA, 2000

2   Serotonin is a neurotransmitter which is produced by certain neurones in the brain. One of its effects is to increase the activity of sensory neurones in the brain. It also usually improves a person's mood and keeps the person awake. **Figure 2** shows a synapse at which serotonin is the neurotransmitter.

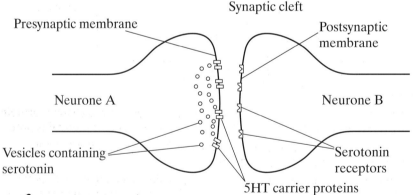

Figure 2

    (a) Explain how the release of the neurotransmitter, serotonin, by neurone **A** would initiate an impulse in neurone **B**.   *(3 marks)*

    (b) The serotonin is normally rapidly reabsorbed from the synaptic cleft by 5HT carrier proteins in the presynaptic membrane.

        Suggest **one** advantage of rapidly reabsorbing the serotonin.   *(1 mark)*

    (c) The active ingredient in the drug, Ecstasy, is MDMA. MDMA blocks the attachment of serotonin molecules to the 5HT carrier proteins.

        (i) Suggest how MDMA may temporarily improve a person's mood.

        (ii) MDMA may cause long-term damage to the 5HT carrier proteins. This leads to a depressed mood. Suggest why.   *(3 marks)*

AQA, 1999

3   **Figure 3** shows the changes in permeability of a neurone membrane to potassium (K^+) and sodium (Na^+) ions during the course of an action potential.

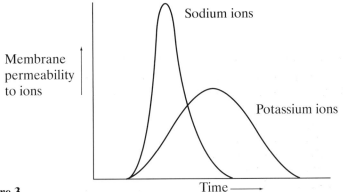

Figure 3

(a) During the course of an action potential, the potential difference across the neurone membrane changes from −70 to +40 and back to −70 mV. Use the information in **Figure 3** to explain what causes these changes.   *(4 marks)*

(b) (i)   What is meant by the 'all-or-nothing' nature of a nerve impulse?

     (ii) Neurones can respond to both strong and weak stimuli. Describe how a neurone conveys information about the strength of a stimulus.   *(3 marks)*

(c) Describe what is meant by *temporal summation* at a synapse.   *(1 mark)*

AQA, 2000

4   (a) **Figure 4** shows a sodium channel protein in the cell surface membrane of a neurone.

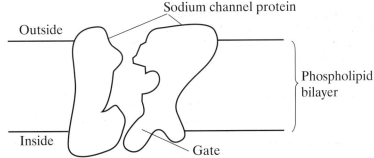

Figure 4

     (i)   A change in potential difference will affect the gate on the channel protein and produce an action potential. Explain how the action potential is produced.

     (ii) Describe what happens to the sodium ions immediately after the passage of an action potential along the neurone.   *(4 marks)*

(b) Local anaesthetics are used by dentists to stop patients feeling pain. **Figure 5** shows how a molecule of a local anaesthetic binds to a sodium channel protein.

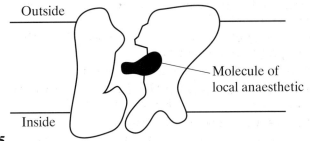

Figure 5

Explain how this local anaesthetic stops the patient feeling pain.   *(3 marks)*

AQA, 2001

6.1 Sympathetic and parasympathetic nervous systems

Learning objectives:

■ What is the autonomic nervous system?

■ What are the effects of the sympathetic and parasympathetic nervous systems?

Specification reference: 3.4.6

You have already learned that the nervous system is subdivided into the **central nervous system (CNS)**, consisting of the brain and spinal cord, and the **peripheral nervous system**. The peripheral nervous system can be divided up into sensory and motor systems. You can see this in Figure 1.

The motor system can be further divided into the **somatic nervous system** and the **autonomic nervous system**. The somatic nervous system carries impulses to skeletal muscles that are under conscious (voluntary) control. The autonomic nervous system carries nerve impulses to glands and muscles that are not under voluntary control.

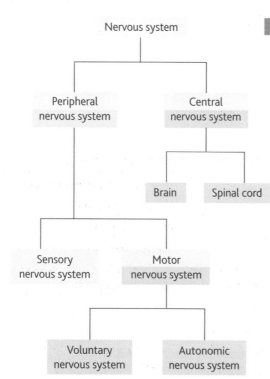

Figure 1 *Organisation of the nervous system*

The autonomic nervous system

The autonomic nervous system is concerned with controlling involuntary activities in the body, such as the beating of the heart, the secretion of sweat and controlling the muscles of the gut. You will remember from AS level that sympathetic and parasympathetic nerves speed up and slow down the heart rate in response to different levels of exercise.

As mentioned above, the autonomic nervous system is subdivided into the sympathetic and parasympathetic systems. Both systems consist of motor neurones, and both systems send impulses to effectors that are not under voluntary control. Both consist of two neurones that connect the brain or spinal cord to effectors. The first neurone synapses with the second neurone at a **ganglion**. You can see this in Figure 2.

Link

You may find it useful to look back at Topic 10.5 in *AS Human Biology* to remind yourself how the heart rate changes in response to exercise.

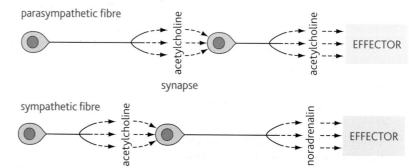

Figure 2 *Comparison of the sympathetic and parasympathetic nervous systems*

There are two main differences between the sympathetic and the parasympathetic nervous systems:

- In the **sympathetic nervous system** the ganglion is very close to the spinal cord, so that the pre-ganglionic neurone is short and the post-ganglionic neurone is long. The **parasympathetic nervous system** is the opposite. The pre-ganglionic neurone is long and the post-ganglionic neurone is short. The ganglion is very close to, or inside the effector.

- The sympathetic nervous system produces the neurotransmitter **noradrenalin** at the synapse with the effector whereas the parasympathetic nervous system produces the neurotransmitter **acetylcholine** at this synapse. For this reason, sympathetic neurones are said to be **adrenergic** while parasympathetic neurones are said to be **cholinergic**.

Application

Atropine

Atropine is a drug that blocks the action of the neurotransmitter acetylcholine. It comes from the plant called 'deadly nightshade', *Atropa belladonna*. Belladonna is Italian for 'beautiful lady'. This plant is named belladonna because women once used this drug to make their pupils bigger. Large pupils are considered as attractive. Nowadays, atropine is used to dilate the pupils before some eye examinations.

1 Explain how atropine makes the pupil bigger.

Effects of the autonomic nervous system

In general, the sympathetic and parasympathetic nervous systems act **antagonistically** on the effectors they innervate. Antagonistic means that they have opposite effects. Some of these effects are described in Table 1. However, you will notice that some effectors have only a sympathetic or parasympathetic supply, not both.

You will notice that the parasympathetic system is active under normal conditions. It is involved with activities to do with 'rest and digest'. The sympathetic nervous system is responsible for reacting to stress, emergencies and danger. You will learn more about this in the next topic.

Table 1 *Summary of the effects of the autonomic nervous system*

Parasympathetic nervous system	Sympathetic nervous system
slows heart rate	speeds up heart rate
dilates arteries	constricts arteries
constricts bronchioles	dilates bronchioles
contracts circular muscles in iris of eye, making pupil constrict	contracts radial muscles in iris of eye, making pupil dilate
stimulates tear production	
stimulates secretion of saliva	
speeds up movement through the gut	slows down movement through the gut
relaxes bladder and anal sphincters	contracts bladder and anal sphincters
causes contraction of the bladder	causes relaxation of the bladder
	contracts erector pili muscles, causing hairs in the skin to 'stand on end'
	increases sweat production

Summary questions

1 What is the main difference between the somatic and the autonomic nervous system?

2 A nerve is a bundle of neurones alongside each other. The nerve bulges where there is a ganglion. Explain why.

6.2 Fight or flight

Learning objectives:

- Where in the body is the hypothalamus?

- What is the function of the hypothalamus?

Specification reference: 3.4.6

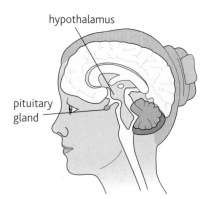

hypothalamus

pituitary gland

Figure 1 *The position of the hypothalamus and the pituitary gland*

The **hypothalamus** is part of the brain. Figure 1 shows you that it is a small area in the floor of the mid-brain. You will notice that it is very close to the pituitary gland. The hypothalamus contains neurosecretory cells which are rather similar to neurones, that link it to the pituitary gland. The pituitary gland is part of the **endocrine system**. This means that it secretes hormones. You will remember from Chapter 1 that the pituitary gland secretes hormones such as FSH and LH that are concerned with the control of reproduction.

Endocrine organs are **ductless** and secrete hormones directly into body fluids, mainly the blood. These hormones travel in the blood and change the activity of cells and organs some distance away.

The hypothalamus

- The hypothalamus receives impulses from the **cerebral cortex**. This is the part of the brain concerned with conscious thought, interpretation of stimuli and memory.
- As a result of information received from the cerebral cortex, the hypothalamus sends out impulses along the sympathetic nerves.
- This brings about changes in effectors, which may be muscles or glands.

For example, the hypothalamus is involved in controlling body temperature. You will learn more about this in Chapter 7.

The adrenal medulla

There are two adrenal glands, one above each kidney. They are endocrine glands. The adrenal glands have an outer region called the adrenal cortex, and an inner region called the adrenal medulla. The main hormone produced by the adrenal medulla is **adrenaline**. Sympathetic nerves going to the adrenal medulla stimulate the release of the hormone adrenaline.

Fight or flight

Physical and mental stress causes the sympathetic nervous system to be more active. It may help you to think how you might feel if you were confronted by a threatening person or a fierce dog. Impulses from the cerebral cortex pass to the hypothalamus, which stimulates the sympathetic nervous system. This brings about changes in the body, including the release of the hormone adrenaline, that prepare you either to face the threat or to run away from it. For this reason it is called the **fight or flight response**.

The hormone adrenaline and the sympathetic nervous system bring about a number of changes as shown in Figure 2. They can be summarised as follows:

- **Effects on the gut and breathing:** The smooth muscles in the gut relax so the diaphragm can be pushed further down into the abdomen. The bronchi become more dilated. This allows more air to enter and leave the lungs so more oxygen can be absorbed into the blood and more carbon dioxide can be removed.

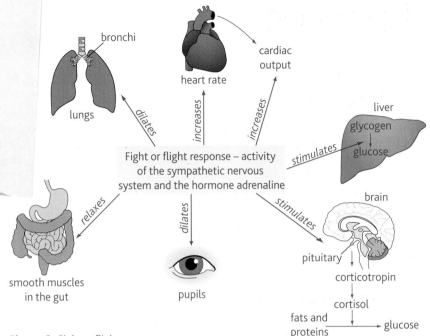

Figure 2 *Fight or flight response*

- **Effects on the blood system:** The heart rate increases. Arterioles in the gut and skin constrict since less blood is needed in these parts of the body, but the arterioles supplying the skeletal muscles dilate. This ensures that an increased supply of oxygenated blood is taken to the skeletal muscles so that the body can be moved quickly.
- **Effects on blood glucose:** Glycogen stored in the liver and muscles is converted to glucose, providing the energy source required for increased muscle contraction. Adrenaline stimulates the release of a hormone, corticotrophin, from the pituitary gland. This stimulates the release of another hormone, cortisol, from the adrenal cortex, which converts fats and proteins to glucose.
- **Effects on the nervous system:** The pupils of the eyes dilate so that visual stimuli may be detected quicker. Adrenaline increases the sensitivity of the nervous system so that stimuli may be detected more rapidly and responses made more rapidly.

How science works

Khat

Khat is a leafy shrub grown in parts of East Africa and the Arabian peninsula. For generations, chewing khat has been a popular social activity in Africa. More recently khat has been used by immigrants into the USA who need to stay awake for long periods. A researcher decided to investigate the effects of khat.

He used five male baboons and gave each of them 250 g of khat juice weekly for 8 weeks. He analysed blood samples taken from the baboons each week. Figure 3 gives some of the results from this investigation.

1. Describe the results of this investigation.
2. What evidence is there that khat reduces stress levels?
3. Discuss the reliability of this investigation.

Summary questions

1. Feeling slightly nervous just before an exam or a sports event can improve your performance. Use the information on this page to explain why.

2. Humans who have very stressful jobs may suffer from constipation or diarrhoea. Explain why.

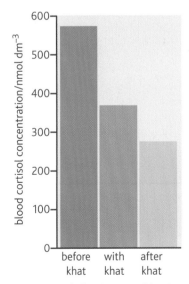

Figure 3 *Graph showing mean blood cortisol levels in baboons before, during and after taking khat*

6.3 Hormones

Learning objectives:

- How do hormones work?

- How is control by hormones different from nervous control?

Specification reference: 3.4.6

You already know that a hormone is a regulating chemical produced by an endocrine gland. It is carried in the blood to its target cells or organs. The target cells have specifically shaped receptor molecules on their plasma membranes. Figure 1 shows the location of the main endocrine glands in a human.

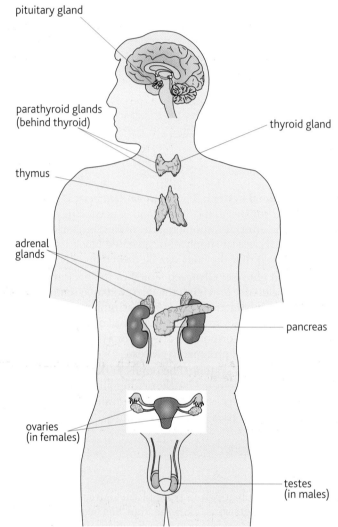

Figure 1 *Location of the main endocrine glands in a human*

Features of hormones

Hormones are usually released in very small concentrations but they may be effective for a long time. They are usually fairly small molecules. Some hormones are proteins or polypeptides, for example insulin or adrenaline. Others are steroids, such as testosterone or cortisol.

Steroid hormones are lipid soluble so they can pass through plasma membranes. Inside the cell they bind to a specific receptor protein that carries it into the nucleus. Inside the nucleus, the hormone–receptor complex binds to a leng... ...NA. This means that a specific gene is either switched on or off. You learned about the effects of testosterone on transcription in Topic 3.2.

Protein hormones bind to receptors in the plasma membrane. The hormone–receptor complex causes changes inside the cell, resulting in the activation of specific enzymes within the cell.

Comparing the nervous system and the endocrine system

Both the nervous system and the endocrine system involve responses being carried out in effectors as a result of the detection of certain stimuli by receptors. In general, the nervous system produces a rapid response and the endocrine system produces a slower, longer lasting response. The two systems are compared in Table 1.

Table 1 *Comparison of the nervous system and the endocrine system*

Endocrine system	Nervous system
involves communication using chemicals called hormones	communication is by nerve impulses but chemical neurotransmitters are used at synapses
transmission involves hormones travelling in the blood system	transmission is by nerve impulses along neurones
transmission is relatively slow	transmission is rapid
stronger stimuli result in a higher concentration of hormone	stronger stimuli result in a higher frequency of action potentials
hormones travel to all parts of the body, but only affect their target organs	nerve impulses only travel to specific parts of the body
the effects of hormones are often widespread	the effects of nervous communication are usually localised
the effect may be temporary and reversible or permanent and irreversible	the effect is temporary and reversible
the response is slow	the response is rapid
the response is usually long-lasting	the response is short-lived

Summary questions

1 Give an example of a temporary and reversible effect of a hormone, and a permanent and irreversible effect of a hormone.

2 Use your knowledge of how hormones work to explain why hormones, especially steroid hormones, do not produce rapid effects.

6.4 Antagonistic muscle action

Learning objectives:

- What is the structure of muscle tissue?

- How do skeletal muscles bring about movement?

- What are the differences between slow-twitch and fast-twitch muscles?

Specification reference: 3.4.6

Skeletal muscle is the muscle tissue attached to the bones of your body. It is also known as voluntary muscle, because you have control over when to contract these muscles and bring about a movement. Another name for this kind of muscle is striated (or striped) muscle because it has a striped appearance when you look at it under the microscope. The meat that you eat is skeletal muscle. You will have noticed that this kind of muscle is made up of many muscle fibres. You can see this in Figure 1.

The muscle fibre is surrounded by a plasma membrane called the **sarcolemma**. The cytoplasm in the muscle fibre, called **sarcoplasm**, contains large numbers of mitochondria. The muscle fibre is not divided up into individual cells but is more like several cells fused together. This means that it has several nuclei all along its length. The muscle fibres appear striped, or striated, because they are made up of bundles of many smaller **myofibrils**, which are also striped.

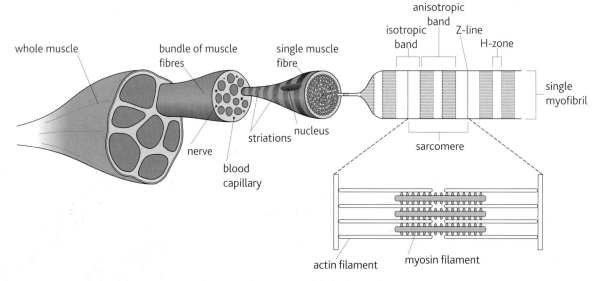

Figure 1 *The structure of skeletal muscle*

Each myofibril is striped in the same way and they line up together, giving a striped appearance. The myofibrils are made up of even smaller protein filaments. You will learn more about this in Topic 6.5.

Antagonistic pairs

Muscles can contract and relax. They can exert a pulling force when they contract but they cannot push. Therefore, to bring about movement an **antagonistic pair** of muscles is needed.

Look at Figure 2. You can see that there is an antagonistic pair of muscles in the upper arm – the biceps and the triceps. When the biceps contracts and the triceps relaxes, the arm bends at the elbow. However, the biceps cannot push the arm straight again. To straighten the arm, the biceps relaxes and the triceps contracts.

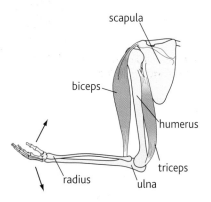

Figure 2 *The muscles in the upper arm*

Muscles that bend a limb, such as the biceps, are called **flexors**. Muscles that straighten a limb, such as the triceps, are called **extensors**. Several different pairs of antagonistic muscles are involved in controlling posture in humans. You will remember from AS level that humans evolved the ability to be bipedal. This involved changes in skeletal and muscle structure.

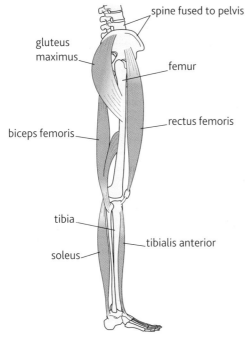

Figure 3 *Some of the muscles in the leg that are used for standing and walking*

Figure 3 shows some of the antagonistic pairs of muscles in the legs involved in standing and walking. Contraction of the soleus muscle pulls the heel up and extends the foot, whereas contraction of the tibialis anterior bends the ankle, raising the foot. To stand upright, both muscles need to contract a little so that they both exert a tension. This is called **isometric contraction**. To keep the leg straight at the knee, the rectus femoris and biceps femoris also need to show isometric contraction. When the gluteus maximus and the rectus femoris both contract isometrically the trunk is kept upright. Only a small number of fibres contract at any time in each of these muscles and the fibres involved change constantly, ensuring that the muscle does not become fatigued.

Slow-twitch and fast-twitch muscle fibres

There are two different kinds of muscle fibres in our skeletal muscles. They are known as slow-twitch and fast-twitch fibres. When we carry out aerobic exercise, the slow-twitch muscle fibres are most active. These muscle fibres are used for endurance sports like distance cycling, distance running or swimming. They look red in colour because they contain a lot of myoglobin. Myoglobin is a red pigment that stores oxygen. The fast-twitch fibres are mainly used for intensive exercise that is carried out for a short period of time, such as sprinting. These produce most of their ATP by anaerobic respiration (which does not require oxygen). They appear whiter in colour as they do not contain much myoglobin.

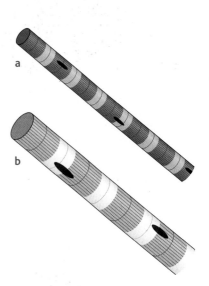

Figure 4 a *Slow-twitch muscle fibre*
b *Fast-twitch muscle fibre*

Most of your muscles contain a mixture of slow-twitch and fast-twitch muscle fibres but there are exceptions. The muscles in your back involved in maintaining posture contain mainly slow-twitch muscle fibres, while the muscles that move your eyes and the muscles in your eyelids are made up mainly of fast-twitch muscle fibres.

Muscle fibres in meat

Meat is made of muscle fibres. If you eat chicken or turkey you will know that it contains both 'red' and 'white' meat. The white meat is found in chicken wings and breasts. It is white because it consists mainly of fast-twitch muscle fibres. Wings are used for brief periods of intense activity. These muscles are very active for a short time and tire quickly. Chickens use their legs for standing and they do this most of the time. These muscles contain a lot of slow-twitch muscle fibres, which can keep working for long periods of time.

The two different kinds of fibre are compared in Table 1.

Table 1 *Slow-twitch and fast-twitch muscle fibres*

Slow-twitch fibres	Fast-twitch fibres
contain large numbers of mitochondria	contain few mitochondria
produce ATP by aerobic respiration	produce ATP by anaerobic respiration
supplied by many capillaries	supplied by fewer capillaries
much smaller diameter than fast-twitch fibres	much larger diameter than slow-twitch fibres
contain a large amount of myoglobin	contain very little myoglobin

Figure 5 shows the relative proportions of the different kinds of fibres found in different people.

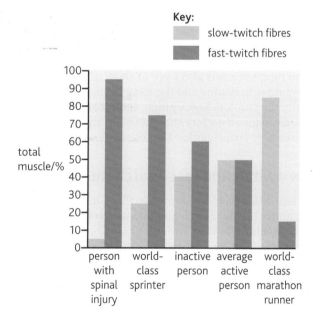

Figure 5 *The relative proportions of slow- and fast-twitch fibres in different people*

1 Explain why muscle fibres contain many mitochondria.

2 Suggest the advantage of having myoglobin in slow-twitch muscle fibres.

3 Explain why a person with spinal injury has a higher proportion of fast-twitch fibres than a world-class sprinter.

6.5 Sliding filaments

Learning objectives:

■ How do muscles contract?

Specification reference: 3.4.6

You will remember from Topic 6.4 that skeletal muscles are made up of muscle fibres. These in turn are made up of many smaller myofibrils. The myofibrils contain bundles of microscopic protein filaments.

There are two kinds of protein filaments:

■ thick filaments, made of the protein **myosin**

■ thin filaments, made of the protein **actin**.

Several myosin molecules lie together in a thick bundle with their bases attached at a point called the M-line. You can see this in Figure 1. The myosin molecules have hook-shaped 'heads' projecting from the bundle. The actin molecules lie together in a long chain, and two chains of actin twist together to form a long, thin filament. The actin filaments attach to each other at a point called the Z-line.

The actin and myosin filaments lie alongside each other and overlap, producing the striped appearance of the myofibril. The darker areas are where the actin and myosin filaments overlap. The distance between two Z-lines is called a **sarcomere**.

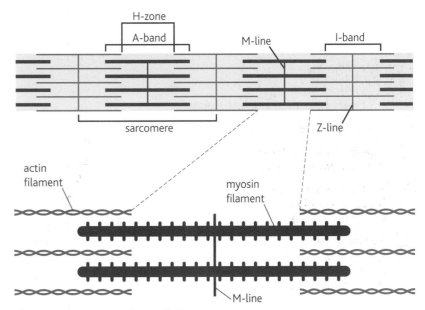

Figure 1 *The structure of a myofibril*

■ Sliding filaments

When muscles contract, the sarcomeres in each myofibril get shorter. This happens because the actin and myosin filaments slide over each other, pulling the Z-lines closer together. Energy from ATP is needed for this to happen. You can see this in Figure 2.

This sequence happens several times, making the actin filament slide past the myosin 'heads'. The two kinds of filaments overlap further, pulling the Z-lines together. Figure 3 shows what happens when a sarcomere contracts.

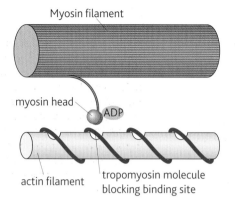

1 Tropomyosin molecule prevents myosin head from attaching to the binding site on the actin molecule.

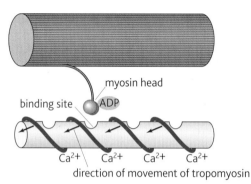

2 Calcium ions released from the endoplasmic reticulum cause the tropomyosin molecule to pull away from the binding sites on the actin molecule.

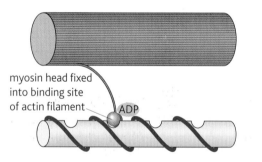

3 Myosin head now attaches to the binding site on the actin filament.

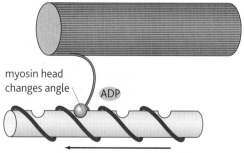

4 Head of myosin changes angle, moving the actin filament along as it does so. The ADP molecule is released.

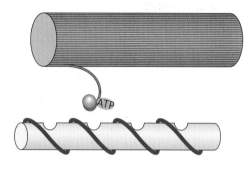

5 ATP molecule fixes to myosin head, causing it to detach from the actin filament.

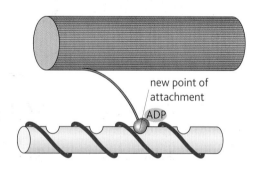

6 Hydrolysis of ATP to ADP by ATPase provides the energy for the myosin head to resume its normal position.

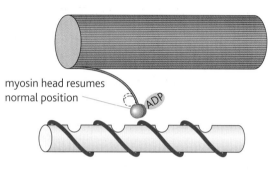

7 Head of myosin reattaches to a binding site further along the actin filament and the cycle is repeated.

Figure 2 *How ATP changes the shape of the myosin 'heads', allowing muscle contraction*

The role of calcium ions

When ATP is present, muscle tissue contracts. This means that there must be an inhibitor present in the tissue to stop it contracting all the time. This inhibitor is called **tropomyosin**. It fits into binding sites on the actin filaments, stopping myosin from binding.

However, when a muscle contracts, calcium ions enter the sarcoplasm. These bind to the protein troponin which acts to pull the tropomyosin out of the way, allowing the myosin 'heads' to bind to the actin binding sites. In relaxed muscles, calcium ions are actively transported out of the sarcoplasm. They are collected together in the membranes of T-tubules in the muscle filaments. T-tubules are infoldings of membrane that run through the muscle fibres at intervals, close to the endoplasmic reticulum.

Stimulating a muscle to contract

Muscles contract when they are stimulated by an action potential from a motor neurone. You will remember from Topic 5.5 that a motor neurone synapses with a muscle at a neuromuscular junction. When the transmitter substance fits into specific receptors on the postsynaptic membrane (the sarcolemma), sodium ions enter the sarcoplasm, depolarising the membrane. This causes calcium ions from the T-tubules to enter the sarcoplasm and muscle contraction occurs.

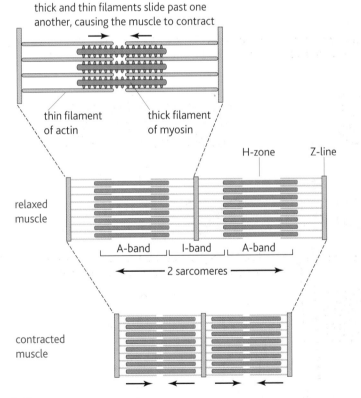

Figure 3 *How a sarcomere contracts*

Link

In Topic 2.8 you learned about the sex-linked condition, Duchenne muscular dystrophy. People with DMD have a faulty allele coding for the protein dystrophin. Dystrophin is a protein in the sarcolemma of the muscle fibre that forms part of the Z-line. Faulty dystrophin means that the actin filaments are not anchored to the sarcolemma as firmly as they should be.

Summary questions

1 Name the proteins present in: a the I-band, b the A-band, c the H-zone.

2 What happens to: a the I-band, b the A-band, c the H-zone when the sarcomere contracts?

3 A short time after a person dies, the muscles of the body contract firmly. The body is said to be in 'rigor mortis'. Use the information on this page to explain how this happens.

AQA Examination-style questions

1   (a)   **Figure 1** shows some of the muscles in a human leg. Describe what happens to the leg when;
   (i)   muscle **A** contracts;
   (ii) muscle **B** contracts.
   (iii) Explain why muscles such as **C** and **D** are arranged in pairs.   *(4 marks)*

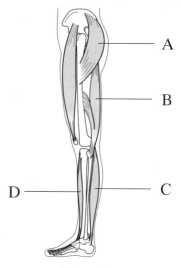

Figure 1

   (b)   **Figure 2** shows part of a skeletal muscle fibre.

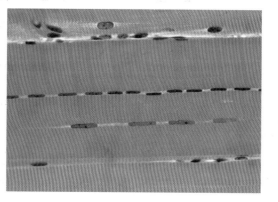

Figure 2

Give **three** differences in structure between a skeletal muscle fibre and an
epithelial cell from the lining of the small intestine.   *(3 marks)*

AQA, 2000

2   (a)   The table summarises features that may be shown by slow (tonic) and fast (twitch)
         muscle fibres.

Structure	A many mitochondria	B few mitochondria
Location	C near the surface	D deeply situated

Which **two** letters from **A, B, C** and **D** correspond to the features shown by slow
muscle fibres?   *(1 mark)*
   (b)   Explain the advantage of possessing **both** types of muscle fibres.   *(2 marks)*

AQA, 2000

3   **Figure 3** shows part of a myofibril as seen through an electron microscope.

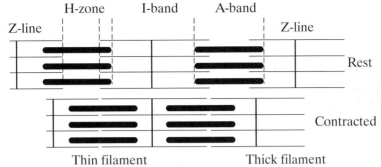

Figure 3

(a)   Name the main protein present in:

(i)   the thick filaments;

(ii) the thin filaments. *(2 marks)*

(b)   Describe the mechanism that brings about the change in position of the filaments when the myofibril contracts. *(4 marks)*

AQA, 2000

4   Myasthenia gravis is a disease which causes muscular weakness. It develops because of an attack by the body's own immune system on neuromuscular junctions. **Figure 4** shows a normal neuromuscular junction and one affected by the disease (myasthenic).

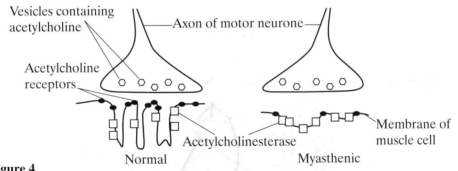

Figure 4

(a)   Describe **two** ways in which a myasthenic neuromuscular junction differs from a normal one and explain how each difference would affect transmissions across the myasthenic neuromuscular junction. *(4 marks)*

(b)   The changes in the neuromuscular junctions in myasthenia gravis result in fewer calcium ions entering muscle fibres. Explain how this reduces interactions between actin and myosin filaments and, thus, the strength of muscle contractions. *(3 marks)*

AQA, 2003

5   (a)   The table shows some differences between slow and fast muscle fibres.

Slow muscle fibres	Fast muscle fibres
Enable sustained muscle contraction to take place	Allow immediate, rapid muscle contractions to take place
Many mitochondria present	Few mitochondria present
Depend mainly on aerobic respiration for the production of ATP	Depend mainly on glycolysis for the production of ATP
Small amounts of glycogen present	Large amounts of glycogen present

(i)   Explain the advantage of having large amounts of glycogen in fast muscle fibres.

(ii) Slow muscle fibres have capillaries in close contact. Explain the advantage of this arrangement. *(4 marks)*

Hypothermia and diabetes – when control fails to work

7.1 Homeostasis and negative feedback

Learning objectives:

■ What is homeostasis and what is negative feedback?

■ Why is it important that the internal environment within the body is regulated?

Specification reference: 3.4.7

Homeostasis

Homeostasis is the regulation of the body's internal environment. It is important that the composition, pH and volume of the blood and tissue fluid are maintained within narrow limits, even if the external environment changes. There are several reasons for this:

■ Enzymes that control the metabolic pathways within cells work most efficiently at their optimum temperature and pH. If conditions vary significantly from the optimum temperature and pH, they start to denature. This also applies to other proteins within the body, such as proteins in cell membranes.

■ If the water potential of the blood and tissue fluid changes significantly, body cells may take in water by osmosis, causing them to swell or even burst. Alternatively, they could lose water by osmosis, stopping them from working efficiently.

Negative feedback

Homeostasis needs a control system that detects stimuli and makes the necessary changes to bring the internal environment back to normal. This control system is shown in Figure 1.

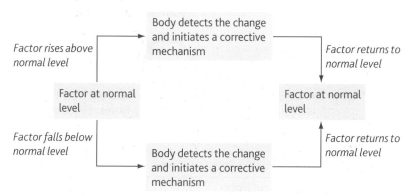

Figure 1 *Negative feedback*

You will see that the body detects the change in the body's internal environment. Hormones and/or nerve impulses are used to bring about a corrective mechanism that returns conditions to the normal level. This mechanism is called **negative feedback**.

Link

Look back at Topic 10.5 of *AS Human Biology* for more detail of how blood pH is regulated.

Controlling respiratory gases

You will remember from Chapter 10 of *AS Human Biology* that chemoreceptors in the carotid bodies, the aortic bodies and the medulla of the brain, monitor blood pH. You should also remember that blood containing more carbon dioxide is slightly acidic, causing blood pH to fall. This happens, for example, when you exercise. The fall in blood pH is detected by chemoreceptors, which send impulses to the cardiovascular and respiratory centres in the medulla of the brain. The respiratory centre in the medulla sends more nerve impulses to the diaphragm and intercostal muscles, causing an increase in the rate and depth of breathing. The cardiovascular centre sends more nerve impulses along the sympathetic nerve to the sino-atrial (SAN) of the heart, causing an increase in heart rate. As a result of these changes, the carbon dioxide concentration in the blood falls back to the normal level.

How science works

Claude Bernard and homeostasis

Claude Bernard (1813–1878), a French physiologist, is considered to be the 'father' of modern experimental physiology. He started his career in Paris, working on the physiology of digestion. He studied internal metabolism and contributed a great deal to our understanding. He was a strong believer in rigorous scientific experimentation as the way to understanding physiology.

Claude Bernard developed the concept of homeostasis. He proposed that 'La fixité du milieu intérieur est la condition de la vie libre'. In English, this means 'the fixity of the internal environment is the condition for free life'. He went on to explain that 'The living body, though it has need of the surrounding environment, is nevertheless relatively independent of it. This independence which the organism has of its external environment, derives from the fact that in the living being, the tissues are in fact withdrawn from direct external influences and are protected by a veritable internal environment which is constituted, in particular, by the fluids circulating in the body'.

Figure 2 *Claude Bernard working in his laboratory*

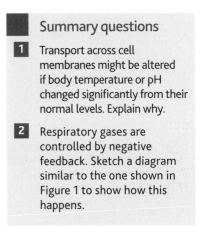

Summary questions

1 Transport across cell membranes might be altered if body temperature or pH changed significantly from their normal levels. Explain why.

2 Respiratory gases are controlled by negative feedback. Sketch a diagram similar to the one shown in Figure 1 to show how this happens.

7.2 Thermoregulation

■ **Link**

Look back at Topic 6.2 to make sure that you know where the hypothalamus is.

There are three ways in which the body gains or loses heat:

■ Radiation – in which heat energy is transferred from one place to another as electromagnetic waves.
■ Conduction – in which heat energy is transferred by the collisions of molecules. Liquids and solids conduct heat better than air.
■ Convection – in which heat is transferred by currents in air or water.

The body also loses heat by evaporation of sweat. When water in sweat turns from liquid into vapour, it takes heat from the body. Heat is also lost in substances that leave the body, such as urine, faeces and exhaled air.

■ Controlling body temperature

Human body temperature is controlled by the hypothalamus in the brain. Sensory receptor cells in the hypothalamus, called **thermoreceptors**, monitor the temperature of the blood flowing through the hypothalamus. This is known as the **core body temperature**. There are also thermoreceptors in the skin which monitor skin temperature. These thermoreceptors send nerve impulses to the hypothalamus.

If the thermoreceptors in the hypothalamus detect a change in body temperature away from the normal level, the hypothalamus brings about corrective mechanisms. You can see these in Figure 1.

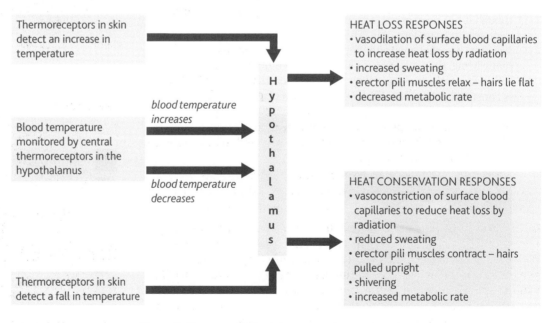

Figure 1 *Thermoregulation in humans*

Heat loss responses

If the thermoreceptors in the hypothalamus detect an increase in body temperature, the hypothalamus corrects this by bringing about **heat loss responses**. These include:

- **Vasodilation**. The hypothalamus sends impulses along the parasympathetic nerves to the skin arterioles, causing smooth muscle in the walls of these arterioles to relax. The shunt vessel constricts, forcing more blood to flow through the surface capillaries. This increases the amount of heat energy lost by radiation. You can see this in Figure 2.
- **Increased rate of sweating**. More sweat is secreted on to the surface of the skin. As the water in the sweat evaporates it removes heat from the body.
- **Pilorelaxation**. This has little effect in modern humans as we do not have hairy skins. However, the parasympathetic nerves going to the erector pili muscles at the base of the skin hairs cause these muscles to relax. This means that the hairs lie flat so that less air is trapped next to the skin. This reduces the insulating effect of the body hair.

Humans also reduce their body temperature by behavioural responses, such as opening a window, taking off a jumper or moving into the shade. The thermoreceptors in the skin and hypothalamus send impulses to the cerebral cortex in the brain. This makes the person aware that s/he is too hot. As a result, s/he does something to help to cool down. People also become more inactive when they are too hot. This helps to reduce body temperature as the rate of respiration in muscle tissue decreases. This means that less heat is produced.

If a person is exposed to high temperatures for a long period of time, the body produces less thyroxine, a hormone that increases metabolic rate. This reduces the person's metabolic rate.

Heat conservation responses

If the thermoreceptors in the hypothalamus detect a decrease in body temperature, the hypothalamus corrects this by bringing about **heat conservation responses**. These include:

- **Vasoconstriction**. The hypothalamus sends impulses along the sympathetic nerves to the skin arterioles, causing smooth muscle in the walls of these arterioles to contract. The shunt vessel relaxes, forcing less blood to flow through the surface capillaries. This reduces the amount of heat energy lost by radiation. Figure 3 illustrates this.
- **Decreased rate of sweating**. Less sweat is secreted on to the surface of the skin, so less heat energy from the body is used to evaporate the water in the sweat into vapour.
- **Piloerection**. Although this has very little effect in humans, the erector pili muscles attached to the hairs in the skin pull the hairs 'on end'. This traps an insulating layer of air, reducing heat loss.

Again, behavioural responses also help to conserve heat, such as putting on a jumper. Shivering also generates heat. This is an involuntary response that occurs when skeletal muscles contract and relax rapidly. This increases the rate of respiration in muscles, so more heat is generated. In addition, the hormone adrenaline is released which increases the metabolic rate. If a person is exposed to cold temperatures for a long time, the body produces more thyroxine. This increases the metabolic rate.

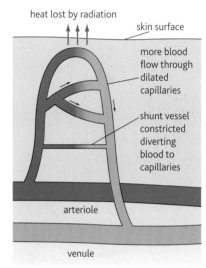

AQA Examiner's tip

Make sure you learn how to describe vasoconstriction and vasodilation properly. Many students think that the blood vessels move, but they do not. The volume of blood flowing through the surface capillaries changes but the capillaries stay in the same place.

Warm conditons

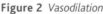

Figure 2 Vasodilation

Cold conditons

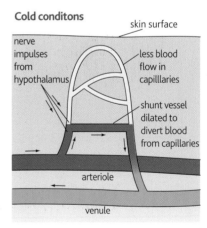

Figure 3 Vasoconstriction

Summary questions

1 Sketch a diagram like the one in Figure 1 in Topic 7.1 to show how body temperature is controlled by negative feedback.

2 When people are feeling cold they tend to curl up their bodies. Explain how this helps them to conserve heat.

7.3 Hypothermia and hyperthermia

Learning objectives:

- What happens when body temperature falls or rises significantly from the normal level?

Specification reference: 3.4.7

Hypothermia

Normal body temperature is about 37 °C. If body temperature falls significantly below this, usually considered to be below 35 °C, then a person is said to be suffering from **hypothermia**. When body temperature falls to this level, the normal heat conservation and heat generation responses of the body are unable to bring the body temperature back up. In other words the normal negative feedback mechanism fails.

As the body temperature falls, the body's rate of metabolism slows down. This happens because molecules have less kinetic energy, so enzyme-controlled reactions happen more slowly. As a result, metabolic reactions such as those involved in respiration slow down. These are the activities that should be producing heat to warm up the body. This is an example of **positive feedback**.

Hypothermia is a risk for elderly people, especially during cold winters. They may be worried about the cost of heating their homes, so they might not switch on the heating for long enough. Also, many elderly people are inactive so that their bodies do not produce much heat in respiration. However, even young and fit people may develop hypothermia if they are exposed to extreme cold. When the air temperature is very cold, heat is lost from the body rapidly. People who engage in outdoor activities, such as climbing or walking, may suffer from hypothermia if they experience difficulties, such as injury or poor weather conditions, that keep them exposed to cold weather for longer than they had anticipated.

When someone is suffering from hypothermia they may not recognise it. They will eventually stop feeling cold, and become relaxed or sleepy. It is important to give the correct first aid treatment at this point. If the person goes more deeply into hypothermia, the heart rate slows, the arterioles dilate and they will start to feel warm. They may even want to remove clothing, saying that they feel too hot. After this, they slip into unconsciousness.

Hyperthermia

Hyperthermia is a condition in which the body temperature rises significantly above normal. Our bodies mainly lose heat by sweating. If someone is dehydrated they may not produce enough sweat. Also, if the air is humid (i.e. saturated with water vapour) then sweat may not evaporate quickly enough to bring body temperature down.

As body temperature rises above the normal level the person starts to feel ill. If body temperature rises above 41 °C the hypothalamus is no longer able to regulate body temperature and it continues to rise. The person will start to feel light-headed and sick. This condition is known as heat stroke.

First aid treatment for hyperthermia is to bring the body temperature down again as soon as possible. This is done by applying cold water to the body and exposing the skin so that the water can evaporate easily. The person should also be given cold drinks if they are able to swallow.

How science works

Fever

Fever is a medical sign that body temperature has risen above its normal level. It is a symptom of disease in which body temperature is raised 1–2 °C above normal. Fever is different from hyperthermia because the body releases substances called **pyrogens**. These change the body's set-point, so that the person feels cold, has an increased heart rate and shivers even though their body temperature is normal. Some scientific studies have suggested that the body recovers more rapidly from an infection if they have a fever. Other scientists are less sure that fever is useful.

Figure 1 *A boy suffering from a fever*

1 Suggest ways in which having a higher than usual body temperature might help the body to fight an infectious disease.

Application

First aid treatment for hypothermia

■ Send for medical help.
■ Prevent further body heat loss and warm the casualty gradually. This is best done by wrapping them in layers of blankets, preferably in a warm room.
■ Put the casualty to bed and keep them well covered, including a hat.
■ Do not try to put the casualty in a bath.
■ Offer the casualty warm drinks.

1 Explain why the best way to warm the person is by wrapping them in blankets and putting them to bed.

2 Explain why the person should be given a hat to wear.

3 Explain why warm drinks will be helpful.

Summary questions

1 There are stories of dead climbers being found naked on a mountain, with their clothes neatly folded nearby. Explain how this might have occurred.

2 People who walk and climb in exposed places are advised to take a survival bag. This is a large plastic bag, big enough for a person to crawl into. Explain how this might help an injured hill walker to survive cold conditions.

7.4 Regulation of blood glucose

Learning objectives:

- How are blood glucose levels regulated?

- What factors affect blood glucose regulation?

Specification reference: 3.4.7

Control of blood glucose levels

In a healthy human, blood glucose levels are normally kept at about 90 mg per 100 cm³ of blood. It is important that blood glucose levels are kept relatively constant because glucose is the most important respiratory substrate in humans. If the blood glucose level changes significantly this will change the water potential of the blood. This means that cells will gain or lose too much water through osmosis.

The receptors that monitor blood glucose levels are found in the **islets of Langerhans** in the pancreas (Figure 1). There are two main kinds of cells here: the α cells which secrete the hormone glucagon and the β cells which secrete the hormone insulin.

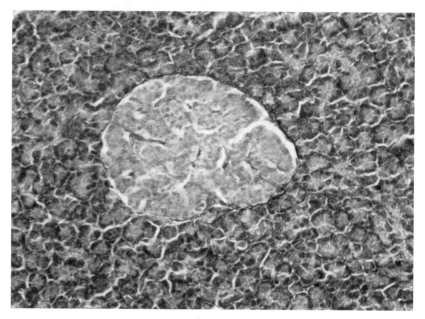

Figure 1 *Section through the pancreas showing an islet of Langerhans*

High blood glucose levels

Blood glucose levels rise when a meal containing carbohydrates is eaten and glucose is absorbed into the blood. This rise is detected by the β cells in the islets of Langerhans, which secrete the hormone insulin. Insulin travels in the blood and fits into specific receptors in the cell membranes of liver and muscle cells. This lowers blood glucose level in several ways:

- Insulin stimulates the uptake of glucose by all respiring cells in the body, especially liver and muscle cells. In these cells insulin causes glucose channels in the plasma membranes to open, allowing more glucose to enter the cells.

- Insulin stimulates an increase in respiration rate so that more glucose is respired.

- Insulin activates enzymes inside liver and muscle cells, increasing the rate at which glucose is converted into glycogen. Glycogen is a polysaccharide made of many glucose units linked together. It is insoluble and compact, so it is an ideal storage molecule. This process is called **glycogenesis**.

■ Insulin activates other enzymes which convert glucose into fatty acids. These are converted into fats which are deposited in adipose tissue (fat storage tissue).

Insulin is secreted until the blood glucose level returns to normal.

Low blood glucose levels

Blood glucose levels can fall when a person has not eaten a meal for some time, especially if they have been very active. This fall in blood glucose level is detected by the α cells in the islets of Langerhans, which respond by secreting the hormone glucagon. Glucagon, like insulin, binds to specific receptor proteins in plasma membranes. This activates enzymes that convert glycogen back to glucose. This process is called **glycogenolysis**. Glucagon also stimulates the conversion of amino acids to glucose which results in more glucose being secreted into the blood.

The way in which insulin and glucagon control blood glucose levels is shown in Figure 2.

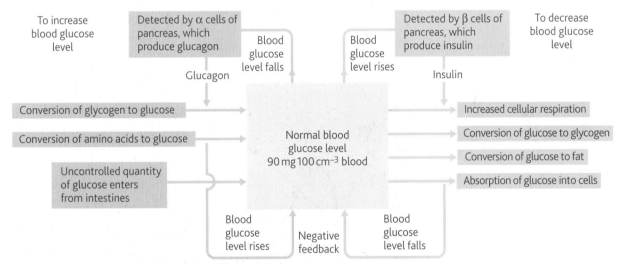

Figure 2 *Summary of the regulation of blood glucose*

The control of blood glucose is an example of negative feedback because a rise in blood glucose level triggers corrective mechanisms that return the blood glucose level back to the normal level. Similarly, when the blood glucose level falls below normal, this triggers corrective mechanisms that bring the blood glucose level back to normal.

Summary questions

1 Sketch a diagram like Figure 1 in Topic 7.1 to show how blood glucose levels are controlled.

2 Explain why a period of physical activity reduces blood glucose levels.

3 Why is glycogen a better storage molecule than glucose?

Learning objectives:

- What is the difference between Type 1 and Type 2 diabetes?

- What are the health implications of diabetes?

- How can diabetes be controlled?

Specification reference: 3.4.7

Type 1 diabetes

Diabetes mellitus is a condition in which people are unable to control their blood glucose levels. There are two types of diabetes. **Type 1 diabetes** occurs early in life. People with Type 1 diabetes do not produce enough insulin. This means that when they have eaten a meal, their blood glucose level can remain very high for a long time. As a result, glucose is present in the urine. A person suffering from Type 1 diabetes may be excessively thirsty and will need to urinate frequently. They will lose weight because body cells do not receive enough glucose. To compensate the body breaks down fats and proteins in body tissues such as muscles. If Type 1 diabetes is left untreated it can be fatal.

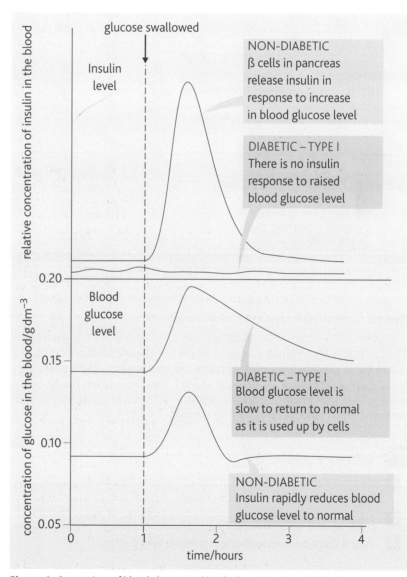

Figure 1 *Comparison of blood glucose and insulin levels in a Type 1 diabetic and a non-diabetic individual after each swallowed a glucose solution*

Type 1 diabetes can be treated with regular insulin injections. It is important that the person injects the correct amount of insulin. If they inject too much insulin, their blood glucose level can become too low (or **hypoglycaemic**) and they may fall into a coma. They could even die. If they have not injected enough insulin or if they have eaten too much sugary food, their blood glucose level may become too high (or **hyperglycaemic**). This can also lead to coma.

Figure 1 shows the changes in blood glucose and insulin levels in a person with Type 1 diabetes and in a non-diabetic person, after they have had a glucose drink. You will notice that the person with diabetes does not produce any insulin, so their blood glucose level rises much higher than in the non-diabetic person. Also, the blood glucose level of the diabetic person stays high for a long time.

■ Type 2 diabetes

Type 2 diabetes usually occurs in older people, especially if they are overweight. However, some young adults and even children are now developing the disease. People with Type 2 diabetes usually produce plenty of insulin, but the liver and muscle cells no longer respond to it. This kind of diabetes is usually treated by exercise and diet. The person is encouraged to lose weight, to exercise regularly and to eat foods with a low glycaemic index.

■ How science works

Diabetes mellitus

The term 'diabetes' was first used in about 250 BC. It is based on a Greek word meaning 'to siphon', because fluid seemed to drain rapidly from an affected person. Thomas Willis, physician to King Charles II, coined the word 'mellitus' in 1674. Mellitus is the Latin word for honey. The urine of a diabetic tasted of honey and at the time, this was the only way for doctors to diagnose the disease.

In 1889, two German doctors showed that dogs would develop diabetes if the pancreas was removed. Later, the Canadians Dr Frederick Banting and his assistant Charles Best, working in the laboratory of Professor John Macleod, tried to isolate an extract from the pancreas that could be used to treat diabetes. They succeeded in 1921.

Banting and Best induced diabetes in several dogs by removing the pancreas. These dogs produced glucose in the urine and had high blood glucose levels. They then injected pancreatic extract into the dogs and showed that this reduced the dogs' blood glucose levels.

A biochemist, Dr John Collip, was employed to purify the pancreatic extract. A young boy, Leonard Thompson, was the first patient to receive insulin treatment. In 1922, aged 14, he was extremely ill and weighed only 4 stone 8 lbs (29 kg). At first the extract only reduced the blood glucose level a little and the boy developed abscesses around the site of the injections. However, Collip worked hard on purifying the extract. Several weeks later, Leonard was injected again and made a remarkable recovery. His blood glucose levels fell, he gained weight and lived for another 13 years. He died from pneumonia at the age of 27.

In 1923, Banting and Macleod were awarded the Nobel Prize for the discovery of insulin. Banting split his prize with Best, and Macleod split his prize with Collip.

Figure 2 *Frederick Banting and Charles Best with one of the dogs they used in their research*

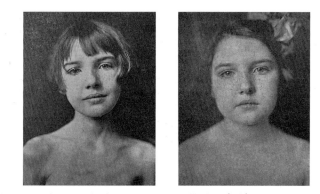

Figure 3 *Photograph of Case VI: young girl before (left) and after insulin treatment*

Testing for diabetes

One test used to diagnose diabetes is the **fasting blood glucose level** test. The person is told not to eat or drink anything, except water, overnight. The blood glucose level is measured in the morning. If this is above $7\,\text{mmol}\,\text{dm}^{-3}$ the person has diabetes.

Another test is the **oral glucose tolerance test**. The person does not eat or drink anything overnight, except water. In the morning, the person drinks a solution containing 75 g glucose. A blood sample is taken 2 hours later. If the blood glucose level is $11.1\,\text{mmol}\,\text{dm}^{-3}$ or higher, the person has diabetes.

1. Why is it important that the person does not eat or drink anything, except water, before these tests are carried out?

2. Explain why a person who is not diabetic will have a blood glucose level below $11.1\,\text{mmol}\,\text{dm}^{-3}$ 2 hours after drinking the glucose solution in the oral glucose tolerance test.

Application

Monitoring blood glucose levels

It is important that a person with diabetes should be able to measure their blood glucose levels regularly if they are to manage their diabetes well. Most blood glucose testing is now done with a biosensor. This kind of device can be used by a health professional or by a person in his or her own home.

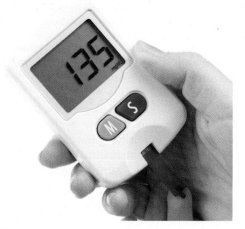

Figure 4 *Blood glucose meter*

A biosensor test strip is placed in the meter. This test strip contains an enzyme, glucose dehydrogenase. The person pricks their finger with a sterile lancet inside a device that looks rather like a pen. A drop of blood is placed on the edge of the test strip. The enzyme glucose dehydrogenase converts any glucose in the blood sample into a substance called gluconolactone. This reaction produces a very small electric current which is picked up by an electrode on the test strip. In turn, this produces a reading for blood glucose concentration on the digital screen of the meter.

1 Explain why this meter will detect only glucose, and not any other substance present in blood.

2 Explain how a biosensor like this can help a person to control their diabetes.

Summary questions

1 Insulin can be made using recombinant DNA technology (see Topic 4.2). Suggest advantages of producing insulin using recombinant DNA technology rather than by purifying pancreatic extract from another animal.

2 Explain how a exercise and b foods with a low glycaemic index can help a person with Type 2 diabetes to control their blood glucose levels.

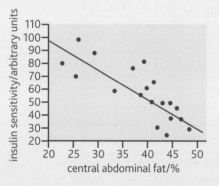

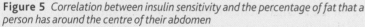

Figure 5 *Correlation between insulin sensitivity and the percentage of fat that a person has around the centre of their abdomen*

3 People with Type 2 diabetes have liver and muscle cells that are less sensitive to insulin than the liver and muscle cells of non-diabetics. Figure 5 shows the correlation between insulin sensitivity and the percentage of fat that a person has around the centre of their abdomen.

a Describe the trend shown by this graph.

b Does this graph provide evidence that people with a lot of central abdominal fat are at greater risk of Type 2 diabetes? Give reasons for your answer.

1   **Figure 1** shows how the concentration of glucose in the blood is regulated.

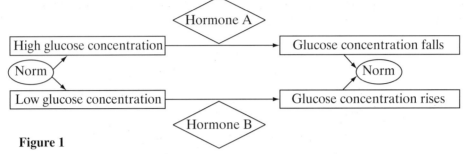

Figure 1

(a) Name:
   (i) hormone **A**;
   (ii) hormone **B**.   *(2 marks)*

(b) Explain how hormone **B** brings about the change shown in the diagram.   *(2 marks)*

(c) Two people fasted for 12 hours and each was then given a drink which contained
    100 g of glucose. Their blood glucose concentration was measured regularly for 3
    hours. The results of the investigation are shown in the table.

Time after glucose drink / minutes	Blood glucose concentration / mg per 100 cm³ of blood	
	Person X	**Person Y**
0	81	90
20	136	131
40	181	142
60	213	89
90	204	79
120	147	74
150	129	86
180	113	86

   (i)   Suggest an explanation for the changes in the blood glucose concentration of
         person **X**.

   (ii) Explain how the concentration of hormone **A** in the blood would vary in
         person **Y** between 0 and 60 minutes.

   (iii) Suggest an explanation for the results shown by person **Y** between 90 and
         180 minutes.   *(4 marks)*

AQA, 1999

2   Size matters for marathon runners. Big athletes produce more heat and find it harder
     to keep cool. Shape matters too; a tall, thin runner has fewer problems keeping cool
     than a short, tubby runner of the same body mass. A 65 kg athlete running a marathon
     in 2 hours 10 minutes in reasonably dry conditions can avoid overheating at air
     temperatures up to 37 °C, but in humid conditions the same level of performance is
     possible only at temperatures below about 17 °C.

(a) Explain how athletes produce heat when they run.   *(2 marks)*

(b) Why does a 'tall, thin runner have fewer problems keeping cool than a short,
    tubby runner of the same body mass'?   *(2 marks)*

(c) Explain why runners are more likely to overheat in humid conditions.   *(3 marks)*

(d) Describe how the body responds to a rise in core body temperature.   *(5 marks)*

AQA, 2001

3 The glucose tolerance test is used in hospitals to assess insulin production. A patient fasts for several hours before swallowing 50 g of glucose in 150 cm^3 water. The concentration of glucose in the patient's blood is then measured immediately and at 30-minute intervals over a period of two to three hours. **Figure 2** shows changes in the blood glucose concentration of three patients who have taken this test.

(a) Explain why it was necessary for the patients to fast for several hours before the test was carried out. *(1 mark)*

(b) Explain the changes which take place in the blood glucose concentration of the non-diabetic person over the period shown in **Figure 2**. *(4 marks)*

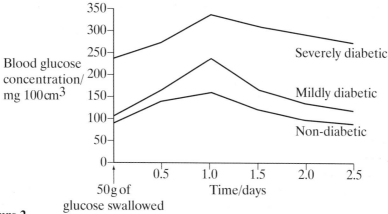

Figure 2

(c) What can be concluded from the results of this test about insulin production in the two diabetic patients? Explain the evidence for your answer. *(3 marks)*

AQA, 2000

4 **Figure 3** shows some important features of homeostatic mechanisms in the body.

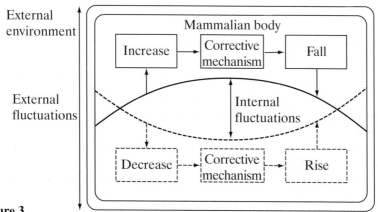

Figure 3

(a) Use the information in the diagram to help explain the importance of a mammal maintaining a constant internal temperature. *(4 marks)*

(b) Explain the role of the hypothalamus and nervous system in the regulation of body temperature. *(5 marks)*

(c) Explain why, in a normal healthy individual, the blood glucose concentration fluctuates very little. *(6 marks)*

AQA, 2000

Unit 4 questions: Bodies and cells in and out of control

1   **Figure 1** shows the nucleus of a sperm entering a human secondary oocyte. Only three of the 23 chromosomes in the oocyte are shown.

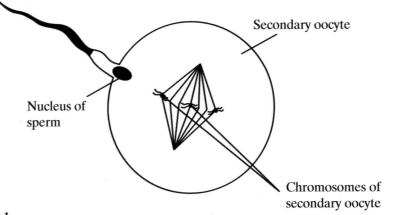

Secondary oocyte

Nucleus of sperm

Chromosomes of secondary oocyte

Figure 1

   (a) Which stage or stages of meiosis must still occur to complete the formation of the nucleus of the ovum? Explain your answer.   *(2 marks)*
   (b) The zygote develops into a blastocyst. What is a blastocyst?   *(2 marks)*
   (c) In the early stage of pregnancy, hCG is found in the urine of the woman. Explain why this is an indicator of pregnancy.   *(2 marks)*

AQA, 2007

2   The graph in **Figure 2** shows the range of blood plasma concentrations of FSH and oestrogen found in females of different ages.

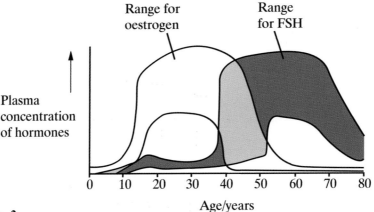

Range for oestrogen   **Range for FSH**

Plasma concentration of hormones

Age/years

Figure 2

   (a) Describe **two** changes in the body caused by the initial increase in oestrogen concentration shown on the graph.   *(1 mark)*
   (b) It is more useful to plot the mean plus and minus one standard deviation than to plot the range of concentrations. Explain why.   *(2 marks)*
   (c) Use data from the graph to explain the change in FSH concentration between the ages of 37 and 50.   *(2 marks)*

AQA, 2007

3   (a) (i)   What is meant by homeostasis?   *(1 mark)*
        (ii) Giving **one** example, explain why homeostasis is important in mammals.   *(2 marks)*
    (b) (i)   Cross-channel swimmers experience a large decrease in external temperature when they enter the water. Describe the processes involved in thermoregulation in response to this large decrease in external temperature.   *(7 marks)*
        (ii) A person swimming in cold water may not be able to maintain their core body temperature and begins to suffer from hypothermia. Explain why a tall, thin swimmer is more likely to suffer from hypothermia than a short, stout swimmer of the same body mass.   *(2 marks)*
    (c) Cross-channel swimmers may suffer from muscle fatigue during which the contraction mechanism is disrupted. One factor thought to contribute to muscle fatigue is a decrease in the availability of calcium ions within muscle fibres. Explain how a decrease in the availability of calcium ions could disrupt the contraction mechanism in muscles.   *(3 marks)*

AQA, 2006

4   IQ test scores have been used as a measure of intelligence. Genetic and environmental factors may both be involved in determining intelligence. In an investigation of families with adopted children, the mean IQ scores of the adopted children was closer to the mean IQ scores of their adoptive parents than to that of their biological parents.
    (a) Explain what the results of this investigation suggest about the importance of genetic and environmental factors in determining intelligence.   *(1 mark)*
    (b) Explain how data from studies of identical twins and non-identical twins could provide further evidence about the genetic control of intelligence   *(4 marks)*

AQA, 2006

5   (a) Describe how the resting potential of a neurone is maintained.   *(2 marks)*
    (b) The resting potential of a neurone is maintained at −70 mV. A metabolic poison was applied to a neurone and the change in the resting potential was measured over several hours. The results are shown in the graph in **Figure 3**.

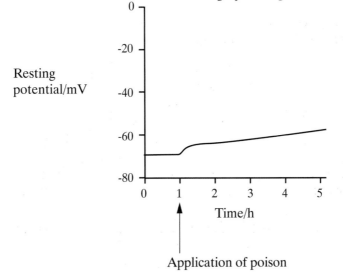

Figure 3

Explain the change in resting potential that takes place after the application of the metabolic poison.   *(4 marks)*

AQA, 2007

6 **Figure 4** shows three cells, **A**, **B** and **C** from the same organism. One of the cells is in the first division of meiosis, one is in the second division of meiosis and one is dividing by mitosis.

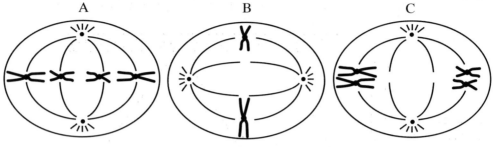

A   B   C

Figure 4

(a) What is the diploid number of chromosomes in the organism from which these cells were taken? *(1 mark)*

(b) Complete the table to show which of the cells, **A**, **B** or **C** is in the first division of meiosis and which is in the second.

Stage of meiosis	Cell
First division	
Second division	

(1 mark)

(c) Explain two ways in which meiosis leads to genetic variation in gametes. *(4 marks)*

AQA, 2007

7 **Figure 5** shows glycolysis and the Krebs cycle.

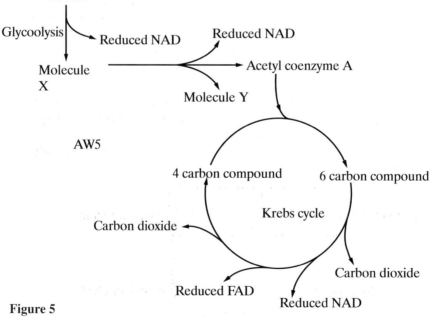

Figure 5

(a) Name
  (i) molecule **X** *(1 mark)*
  (ii) molecule **Y** *(1 mark)*

(b) Where, in a cell, does glycolysis occur? *(1 mark)*

(c) High concentrations of ATP inhibit an enzyme involved in glycolysis.
  (i) Describe how inhibition of glycolysis will affect the production of ATP by the electron transfer chain *(1 mark)*
  (ii) Explain this effect *(3 marks)*

AQA, 2007

8   Some people have red-green colour blindness. This may be caused by a mutant allele that results in the failure to produce a light-sensitive pigment in one type of cone cell. The graph in **Figure 6** shows the absorption spectra of the pigments from the cone cells of a person with this form of colour blindness.

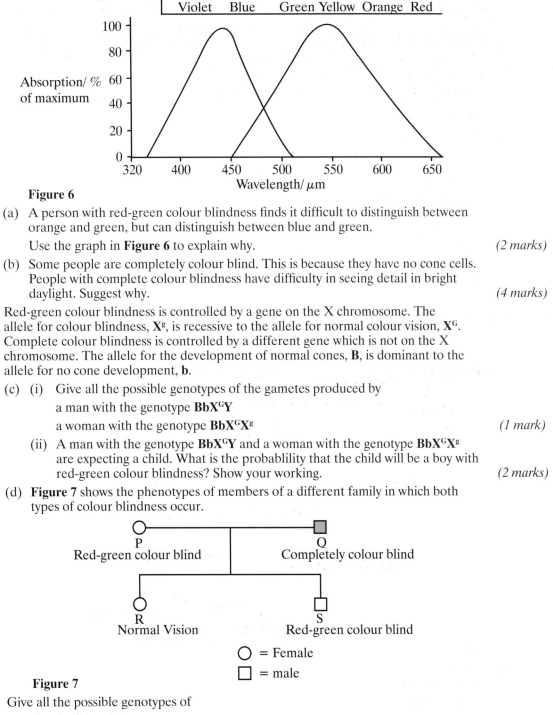

Figure 6

(a)   A person with red-green colour blindness finds it difficult to distinguish between orange and green, but can distinguish between blue and green.
      Use the graph in **Figure 6** to explain why.   *(2 marks)*

(b)   Some people are completely colour blind. This is because they have no cone cells. People with complete colour blindness have difficulty in seeing detail in bright daylight. Suggest why.   *(4 marks)*

Red-green colour blindness is controlled by a gene on the X chromosome. The allele for colour blindness, X^g, is recessive to the allele for normal colour vision, X^G. Complete colour blindness is controlled by a different gene which is not on the X chromosome. The allele for the development of normal cones, **B**, is dominant to the allele for no cone development, **b**.

(c)   (i)   Give all the possible genotypes of the gametes produced by
            a man with the genotype BbX^GY
            a woman with the genotype BbX^GX^g *(1 mark)*

      (ii) A man with the genotype BbX^GY and a woman with the genotype BbX^GX^g are expecting a child. What is the probablility that the child will be a boy with red-green colour blindness? Show your working.   *(2 marks)*

(d)   **Figure 7** shows the phenotypes of members of a different family in which both types of colour blindness occur.

Figure 7

Give all the possible genotypes of
      (i)   individual **P**
      (ii) individual **Q**
      (iii) individual **S**   *(4 marks)*

(e)   Complete colour blindness occurs in the same frequency in males and females, but red-green colour blindness occurs more frequently in males. Explain why.   *(2 marks)*

AQA, 2007

The air we breathe, the water we drink, the food we eat

Chapters in this unit

Human populations exist within communities and ecosystems, and depend upon other organisms for survival. However, humans are changing the ecosystems that they share with populations of other organisms. There are conflicts of interest between human demands and conservation of ecosystems. This Unit will explore how human populations integrate with others.

In Chapter 8 you will learn what effect human populations have on evolution. Chapter 9 looks at how human impact is affecting the stability of populations, communities and ecosystems and altering the selection pressures on populations of organisms.

The environmental changes produced by human activities are also affecting human health. In Chapter 10 you will learn how changes to diet and lifestyle may have contributed to the increase in a range of allergies and how the pollution of air and water can also lead to illness.

Human activities can damage ecosystems and create new ones and scientists must study ecosystems to monitor environmental changes. You will learn in Chapter 11 about the techniques used to study ecosystems and how waste land can provide sites where communities of plants and animals can survive. Human populations produce large quantities of waste that can damage ecosystems. You will learn about waste management schemes that aim to make the disposal of waste sustainable.

When fossil fuels are burnt they produce greenhouse gases. In Chapter 12 you will learn how the carbon footprint provides a measure of the impact that human activities have on the amount of greenhouse gases produced. You will learn that greenhouse gases have an effect on climate change and how this climate change is affecting the distribution of plants and animals. Chapter 12 also looks at how plants can reduce the impact of the use of fossil fuels.

The final chapter of this unit looks at humans as ecosystems. You will learn about the communities of microorganisms that live on the skin and in the gut and how human actions can change these communities and affect their stability.

What you already know

Whilst the material in this unit is intended to be self-explanatory, there is certain information from GCSE and the AS Human Biology course that will prove very helpful to the understanding of this unit. Knowledge of the following elements of GCSE and the AS Human Biology course will be of assistance:

- ☐ Living organisms are interdependent and adapted to their environment.

- ☐ A species is a group of similar organisms that can interbreed to produce fertile offspring.

- ☐ There are differences between different species of plants and animals and between individuals of the same species.

- ☐ Particular genes or mutations in the genes of plants or animals may give them characteristics that enable them to survive better. Over time this may result in the evolution of an entirely new species.

- ☐ Humans often upset the balance of different populations in natural ecosystems, or change the environment so that some species find it difficult to survive.

- ☐ The landscape and ecosystems that are present in the UK today are largely the result of human activities.

- ☐ Green plants use light energy to make their own food during photosynthesis. They make this food using carbon dioxide from the air and water from the soil.

- ☐ By observing the numbers and sizes of the organisms in food chains we can find out what happens to energy as it passes along the food chain.

- ☐ Microbes play an important part in decomposing dead plants and animals. The same material is recycled over and over.

- ☐ During aerobic respiration chemical reactions occur which use glucose and oxygen and release energy.

- ☐ Our bodies provide an excellent environment for many microbes

- ☐ Many strains of bacteria, including MRSA, have developed resistance to antibiotics as a result of natural selection.

8.1 Alleles and variation

Learning objectives:

■ What can change the frequency of an allele within a population?

■ What produces phenotypic variation?

Specification reference: 3.5.1

In Chapter 2 you learnt about mutations. A **mutation** is any change to the quantity or structure of an organism's DNA. A mutation that changes the sequence of bases in a gene results in a different form of that gene, called an **allele**. If the mutation occurs in a gamete, the allele will be passed on to the offspring. The **phenotypes** of the individuals that inherit this allele may be changed, adding to the **phenotypic variation** within the **population**. If the individuals that inherit the allele are more likely to survive and reproduce than individuals without the allele, then after several generations the frequency of that allele will increase within the population. You have already seen an example of this when you looked at sickle-cell anaemia in Topic 2.6. Evolution involves a change in the **allele frequency** in a population.

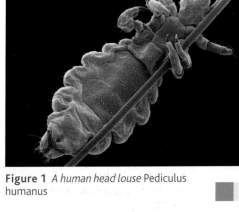

Figure 1 shows a head louse, *Pediculus humanus*. Head lice are small, wingless insects that are human **parasites**. They cling to hairs on the head where they lay their eggs and feed on blood which they take from the scalp. Head louse infestations are treated with chemical shampoos that contain insecticides, but some head lice are not being killed by these treatments – they are resistant to them.

Figure 1 *A human head louse* Pediculus humanus

How science works

Head lice forming a resistance movement

Recent research carried out in Wales suggests that 80% of head lice appear to be resistant to common chemicals used as insecticide treatments against them. The chemicals, called pyrethroids, are effective against non-resistant head lice because they bind to sodium channels in neurones and disrupt nerve impulses.

Scientists from the Liverpool School of Tropical Medicine in the UK screened 3000 children selected randomly from 31 schools across Wales. They found that 8% of the children had head lice.

The researchers sent 316 of the head lice obtained from infested children for genetic analysis and found that 82% tested positive for pyrethroid-resistance alleles. These alleles code for a sodium channel protein that will not bind to pyrethroid, so nerve impulses are not disrupted.

Resistance was greatest in areas that had the most use of pyrethroid treatments for head lice.

In southern parts of Wales more than 11% of head lice were found to be homozygous for the pyrethroid-resistance allele. But in northern Wales, where pyrethroid treatments had been discouraged for 5 years preceding the study, only 3% of the head lice were homozygous for this allele.

Resistance to pyrethroids has also been documented in other countries, such as Israel.

© *New Scientist Magazine*

1 What is a parasite?

2 Calculate the number of children in the sample who were found to have head lice.

3 Why was the screening carried out randomly?

4 How has the allele for resistance to the pyrethroid insecticide arisen?

5 Suggest a reason for the higher incidence of head lice that are homozygous for the pyrethroid-resistance allele in southern Wales, compared to northern Wales.

6 How could local health authorities discourage the use of pyrethroid treatments against head lice, yet still keep the infestation rates low?

Within a population, the phenotypic variation that exists is the result of **genetic factors** (e.g. the presence of certain alleles which would have originally arisen within the population by mutation), **environmental factors** (e.g. food availability) or a combination of both. In certain environments, individuals that inherit a mutant allele may be at a **selective advantage**. This means they are more likely to survive and produce offspring that also carry the mutant allele than individuals without the mutant allele. This leads to an increase in the **frequency** of the mutant **allele** in the population – the population is **evolving**.

Summary questions

1 Give definitions for the following terms: population, phenotypic variation, mutation, allele frequency, evolution.

2 Apart from the examples given in the text, give **two** genetic and **two** environmental factors that could contribute to phenotypic variation in a population of head lice.

3 Warfarin is a rat poison that was introduced into Britain in 1953. In 1960, a population of rats near Welshpool was found to be resistant to warfarin. Within a few years, warfarin-resistant rats were found in many different areas. Use your knowledge of alleles and natural selection to explain how this might have happened.

AQA Examiner's tip

Populations evolve, not individual organisms. You should be able to explain why.

8.2 Selection

Learning objectives:

- Why are some members of a population more likely to survive, reproduce and pass on their genes to the next generation?

- What is a selection pressure?

Specification reference: 3.5.1

Link

Look back at Topic 4.4 of *AS Human Biology* to remind yourself how antibiotics work.

Living things produce more offspring than their environment can support. When large numbers of offspring are produced within a population, **intraspecific competition** occurs. **Phenotypic variation** within a population will mean that some offspring will die or will not reproduce. This may be because they are more susceptible to disease, less able to escape predators, less able to find a mate or unable to obtain sufficient food. As a result, the offspring best suited to their environment are most likely to survive, reproduce and pass on their alleles to the next generation. This is the basis of **natural selection**, and the survivors are said to have a **selective advantage**.

Selection and resistance to antibiotics in bacterial populations

Microorganisms such as bacteria reproduce very rapidly, sometimes dividing by **binary fission** as often as every 20 minutes. This means that changes in allele frequency within a population of bacteria can be observed taking place over a relatively short time.

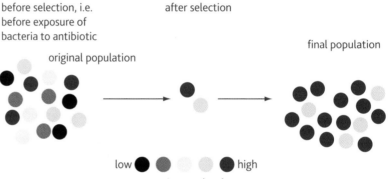

Figure 1 *How the use of an antibiotic can increase the level of resistance within a bacterial population*

Look at Figure 1. It shows how a bacterial population in a person may respond to treatment with a particular antibiotic. You will remember from AS level that many bacteria carry genes for resistance to certain antibiotics on plasmids. The bacteria vary in their susceptibility to a particular antibiotic. The most susceptible bacteria die first when the population is exposed to the antibiotic. Gradually, the proportion of more resistant bacteria in the population increases. The bacteria that can survive exposure to the antibiotic are able to reproduce and pass on their genes for resistance to their offspring. The antibiotic is exerting a **selection pressure** on the bacterial population and the frequency of the gene for resistance to it is increasing within the population. If a person completes their course of prescribed antibiotics, the chance of any bacteria surviving is small. Their immune system can destroy the remaining bacterial cells. However, if a person stops the antibiotic treatment early, a high proportion of the surviving bacteria will be resistant. Anyone who becomes infected with these more resistant bacteria will need stronger doses of that antibiotic, or another type of antibiotic, to combat the infection. If incomplete antibiotic treatment is repeated often enough, then after many generations, all of the bacteria in the population will have evolved complete antibiotic resistance. You will remember from AS level that this has happened within some populations of the bacterium *Staphylococcus aureus*, producing MRSA (methicillin-resistant *Staphylococcus aureus*).

In the absence of antibiotics, the proportion of resistance genes in the population tends to decrease to a low level. This is because within the population, the resistant bacteria tend to be at a disadvantage relative to those bacteria that are not resistant to antibiotics.

How science works

Antibiotic prescriptions

On the 23rd July 2008 doctors were told to no longer prescribe antibiotics to all patients with colds, coughs and sore throats. The new guidelines were put in place by the National Institute for Clinical Excellence (NICE). NICE is an independent watchdog responsible for providing national guidance on healthcare and they believe that most colds and sore throats would clear up by themselves, and that antibiotics were not necessary. They advised that antibiotics should only be prescribed when symptoms persist, or to very young or elderly patients.

The main reason for these guidelines being put in place is that many experts believe harmful bacteria may build up a resistance to antibiotics, which could lead to the spread of 'superbugs'. There are also concerns about how much antibiotics are costing the National Health Service.

1 Colds are caused by viruses. Explain why antibiotics will not help to treat colds.

2 Explain how overuse of antibiotics has led to the spread of MRSA.

3 A person with a sore throat infection, caused by bacteria, may recover within a few days even if they are not given antibiotics. Explain how.

4 Suggest why NICE recommends that some people, such as the very young or elderly, should be given antibiotics.

How science works

Birth weight in humans

Studies carried out during the 20th century indicated that there was selection for an optimum human birth weight. This is summarised in Figure 2. Babies with a birth weight below the optimum were more susceptible to disease and less likely to survive. Heavier birth weight babies were more likely to suffer injury as a result of a difficult delivery and this reduced their chances of survival. This selection for optimum birth weight has probably acted on populations of humans and their ancestors for over a million years. However, in more economically developed countries such as Britain, the selection pressure for an optimum birth weight has almost disappeared.

1 Suggest why selection for optimum birth weight may have acted on populations of humans and their ancestors for over a million years.

2 What may have caused the selection pressure for an optimum birth weight to almost disappear in modern Britain?

Summary questions

1 Genes for antibiotic resistance can be transferred from one bacterial cell to another by means of plasmids. What is a plasmid?

2 Why is it important that a person completes the course of antibiotics that they are prescribed?

3 Why is the use of antibiotics for treatment of minor infections discouraged?

4 Explain why the proportion of genes for resistance to a particular antibiotic will decrease to a low level in a population if the use of that antibiotic is discontinued.

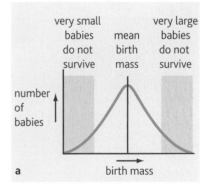

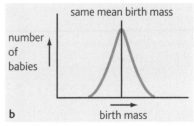

Figure 2 *Selection for an optimum birth weight in humans* **a** *before selection,* **b** *after selection*

8.3 Isolation and the formation of new species

Learning objectives:

- How can the reproductive isolation of populations result in the formation of new species?

- What is the difference between allopatric speciation and sympatric speciation?

Specification reference: 3.5.1

Link

Look back at Topic 9.1 of *AS Human Biology* to remind yourself how living things are classified and named.

Hint

A gene pool is all the genetic information (genes) present within a population at a given time.

Speciation

You will recall from AS level that a **species** is a group of similar organisms that can interbreed to produce fertile offspring.

Speciation is the evolution of a new species from an existing one. New species may arise in two different ways:

- Individuals from two different species cross-fertilise and produce a fertile hybrid. This is thought to have happened in the case of the grass species, *Spartina anglica*.

- Groups within the population become **isolated** from one another and cannot interbreed. This is called **reproductive isolation**. **Gene flow** amongst the separate groups is prevented and the population's single **gene pool** is now split into several separate gene pools. **Allele frequencies** within each gene pool may change because **selection pressures** acting on the **phenotypes** of the organisms may be different. The separate gene pools may become so different from each other that successful interbreeding of the separate groups is no longer possible. Each group is now a different species with its own gene pool. This type of speciation has two main forms: **allopatric speciation** and **sympatric speciation**.

Application and How science works

Spartina alterniflora

The grass *Spartina alterniflora*, a native of North America, was recorded in the United Kingdom at Southampton Water in 1829. The native *Spartina maritima* interbred with *Spartina alterniflora* to produce a sterile male hybrid called *Spartina x townsendii*, which was first identified at Hythe on the south coast of Britain in 1870. By 1880 *Spartina x townsendii* had colonised many South Coast estuaries, spreading by rhizomes (roots). A chromosomal mutation in *Spartina x townsendii* gave rise to *Spartina anglica*, which was recognised in 1968 as a separate species by the scientist C.E. Hubbard. *Spartina anglica* was able to spread and reproduce from seed dispersal, and this allowed populations of this species to become established round the coast of Britain.

1. Why did *Spartina x townsendii* spread by rhizomes (roots)?

2. Why is *Spartina anglica* defined as a new species?

3. *Spartina* species all look very similar. What evidence may the scientist C. E. Hubbard have used to support his hypothesis that *Spartina anglica* was a separate species from *Spartina x townsendii*, *Spartina alterniflora* and *Spartina maritima*?

Allopatric speciation

Allopatric speciation occurs when populations are prevented from interbreeding because they become **geographically isolated**. You learned about speciation involving geographical isolation at AS level. Figure 1 will remind you how geographical isolation can result in the formation of a new species.

1 Species X occupies a forest area. Individuals within the forest form a single gene pool and freely interbreed.

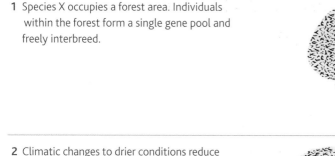

Species X lives and breeds in the forest

Forest

2 Climatic changes to drier conditions reduce the size of the forest to two isolated regions. The distance between the two regions is too great for the two groups of species X to cross to each other.

Forest A

Group X_1

Arid grassland

Forest B

Group X_2

3 Further climatic changes result in one region (Forest A) becoming colder and wetter. Selection pressures in the two isolated regions of forest are now different. Natural selection results in survival of group X_1 in forest A and group X_2 in forest B.

Colder and wetter

Warmer and drier

Forest A

Group X_1

Arid grassland

Forest B

Group X_2

4 The differing selection pressures in the two areas result in the gene pools of the two groups becoming very different. There is no gene flow between the two groups.

Colder and wetter

Warmer and drier

Forest A

Group X_1

Forest B

Group X_2

5 A return to the original climatic conditions results in regrowth of forest. Forests A and B merge and the two groups of species are reunited. The two groups are no longer capable of interbreeding. They are now two species, Y and Z, each with its own gene pool.

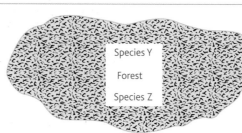

Species Y

Forest

Species Z

Figure 1 *Allopatric speciation as a result of geographical isolation*

Sympatric speciation

This occurs when populations living together become **reproductively isolated**. This isolation may take place:

■ before mating occurs, thus preventing the exchange of gametes (**prezygotic mechanisms**).

■ after mating has taken place, thus preventing the development of the zygotes into offspring (**postzygotic mechanisms**).

Hint

■ The word 'allopatric' is derived from 'allo' meaning other and 'patric' meaning place or homeland.

■ The word 'sympatric' is derived from 'sym' meaning same and 'patric' meaning place or homeland.

Table 1 summarises the different forms of reproductive isolation mechanisms.

Table 1 *The different forms of reproductive isolation mechanisms*

Time of isolation	Type of variation	Explanation of isolation
Pre-mating (prezygotic)	Geographical	Populations are isolated by physical barriers such as oceans, mountain ranges, rivers, etc.
	Ecological	Populations inhabit different habitats within the same area and so individuals rarely meet
	Temporal	The breeding seasons of each population do not coincide and so they do not interbreed
	Behavioural	Mating is often preceded by courtship, which is stimulated by the colour or markings of the opposite sex, the call or particular actions of a mate. Any variations in these patterns may prevent mating, e.g. if a female stickleback does not respond appropriately to the actions of the male, he ceases to court her
	Mechanical	Anatomical differences may prevent mating occurring, e.g. it may be physically impossible for the penis to enter the vagina in mammals
Post-mating but prezygotic	Gametic	The gametes may be prevented from meeting due to genetic or biochemical incompatibility. For instance, some pollen grains fail to germinate or grow when they land on a stigma of different genetic makeup. Some sperm are destroyed by chemicals in the female reproductive tract
Post-mating (postzygotic)	Hybrid sterility	Hybrids formed from the fusion of gametes from different species are often sterile because they cannot produce gametes. For example, in a cross between a horse ($2n = 64$) and a donkey ($2n = 62$) the resultant mule has 63 chromosomes. It is impossible for these chromosomes to pair up during meiosis and so no gametes are formed and the mule is sterile
	Hybrid inviability	Despite fertilisation taking place, further development does not occur or fatal abnormalities arise in early growth. As the offspring do not reach sexual maturity, breeding does not occur
	Hybrid breakdown	The first generation of hybrids is fertile but the second generation fails to develop or, if it does, it is sterile

Summary questions

1 Define the following terms: gene pool, geographic isolation, prezygotic isolation mechanism, postzygotic isolation mechanism.

2 Look at Table 1. In the explanation of hybrid sterility, it states that because a mule has 63 chromosomes, it is impossible for these chromosomes to pair up during meiosis, so no gametes are formed and the mule is sterile. Explain this statement.

How science works

Butterflies and evolution

Recent studies of the butterfly genus *Agrodiaetus* have provided scientists with interesting information on evolution. *Agrodiaetus* butterflies have a wide-ranging habitat and different species can be found large distances apart. Scientists found that closely related species of *Agrodiaetus* that have habitats far apart can look very similar. They also found that species living close together tended to look very different. Scientists believe that the reason for this striking difference in looks is to discourage breeding between the species. Offspring produced from mating of the closely related species tend to be weak and will not reproduce. The result of this is that newer species of *Agrodiaetus* will be genetically isolated from other closely related species.

1 Name the type of speciation that is isolating the gene pools of closely related species of *Agrodiaetus* that are living side-by-side.

2 Which form of reproductive isolation mechanism do you think is operating in this example?

3 Describe how distinctive wing colouration in related species, or sub species, living side-by-side could evolve by natural selection. Include the terms: phenotypic variation, allele frequency, genetic isolation in your answer.

8.4 Human activities and selection

Specification reference: 3.5.1

Human activity is having a major effect on most of the ecosystems found on Earth. Populations living within these ecosystems are exposed to altered selection pressures as a result of human activity and this leads to differing survival and reproduction within the populations. This in turn affects the frequency of alleles within the gene pools of these populations. Humans are therefore affecting the evolution of populations and species.

Learning objectives:

- If human activities are altering the environments of many living organisms, are they changing the selection pressures acting on populations?
- Could this affect the evolution of populations and species?

Hint

Organisms do not decide to evolve. Changes to a population's environment can alter the selection pressures acting on individuals within the population. Populations adapt to these changes by evolving.

Figure 1 *Great tits in cities sing at a higher frequency than those in woodland*

How science works

City song birds

Animals adapt to their surroundings in many ways. Species of animals that live in the countryside may have very different characteristics to members of the same species living in an urban environment. A good example of this is the difference in behaviour of birds living in the city to the behaviour of those living in the countryside. In recent years, studies have been carried out on great tits living in both cities and rural environments. The noise in city environments can cover up bird songs. These songs are vital for communication and attracting mates. Great tits have responded to this by increasing the frequency and speed of their songs. This means their song can be heard over the low frequency noises of their environment. The research also suggests that great tits can pick up new songs from neighbours in new environments.

1 In a city environment, why might great tits with shorter, more high-pitched songs attract mates more successfully than great tits with lower pitched, longer songs?

2 Explain how natural selection may be acting on the great tits in cities.

3 If these changes in bird song represent the first step in speciation, use Table 1 on page 166 to help you decide which type of reproductive isolating mechanism is affecting the great tits studied.

4 Suggest how scientists could investigate whether the alteration in great-tit song in city populations is the result of changes in allele frequency or the environment or a mixture of both.

Summary questions

1 Can populations evolve without forming new species?

2 Why can pest species, such as head lice, survive, even though humans constantly try to eradicate their populations, while some large mammals, such as the tiger, are threatened by extinction?

3 In the UK, many animals, including birds and frogs, now breed much earlier in the spring than they did 20 years ago. Suggest a possible cause of this change in behaviour. In your answer, include the terms: human impact, phenotypic variation, natural selection, allele frequency and gene pool.

1 Some forms of clover are cyanogenic. They produce poisonous cyanide when their tissues are damaged. The production of hydrogen cyanide takes place in two steps, each of which is catalysed by an enzyme produced by a different gene.

Dominant allele A   Dominant allele B

↓   ↓

Enzyme A   Enzyme B

Substrate ⟶ Cyanogenic glucoside ⟶ Hydrogen cyanide

(a) What are the phenotypes (cyanogenic or non-cyanogenic) of the following genotypes? Explain your answer.

   (i) AABb,

   (ii) aaBB.   *(3 marks)*

(b) Two clover plants, each with the genotype AaBb were crossed.

   (i) Use a genetic diagram to show the genotypes of the offspring.

   (ii) What is the expected ratio of cyanogenic to non-cyanogenic plants in the offspring?   *(3 marks)*

(c) Cyanogenic clover plants are toxic to slugs. Slugs are common at low altitude but not at high altitude where the climate is colder. When the cells of cyanogenic clover plants freeze, the hydrogen cyanide that is released kills the cells and as a result the plant dies. The frequencies of the alleles for cyanogenesis are different in clover plants growing at low and at high altitudes.

   (i) How would you expect the allele frequencies to differ?

   (ii) Use the information given to explain how the different allele frequencies might be the result of natural selection.   *(6 marks)*

AQA, 2000

2 **Figure 1** shows the range in wing length of a species of small bird on different islands to the north of Great Britain.

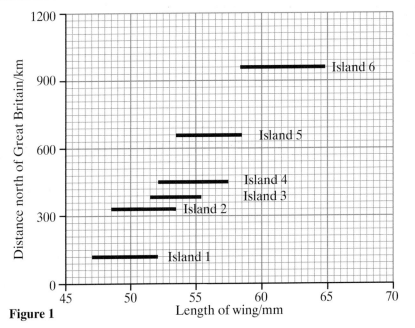

Figure 1

(a) On which islands might a bird with wings 55mm long live? *(1 mark)*

(b) What would be the mid-point of range of wing length expected on an island that is 750 km north of Great Britain? *(1 mark)*

(c) In this species wing length increases with body size. Suggest how this trend in body size might help the birds to survive. *(2 marks)*

(d) Explain how selection may have resulted in an increase in body size. *(3 marks)*

AQA, 2001

3   New species may arise from populations of existing species. The process is called speciation.

(a) What is a species? *(2 marks)*

(b) Explain the difference between allopatric speciation and sympatric speciation. *(3 marks)*

AQA, 2006

4   The apple maggot fly (*Rhagoletis pomonella*) was found only on hawthorn trees in North America. In the 19th century, it spread as a pest to apple trees introduced into North America by farmers. The flies attacking the apple trees are now genetically different from the files that attack hawthorn trees. It is thought that a new species of fly may be evolving.

(a) Suggest how the flies that feed on apple trees could evolve from those that fed on hawthorn trees. *(3 marks)*

(b) Suggest **one** way in which scientists could find out whether the flies from apple trees were a different species from those found on hawthorn trees. *(2 marks)*

AQA, 2001

5   (You may need to read Chapter 9 before you attempt this question)

One of the most valuable crops planted by the Forestry Commission is spruce because its yield of timber is high. Early trial plantings showed that spruce trees grew very slowly when planted on land on which heather was also growing.

(a) Name the type of competition shown between spruce and heather. *(1 mark)*

(b) Give **two** resources for which spruce and heather are likely to be competing. *(1 mark)*

(c) Further trial plantings on land dominated by heather showed that the growth of spruce was greatly assisted by planting another tree species at the same time. This use of a 'nurse' crop is now standard practice. The table shows the results of some of these trial plantings.

Species planted	Height of spruce after 15 years / metres
Spruce, heather Japanese larch	4.5
Spruce, heather Scots pine	3.1
Spruce, heather and Corsican pine	3.5
Spruce and heather	2.0

(i) Young spruce trees were 50 cm high when planted. Calculate the difference in the rate of growth when these trees were grown with Japanese larch compared to the control. Show your working. *(2 marks)*

(ii) Suggest one way in which a 'nurse' crop may aid the growth of spruce trees. *(1 mark)*

AQA, 1999

People change communities

9.1 Ecosystems and the stability of populations

Learning objectives:

- What is meant by the terms: population, niche, community and ecosystem?

- Why may a population become larger or smaller?

Specification reference: 3.5.2

Ecosystems

In Chapter 8 you learned that a **population** is all the freely interbreeding individuals of the same species occupying the same place at the same time. For example, all the oak trees growing in a woodland form the oak tree population, while the woodlice living under dead branches and in the leaf litter in the same woodland form the woodlouse population. These populations depend on other populations of plants and animals for their survival. The name given to all the populations of different organisms living and interacting in a particular place at a particular time is a **community**. In an oak woodland, the oak trees are part of its plant community and the woodlice belong to the woodland's animal community. Dynamic (changing) feeding relationships exist within a community. For example, caterpillars can only feed on the leaves of the oak trees at the times of the year when leaves are present. The interaction of the community (which is made up of living things, called the **biotic** component) with the physical environment (which consists of the non-living, **abiotic** component, such as the soil or the climate) in a specific area is called an **ecosystem**. Ecosystems vary in size from the very small, such as a pond or a wall, to much larger examples such as woodlands, moorlands or lakes. An ecosystem is more or less self-contained in terms of energy flow through it and the cycling of essential nutrients such as nitrogen and carbon within it. If these processes are disrupted, the community is affected and the ecosystem becomes unstable.

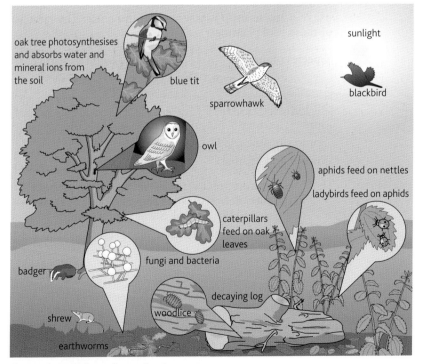

Figure 1 *Part of an oak woodland ecosystem*

Look at Figure 1. It shows some of the organisms that live together in an oak woodland. The place where an organism lives is called its **habitat**. An oak woodland contains many different kinds of habitat and this is why oak woodlands have great **biodiversity**. Populations of many different species live there, each having their own ecological role within the woodland community. The ecological role of a species within its community is called its **ecological niche** and consists of all the environmental conditions and resources needed for the survival of a species within an ecosystem. If a new species is introduced to an ecosystem, its niche may overlap with that of an existing species and this can affect the stability of the ecosystem and all the populations of living things within it.

How science works

The harlequin ladybird

The harlequin ladybird is a native of Eastern Asia. It has been introduced into many countries as a biological control agent against aphid and scale insect infestations in greenhouses, crops and gardens. Populations of harlequin ladybirds are now widespread in North America and many European countries and the species is also present in parts of South America and Africa. The harlequin ladybird arrived in Britain in 2004. Some entered the country in flowers imported from Europe, others in packing cases from Canada and some probably flew across the English Channel. Harlequin ladybirds are very effective predators of aphids and have a wider food range and habitat than most other British ladybird species. This allows the harlequin ladybirds to easily out-compete native ladybirds that feed on aphids, such as the 7-spot ladybird. If aphids are scarce, harlequin ladybirds will also eat the eggs, larvae and pupae of native ladybird species, and the eggs and caterpillars of butterflies and moths. Harlequin ladybirds have a longer reproductive period than most British ladybirds and can disperse over long distances.

Populations of harlequin ladybirds are rapidly becoming established throughout Britain and scientists are asking people to report sightings of the harlequin ladybird to them so that they can monitor the spread of this species. They are very worried about the effect this introduced species is having on native British ladybirds.

Figure 2 *7-spot ladybirds*

Figure 3 *Harlequin ladybirds*

1 Why are harlequin ladybirds able to out-compete 7-spot ladybirds?

2 Suggest **three** reasons for the rapid spread of harlequin ladybirds in Britain since their introduction in 2004.

3 Should scientists try to eradicate the harlequin ladybird from Britain? Suggest **two** reasons why they should and **two** reasons why they should not.

Summary questions

1 How do the caterpillars that feed off oak leaves survive at times of the year when there are no leaves on the oak trees?

2 Do woodlice have plenty of food all year round? How could you investigate this?

3 What does biodiversity mean?

4 Explain why an oak woodland contains so many different ecological niches.

9.2 Population growth

Learning objectives:

■ Why do populations stop growing?

■ Do you know what the term carrying capacity means?

■ What is the difference between interspecific competition and intraspecific competition?

Specification reference: 3.5.2

Populations do not stay exactly the same size all the time. The size of a population depends upon its birth rate, its death rate and migration into and out of the population. These are in turn affected by food supply, ability to reproduce, predation and disease. In a stable ecosystem where these factors are balanced, the size of a population of plants or animals remains fairly constant. **Limiting factors** keep the population at the maximum size that can be sustained by the ecosystem. This population size is called the **carrying capacity** and the limiting factors that determine its size are collectively called **environmental resistance**. Look at Figure 1. It shows a population growth curve for a natural population that has reached its carrying capacity

You can see that the growth of the population has three clear phases. In the lag phase, growth is slow because the size of the population is so small that even a doubling of the population does not produce a great increase in population numbers. In the exponential (log) phase, the growth rate of the population is at its maximum because food, water, shelter, potential mates and other factors essential for survival and reproduction are abundant. In the third phase, called the stationary phase, the population size remains fairly constant, held at the carrying capacity by limiting factors such as food supply. There are some small fluctuations above and below the carrying capacity due to variations in limiting factors but overall, the size of the population remains fairly constant.

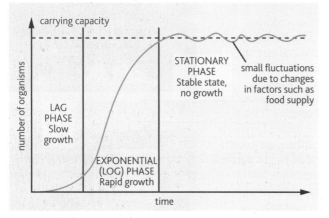

Figure 1 *Population growth curve*

Table 1 *Limiting factors that influence the carrying capacity of populations*

Factors that increase the carrying capacity	Factors that decrease the carrying capacity
Plentiful food/nutrient supply	Inadequate food/nutrient supply
Adequate supply of water	Inadequate supply of water
Favourable intensity and duration of light	Light of inappropriate intensity and duration
Fewer predators	Increased number of predators
Effective means of avoiding predation, e.g. camouflage, escape mechanisms	Ineffective means of avoiding predation
Effective means of resisting disease	Inability to resist disease
Favourable climatic conditions, e.g. suitable temperature, humidity, rainfall	Unfavourable climatic conditions, e.g. extremes of temperature, humidity, rainfall
Constant abiotic conditions, e.g. pH, water potential, climate	Fluctuating abiotic conditions
High reproductive rate	Low reproductive rate

There are two types of limiting factor:

■ Density-dependent factors have effects that depend on population density. They have a greater effect when the population is large and individuals are living close together. For example, it is easier for a disease to spread from one organism to another when the organisms are living close together. Density-dependent factors are usually **biotic** ones such as food supply, predation and disease. The larger the population, the greater the effect.

- Density-independent factors are **abiotic** factors such as temperature, rainfall, light intensity and fire. They have similar effects regardless of the population size/density.

Competition

Density-dependent limiting factors result in competition between individuals for limited resources such as water, food and territories. If the competition is between members of **different species**, it is called **interspecific competition.** In Topic 9.1 you saw that newly introduced harlequin ladybirds are competing with native British 7-spot ladybirds for aphids. Competition for limited resources between members of the **same species** is called **intraspecific competition**. In Chapter 8 you saw how natural selection can change the frequency of alleles within a population. Intraspecific competition is the driving force behind natural selection. Individuals with phentoypes that give them a competitive advantage are most likely to win the competition and survive, reproduce and pass on their alleles to their offspring.

How science works

Chiffchaffs change their migration pattern

Chiffchaffs are small birds that migrate to Britain each spring. They set up territories, find a mate, build nests, lay eggs and raise their young. Chiffchaffs are among the first birds to arrive back in Britain, after spending the winter around the Mediterranean and in Senegal on the west coast of Africa. Migration north starts in February, with the main movement in March and April. Making this long journey north is very hazardous for small birds like the chiffchaff, but there is a spring surplus of insect food available in Britain and long hours of daylight in which to search for it. Their young are more likely to survive if born in Britain than in their wintering grounds because there is less competition for food. Scientists have kept records of the dates and numbers of migrating birds in Britain, including chiffchaffs, for many years. They have discovered that while many chiffchaffs still make the long journey south for the winter, some are staying in Britain and not migrating at all. Milder winters in Britain may mean that there is plenty of food to support some chiffchaff populations, without the need to make the dangerous journey south for the winter.

Figure 2 *Chiffchaff*

Hint

Don't forget the difference between interspecific and intraspecific competition. Try to think of an easy way to remember them. Here's one way: Int**er**specific – diff**er**ent species. Intr**a**specific – s**a**me species.

AQA Examiner's tip

You may be given data in an exam question that suggests that organisms are competing with each other. You not only need to state what kind of competition it is – whether interspecific or intraspecific – but also suggest what resources they might be competing for.

1. Why is migration so hazardous for a small bird like the chiffchaff?

2. Many insects survive the winter as pupae, for example some butterflies and moths spend the winter as chrysalises. Milder winters and warmer springs are encouraging these pupae to develop into adults earlier in the year.

   a How may this give the chiffchaffs that spend the winter in Britain a competitive advantage over chiffchaffs that migrate back to Britain in the spring?

   b How will this change in the migration pattern of the chiffchaff affect other species of insect-eating birds that share their ecosystems?

3. Do you think that the selection pressure acting on the chiffchaffs is changing? Explain your answer using the terms selective advantage, allele frequency and population.

Summary questions

1. Explain how predators can help to keep a population at its carrying capacity.

2. Using the woodland community as an example, give **three** effects of human impact on woodland ecosystems. For each effect, say how the growth of populations within the woodland will be affected.

9.3 Winners and losers

Learning objectives:

- How can the introduction of a new species change the stability of populations in a community?

- Should we try to control the size of populations of introduced species?

Specification reference: 3.5.2

Figure 1 *Field poppy growing in a wheat field*

Some of Britain's wild plants and animals have lived together for many thousands of years. During this time, stable relationships have been established between the populations of these species in their communities. However, many of the species that live in Britain today have been introduced by humans. Some species, like the harlequin ladybird, are recent, accidental introductions. Other accidental arrivals are more ancient, such as the field poppy, whose seeds probably arrived in Britain with the grain traded with Mediterranean countries during the Neolithic period (about 7000 years ago).

Some species have been introduced deliberately, and many of these are domesticated species used as food sources, such as cattle, sheep, goats, wheat and potatoes. Rabbits were introduced to Britain following the Roman invasion about 2000 years ago as a source of food, but now live wild throughout Britain. At AS level, you learnt about the domestication of plants and animals in the Middle East during Neolithic times. Dogs probably arrived in Britain during Neolithic times with human settlers who valued them for their ability to herd animals and their companionship. The introduced species have found ecological niches in many different types of ecosystem, changing the species' interactions and affecting the size and distribution of many populations of native species.

Application and How science works

Reds versus greys

Red Squirrels are native to the British Isles. Grey squirrels were introduced to Britain from North America at the end of the 19th century. They were originally kept in cages in private collections, but following their release into the wild they have become so successful that they now threaten the survival of Britain's red squirrel populations. Look at Figure 2. The grey squirrel is much larger than the red squirrel. Both types of squirrel are adapted to feed in mixed woodland but the larger grey squirrels can put on a lot more body fat than red squirrels. This gives them a better chance of surviving, especially in the cold winter months. As a result, grey squirrels can displace a red squirrel population within 15 years.

The grey squirrel is now widespread in England and Wales, and has established populations in parks and gardens as well as large woods and forests. Meanwhile, the only habitat where the native red squirrel seems better adapted for survival than the grey is in large areas of coniferous woodland such as those found in Scotland, Northumberland and Cumbria. Here, the small seeds of the conifers provide enough energy for the red squirrels, but not for the larger, grey squirrels. Scientists hope that the remaining populations of red squirrels in mainland Britain will survive in red squirrel reserves. These have been created in coniferous forests in places like Formby on Merseyside, where the grey squirrels find it difficult to take over. In an effort to prevent grey squirrels from becoming established in red squirrel reserves, the Forestry Commission is developing a new squirrel feeder designed to poison grey but not red squirrels. If the Government gives permission for the feeder to be used, it will be a major step forward in controlling the size of grey squirrel populations.

1 What type of competition is taking place in this example and what are the squirrels competing for?

2 Why are scientists considering controlling the size of grey squirrel populations in Britain?

3 Is it wrong to try to kill one species if it is threatening the survival of another? Give a reason for your answer.

4 Do you think that killing the grey squirrels is the only way to save the red squirrel in the UK?

5 What effects might a large population of grey squirrels have on the other species living in a woodland ecosystem?

Figure 2 **a** *Red squirrel and* **b** *Grey squirrel*

Application and How science works

A growing concern

Japanese Knotweed is a plant that was introduced to European parks and gardens from Japan, Taiwan and China in the early 19th century. Like many introduced species, it has escaped from its garden habitat and has become a real pest throughout Britain. In its native Asian habitats, natural pests and diseases usually limit the size of Japanese Knotweed populations. In Britain, Japanese Knotweed doesn't appear to have any natural enemies and once it has established itself in a habitat, it is very difficult to get rid of it. A piece of root as small as 0.8 grams can quickly grow to form a new plant which rapidly out-competes other plants for light, water and nutrients. Japanese Knotweed doesn't just threaten native wildlife. Buildings and hard surfaces such as flood defences can be damaged by the plant, and drainage channels can become clogged. Estimates of how much it would cost to try to eradicate Japanese Knotweed from Britain are in excess of £1.5 million.

Japanese Knotweed poses such a threat to the survival of native species that it is an offence to plant this species or cause it to grow in the wild. Anyone convicted of such an act may face a fine of £5000 and/or 6 months' imprisonment!

Figure 3 *Japanese Knotweed*

1 Should so much of taxpayers' money be spent on trying to get rid of this plant? Justify your answer.

2 Does it matter if Japanese Knotweed threatens the survival of native species?

Summary questions

1 Populations of native British species have had thousands of years to establish stable relationships in their communities. Many of these are feeding relationships. For example, herbivores such as snails feed on producers such as grassland plants. Describe **three** other types of feeding relationships that may be present within a community and try to give an example of each one.

2 If a newly introduced species doesn't seem to affect the stability of the ecosystem it moves in to, what can you conclude about its ecological niche?

9.4 Urban wildlife

Learning objectives:

■ Why does the introduction of one new species often change a whole ecosystem?

■ Why have populations of some species grown as our cities grow?

■ Can wild plants and animals survive in urban environments?

Specification reference: 3.5.2

Cats

Domestic cats are thought to be descended from wild cats that were probably attracted to the earliest agricultural settlements in the Middle East. Here, they would hunt the rodents that infested the grain stores, and because of this they would have been highly valued by the first farmers. The cat's ability to track down and kill rodent pests makes it a well-respected human companion to this day.

How science works

Felis catus on the prowl!

The cat *Felis catus* is Britain's most abundant carnivore, and because there are so many of them, domestic and feral (reverted back to wild from domestic) cats are a major predator of wild animals in Great Britain. In 1997, in an attempt to find out the impact of the domestic cat on Britain's wildlife, scientists carried out a survey where they asked cat owners to record in as much detail as possible the prey items brought home by each cat in their ownership. A questionnaire survey was carried out between 1st April and 31st August 1997. The survey found that a total of 14 370 prey items were brought home by 986 cats living in 618 households. Mammals made up 69% of the items, birds 24%, amphibians 4%, reptiles 1%, fish <1%, invertebrates 1% and unidentified items 1%. A minimum of 44 species of wild bird, 20 species of wild mammal, 3 species of reptile and 3 species of amphibian were recorded. Gardens are becoming increasingly important for wildlife as more and more open countryside is developed for housing, industry and transport. Is the potential of the garden as a refuge for wildlife being reduced by the resident cats?

1 Calculate the actual numbers of mammals, birds, amphibians and reptiles that were brought home by the cats in this survey. Present your results in a table to the nearest whole number and show how you have made your calculations.

2 Present the percentage data obtained in the survey as a fully labelled bar chart or a pie chart.

3 About 9 million cats are thought to live in Britain. If the cats in the survey are typical of all cats in Britain, how many mammals, birds, reptiles and amphibians were captured and brought home by all the cats living in Britain during the survey period? Show your calculations.

4 Do you think the data collected is an overestimate of the numbers of animals caught by cats during this time period or an underestimate? Give a reason for your answer.

5 These results were obtained using a questionnaire. Suggest another way of gathering similar data. Do you think your suggested method is more or less valid than the questionnaire used in this study? Give reasons for your answer.

Figure 1 *The domestic cat*

Pigeons

Pigeons were domesticated by humans thousands of years ago as a source of food and fertiliser. Feral pigeons can now be found in cities worldwide. Their ancestors are the rock doves that inhabit rocky cliffs. Feral pigeons occupy the same niche in the urban environment. They perch and nest on ledges, roofs and windowsills and are often found in flocks, feeding by day and roosting by night. Their natural diet is grain and seeds, but they also scavenge food discarded on urban streets. When food is readily available, feral pigeons can breed up to six times a year, allowing rapid population growth. Sometimes, large numbers of pigeons are attracted into an area because people enjoy feeding them. They will also visit bird tables that offer grain and seed. However, large numbers of pigeons living in an urban environment can cause problems. In wet weather, the accumulation of their slippery droppings on pavements can pose hazards to pedestrians. Their droppings are also acidic and can corrode brick- and stonework. Gutters and drains blocked by droppings, nest material and dead birds can overflow, causing water damage to buildings; and dead pigeons in uncovered water tanks can contaminate the water supply.

How science works

Flock evicted from favourite feeding ground

London's Trafalgar Square is world-famous for its feral pigeons, but in an attempt to discourage the large flocks of pigeons that traditionally feed there, bird-food sellers have been banned from trading in the square. The square is now regularly patrolled by a falconer who flies Harris hawks to scare the pigeons away from the area, and throwing food to the birds has become an offence. Since the deterrents were introduced, it is estimated that the number of pigeons found in the square at any one time has fallen from 4000 to around 200 and it is claimed that the use of the hawks has led to a saving of £140,000 a year on cleaning up feathers and bird excrement. However, groups campaigning for the welfare of the Trafalgar Square pigeons continue to feed the birds, despite the threat of legal action.

Figure 2 *Feral pigeons in London*

1 The pigeons in Trafalgar Square have been a tourist attraction in their own right for many years. Why do you think the local council has now decided to reduce the numbers?

2 Opponents of the scheme claim that the new laws will cause pigeons to starve to death if they are not fed in Trafalgar Square. Do you agree with this point of view? Give a reason for your answer.

3 The Local Authority claims that using a bird of prey to scare the pigeons is a humane way to reduce the size of the Trafalgar Square pigeon population. Do you agree? Give reasons for your answer.

Foxes

Foxes first moved into our cities from the countryside in the 1930s. Large areas of suburban housing were built in the period leading up to World War II. The relatively large gardens of these well-spaced houses provided an ideal habitat for foxes and they quickly increased in numbers. From these new suburbs, foxes then colonised other urban areas. Many cities now have urban foxes. For most towns and cities the fox population reached its carrying capacity many years ago and the population is stable, with no significant increases or decreases. There are only a few cities where fox numbers are still increasing and these are ones that have only recently been colonised. Urban foxes have a varied diet that includes earthworms, insects, fruit and vegetables and a wide variety of both domestic and wild birds and mammals. Most of the birds they eat are feral pigeons and small garden birds, and the most frequently eaten mammals are generally field voles that are abundant on allotments, railway lines and other grassy areas.

Figure 3 *A red fox feeding her cub*

Rats

Wherever there are humans, rats will not be far away! Rats live successfully throughout the UK, in both urban and rural environments. Rats are quick to learn and so will exploit any easily accessible food source. It is the brown, or Norway rat *Rattus norvegicus*, introduced to Britain by shipping during the 18th century, that is most commonly found in the UK. Brown rats tend to infest the areas around a building rather than the building itself. They can create extensive burrow systems but will take advantage of artificial tunnels, such as sewers. They cause widespread contamination of their habitat with droppings, urine and hairs. They also carry many diseases and parasites that are potentially harmful to humans and animals. Until the arrival of the brown rat, the black rat *Rattus rattus* was the UK's only resident rat species. The black rat is now rare in the UK and is confined to a few port towns and a few offshore islands.

Summary questions

1. Why do you think that foxes and pigeons have become so successfully established in towns and cities?

2. There is evidence to suggest that other species that usually live in farmland and woodland habitats, such as the wood pigeon, are establishing populations in towns and cities. What may be causing this migration?

9.5 Genetically modified organisms (GMOs)

Learning objectives:

■ What are GMOs?

■ Why do we need large-scale use of GM crops?

■ How can the environmental impact of GM crops be measured?

■ What are the ethical issues associated with the release of GMOs into the environment?

Specification reference: 3.5.2

Link

Refer back to Topic 4.4 to check your understanding of how selective breeding and gene technology work.

GM crops

Humans have improved crop plants through selective breeding for many thousands of years. In 1977, this time-consuming process was speeded up when it was discovered that a bacterium called *Agrobacterium tumefasciens* could be used to introduce foreign genes into plant cells, producing **genetically modified** (**GM**) crop plants. Using this bacterium and several other techniques for gene transfer, scientists have since produced many different GM crops.

Many GM crops are modified to be resistant to pests, disease or herbicides. These include commercially important crops such as soya, wheat, corn (maize), oilseed rape, cotton and sugar beet. One of the most commonly inserted genes is the bacterial gene, **Bt**. It allows the crop plants to produce a toxin that kills insect pests but is harmless to humans. Other crop plants have been engineered to increase shelf life, improve flavour, increase hardiness, increase nutrient content or to be free from certain chemicals that may cause allergies.

Supporters of GM technology argue that genetically engineered crops will provide solutions to a number of global issues. They could help to protect the environment by minimising pesticide use, they could improve nutrition, and by flourishing in conditions where other crops would fail or would have reduced yields, they could help to alleviate world hunger.

Critics of GM technology fear that herbicide-resistant GM crops could become 'superweeds', or that they could accidentally breed with wild plants or other crops, genetically polluting the environment. Large numbers of field trials, carried out by the UK Government and others, reveal that gene transfer does occur. Many scientists agree that the widespread insertion of insecticide genes into crop plants, such as the Bt gene, will also increase the rate of evolution of insecticide-resistant pests.

GM crops and health

There are also fears that GM crops could have adverse health effects. If GM crops are eaten by humans, there are worries that the antibiotic resistance marker genes that they contain may be taken up by bacteria living in the human gut. This could produce populations of human gut bacteria that are resistant to certain antibiotics.

GM crops have been engineered to produce drugs and vaccines. This process is called **pharming**. Supporters of the technology say the use of these plants will allow the cheap production of new medicines. Critics worry that these plants may cross-breed with varieties of food crops and this could lead to contamination of food supplies and possible dangers to health.

Figure 1 *GM maize growing in a field trial in Lincolnshire, UK*

GMO regulation

Currently all releases of genetically modified organisms (GMOs) into the environment are strictly regulated and require consent. Advice on whether a release should be given consent is provided by the Advisory Committee on Releases to the Environment (ACRE). When making an application, an Environmental Impact Assessment (EIA) must be

Figure 2 *Ingredients, listing GM soya*

submitted. An EIA is a detailed report on the environmental impact the project would have. The purpose of the assessment is to make sure the impact on the environment is considered fully. In the European Union, an EIA must cover several areas, including a description of the project, description of impact and details of any weaknesses in knowledge.

In the UK there are also regulations on how GM food is labelled. If any genetically modified ingredients are used in food it must be stated on the packaging.

■ How science works

Farm-scale evaluations

As part of its scientific review of GM crops, the UK Government carried out farm-scale evaluations (FSEs). These were set up in 1999 to investigate whether farming using GM herbicide-tolerant crops is any more harmful to plants and animals than conventional farming. In the first part of the trial, three crops of potential interest to UK farmers were studied: spring oilseed rape, beet and maize. The GM crop varieties have been modified to tolerate broad-spectrum herbicides that kill most plants, including conventional crops. This enables farmers to use new weed control strategies. Results for the first part of the trial were published in October 2003. In general, the farm-scale evaluations found the following:

■ Growing GM beet and spring rape was worse for many groups of wildlife than growing conventional beet and spring rape. The herbicide used on the GM crops controlled weeds more effectively so there were fewer weeds, with fewer insects and less weed seeds (which are important in the diets of birds).

■ Growing GM maize was better for many groups of wildlife than conventional maize. Around GM maize crops there were more weeds, more butterflies and bees at certain times of year and more weed seeds.

■ The differences arise because these GM crops gave the farmers new options for weed control, allowing them to use different herbicides and apply them differently than they would for non-GM crops.

The study concluded that growing such GM crops could impact on wider farmland biodiversity, but that other factors (e.g. the amount of land cultivated, how it is cultivated and how crop rotations are managed) will also be important in determining the overall environmental impact of GM crops.

1 The FSEs lasted 4 years. Why do you think so much time was needed to carry out the evaluations?

2 Why was it important that each GM variety of a particular crop was trialled against a non-GM variety of the same crop?

3 Based on the results of this trial, what advice would you give to farmers who wanted to encourage wildlife on their farms?

Public reaction in the UK, coupled with the results from the farm-scale evaluations, means it is unlikely that GM crops will be grown in the UK in the next few years. Elsewhere it is a different story. Farmers planted 81 million hectares (or 200 million acres) of GM crops worldwide in 2004, up from 67·7 million hectares (or 167 million acres) in 2003.

How science works

Security for GM crop trials?

In 2008, only two major GM crop trials were approved in the UK. One of these trials was disrupted when protesters destroyed crops at the trial site. Currently, the location of the trial sites must be made public. Some scientists think that the GM crop trials should be conducted at secure, secret locations or at specialist high-security facilities. They believe the disruption from protesters and activists is greatly hampering their research into how GM crops can help with worldwide problems. The research could potentially help bring the cost and environmental impact of farming down. It could also lead to discoveries of crops that can help relieve world hunger and survive drought. However, some environmental groups think the research is just a way for big companies to make money and patent crop strains.

Figure 3 *Ripening GM soya bean pods*

1. Why do the scientists want to carry out field trials of GM crops?
2. Why do you think the protestors have vandalised the field trials of GM crops?
3. The EU requires a six-figure grid reference of any GM crop trial site to be made public. Suggest why.
4. Suggest **one** reason why no GM crops are grown commercially in Britain.

Genetically modified animals

The most common GMOs are crop plants, but the technology has been applied to all forms of life. Genetically modified animals with particular genes added or 'knocked out' are very important in medical research. Any animal whose genetic composition has been altered by the **addition** of foreign DNA is said to be **transgenic**. Transgenic mice can be designed to develop conditions such as Alzheimer's disease by incorporating known human disease genes into their DNA. 'Knockout' mice have had certain genes 'knocked out' or disabled. For example, p53 knockout mice have had their p53 tumour-suppressing gene disabled. Transgenic and 'knockout' mice allow scientists to learn more about human diseases. There are potentially many uses for genetically modified animals but there are many ethical concerns, including the possible hazards associated with their release into the environment.

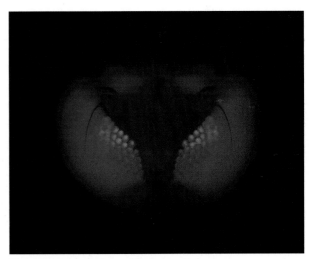

Figure 4 *Genetically modified mosquito's eyes*

Figure 4 shows the eyes of a mosquito (*Anopheles stephensi*) glowing red under ultraviolet light. The gene for DsRed, a red fluorescent protein, has been introduced into the mosquito genome from the coral *Discosoma* sp. Placing a gene from one organism into the genome of another is called transgenics. The red glow shows that the DsRed gene has been successfully introduced. This raises hopes that a gene could be introduced that would make the mosquitoes unable to carry the *Plasmodium* sp. protozoa that cause malaria, which would save millions of lives.

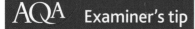

Examiner's tip

In an exam you may be given an example of a GM organism and you could be asked to evaluate its benefits. This means you will need to consider the pros and cons of using the GM organism.

■ Application and How science works

Genetically modified mosquitoes

The spread of malaria by mosquitoes is a huge problem in developing countries. The spread of the disease, however, could be reduced by releasing genetically modified mosquitoes into the wild.

Scientists in the UK have been conducting research into the genetic modification of mosquitoes. The British scientists have created genetically modified male mosquitoes that have glowing testicles. The mosquitoes can be easily identified and sterilised. The sterile males would then be released into the wild. They would mate with female mosquitoes but no fertilised eggs would be laid.

American scientists have genetically engineered a species of mosquito that cannot transmit malaria. The species has been engineered to be resistant to the *Plasmodium* organism that causes the infection. These mosquitoes could be released into areas where malaria is common. The modified mosquitoes have a gene that makes their eyes glow green or red so they can be identified. The species of mosquito involved in the research is not the species that is most harmful to humans. This is also the case with the parasite that was used. *Plasmodium bergehei* was used in the experiment. It is considered to be a good model for *Plasmodium falciparum*.

1 Why would the American scientists introduce large numbers of GM mosquitoes into areas where malaria is common?

2 Why is it important that the American scientists can easily distinguish the GM mosquitoes from the wild insects?

3 Suggest why the researchers used a species of the malaria parasite that does not cause serious harm to humans in their study.

4 Why may some people favour the introduction of sterile male GM mosquitoes, rather than the GM mosquitoes that are resistant to the *Plasmodium* parasite?

Summary questions

1 In an effort to prevent the evolution of populations of pests that are resistant to the **Bt** toxin, conventional crops are often grown alongside fields containing their GM counterparts. The conventional crop is a refuge for the pest species. Using the terms mutation, allele frequency and selection pressure explain why this is done.

2 Produce an information leaflet about GM soya and GM maize. Your leaflet should include information about where and why GM soya and maize crops are grown, the advantages and disadvantages of growing these crops and where people can find out more information.

3 Genetically modified bacteria are already widely used in the production of pharmaceutical products such as human insulin and human growth hormone. Why are there fewer concerns about the safe and ethical use of these organisms compared to other organisms such as mosquitoes?

4 Organise a debate within your class. The title of the debate is 'This house believes that GM crops will play a vital part in sustaining the population of the world during the 21st century'. You will need a chairperson and six speakers: a GM crop research scientist; an anti-GM protestor; an organic farmer from the UK; an intensive farmer from Argentina; a subsistence farmer from Ghana (who only grows enough food to support his family); a UK adult from a family with a low income. They must each prepare a speech of about 500 words to put forward their view on the statement. Their speech must conclude with a statement that either agrees or disagrees with the motion being debated. Each speaker could work with a group to research and prepare their speech. At the end of all the speeches, the chairperson organises a class vote of those for the motion (who do believe that GM crops will play a vital part in sustaining the population of the world during the 21st century) versus those who do not.

1   In organic farming chemical weed-killers are not used on crops. Weeds have to be kept down manually. The number of weeds that can grow and the number of different species vary with the type of crop. **Figure 1** show the weeds in barley and potato crops in two fields on an organic farm.

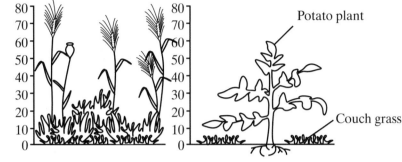

Figure 1

(a) Use information from **Figure 1** to suggest one explanation for the fact that there are more weeds in the barley crop than in the potato crop. *(2 marks)*

(b) One of the weeds in the barley fields is couch grass. In a laboratory investigation, scientists found that when couch grass and barley were germinated at the same time and the seedlings were grown close together during their development, the yield from the barley plants was the same as if they were grown alone. If couch grass was sown 14 days before barley, the yield from the barley plants was significantly smaller than when they were grown alone.

One hypothesis to explain these results is that older couch grass roots produce a substance that inhibits the growth of younger barley roots.

(i) Briefly describe how this hypothesis might be tested.

(ii) Suggest **one** alternative hypothesis to explain the result. *(3 marks)*

AQA, 1999

2   An investigation was carried out on competition using two species of *Desmodium*, which are herbaceous plants. The following plots were set up for each of the species, *D. glutinosum* and *D. nudiflorum*, resulting in a total of six plots.

Plot **A** – small individuals were planted 10 cm from large individuals of the same species.

Plot **B** – small individuals of one species were planted 10 cm from large individuals of the other species.

Plot **C** – small individuals of each species were planted at least 3 metres from any other *Desmodium* plant.

The total increase in leaf length on each of the small plants was measured after four weeks. The results are shown in **Figure 2**.

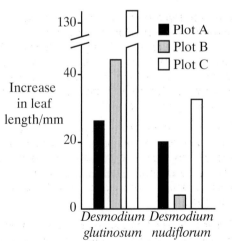

Figure 2

(a) Name the type of competition in:
    (i)   Plot **A**;
    (ii) Plot **B**.   *(1 mark)*
(b) Explain how plot **C** was used as a control.   *(1 mark)*
(c) Use **Figure 2** to explain how competition affects the growth of these two species.   *(3 marks)*

AQA, 1999

3   Barnacles are animals that live on rocky shores. The adults are fixed to the surface of
    rocks and do not move. The young larvae can swim freely in the sea. As they get older
    the larvae settle and attach themselves to a rock surface. Here they develop into adults
    which feed on microscopic plants and animals in the sea when the tide is in. In Britain
    two species of barnacle, *Chthamalus stellatus* and *Balanus balanoides*, commonly occur
    together on the same rocky shore.

    Figure 3 shows the typical distribution of the two in relation to the tide levels on a rocky shore.

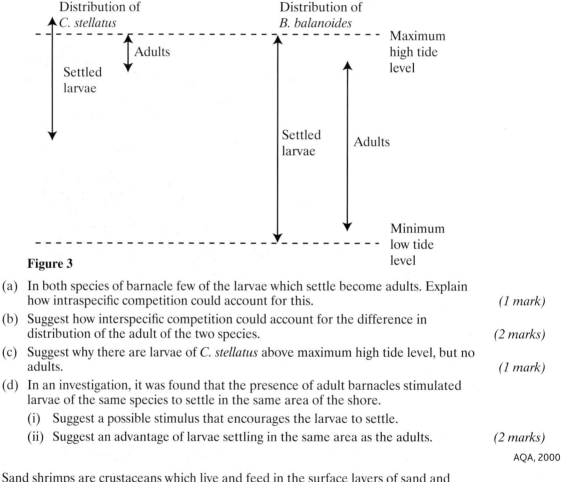

Figure 3

(a) In both species of barnacle few of the larvae which settle become adults. Explain
    how intraspecific competition could account for this.   *(1 mark)*
(b) Suggest how interspecific competition could account for the difference in
    distribution of the adult of the two species.   *(2 marks)*
(c) Suggest why there are larvae of *C. stellatus* above maximum high tide level, but no
    adults.   *(1 mark)*
(d) In an investigation, it was found that the presence of adult barnacles stimulated
    larvae of the same species to settle in the same area of the shore.
    (i)   Suggest a possible stimulus that encourages the larvae to settle.
    (ii) Suggest an advantage of larvae settling in the same area as the adults.   *(2 marks)*

AQA, 2000

4   Sand shrimps are crustaceans which live and feed in the surface layers of sand and
    mud in estuaries. **Figure 4** shows the range of temperature and salt concentration in
    which these sand shrimps can survive.

    Figure 5 shows a sketch map of a tropical estuary where sand shrimps live. The salt
    concentration varies as the tide goes in and out, and as different volumes of sea water
    mix with the fresh water from the river. Below the map is a graph which shows the
    maximum and minimum salt concentration at different points along the estuary.

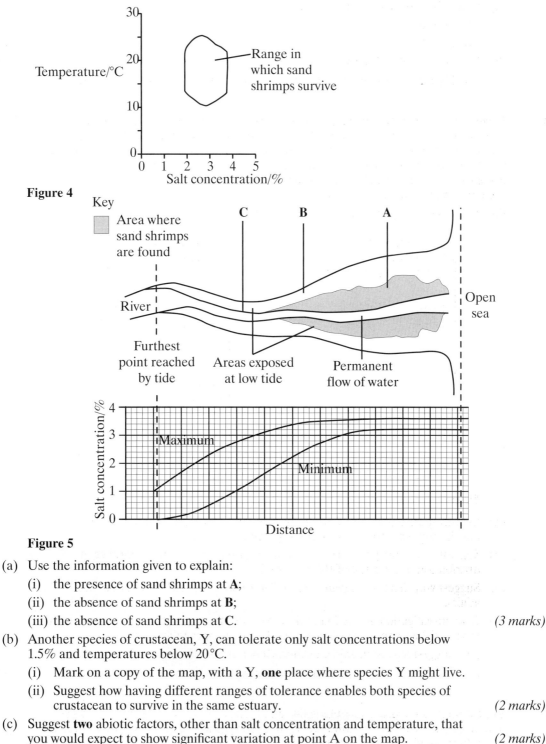

Figure 4

Figure 5

(a) Use the information given to explain:
  (i) the presence of sand shrimps at **A**;
  (ii) the absence of sand shrimps at **B**;
  (iii) the absence of sand shrimps at **C**. *(3 marks)*

(b) Another species of crustacean, **Y**, can tolerate only salt concentrations below 1.5% and temperatures below 20 °C.
  (i) Mark on a copy of the map, with a **Y**, **one** place where species Y might live.
  (ii) Suggest how having different ranges of tolerance enables both species of crustacean to survive in the same estuary. *(2 marks)*

(c) Suggest **two** abiotic factors, other than salt concentration and temperature, that you would expect to show significant variation at point A on the map. *(2 marks)*

AQA, 2000

5   Genetic engineering has made it possible to transfer genes from one species to another. For example, a gene that gives resistance to herbicide and another gene which gives resistance to insect attack have been transferred into maize. Some people think that this technology will improve agriculture. However some people think that there will be great long-term dangers in growing crops of this maize.

Evaluate both of these viewpoints. *(6 marks)*

AQA, 2000

10.1 Diet, crops and food allergies

Allergies occur when a person's immune system responds to substances in the environment called **allergens**. Allergens contain proteins that are harmless to most people, but can produce unpleasant or even potentially life-threatening symptoms in people with allergies. In the UK there has been a dramatic increase in the reported cases of allergy over the past 50 years. This increase has also been experienced in other more economically developed countries in Western Europe, the United States, Canada, Australia and New Zealand. However, allergies are uncommon in less economically developed countries in Africa and the Middle East.

Research has shown that the development of an allergic disorder depends on both genetic and environmental factors. Studies have found that the genetic background of the UK population has not changed significantly enough in the past 50 years to explain the dramatic rise in the number of reported cases of allergy. The increase in reported allergic conditions in the UK over the second half of the 20th century must therefore be due to environmental factors. Some of the environmental factors thought to contribute to the development or worsening of allergic disorders include: diet, exposure to allergens, atmospheric pollution, tobacco smoke and lack of exposure to infections early in life. Human impact has certainly brought about great changes in these areas in the past 50 years.

Our diet has changed considerably in this time and we now consume a great diversity of foods containing many different ingredients. Some of these ingredients have the potential to cause allergies.

Hint

At AS Level you looked at how food labelling can help people to choose a healthy, balanced diet. Food labelling is very important to people with food allergies because the consequences of eating a food they are allergic to could be very serious. European directives state that there are 14 food allergens that have to be listed whenever they, or ingredients made from them, are used at any level in pre-packed foods, including alcoholic drinks. The list consists of cereals containing gluten, crustaceans, molluscs, eggs, fish, peanuts, nuts, soybeans, milk, celery, mustard, sesame, lupin and sulfur dioxide at levels above $10\,mg\,kg^{-1}$, or $10\,mg\,dm^{-3}$, expressed as SO_2.

Figure 1 *Food allergy advice. A label on the packaging of this chocolate warns that it contains soya and possibly traces of nuts, which may cause an allergic reaction*

Figure 2 *Peanuts, shelled and un-shelled*

How science works

Peanut allergy trials

Four hundred and eighty babies will be involved in a 7-year trial to identify factors that put people at risk of developing a peanut allergy in childhood.

Peanut allergy poses a serious threat to the health of affected individuals and cases among children in the UK have doubled in the past 10 years, with one in 70 school children now affected. It was thought that a child's chances of developing a peanut allergy were increased by the consumption of peanuts early in childhood. However, more recent evidence suggests that children who are introduced to peanuts in the diet earlier in infancy may be protected against developing a peanut allergy.

The babies recruited into the trial must be between 4 and 11 months old, known to have eczema and to suffer from an egg allergy. Babies with both these conditions have an increased risk of becoming allergic to peanuts by the time they reach school age. To investigate whether the inclusion of peanuts in the diets of infants changes their risk of developing a peanut allergy, half the babies recruited to the trial will not be allowed any products containing peanuts, and the other half will receive a peanut snack regularly from 4 months of age for 3 years. The health of the children will be carefully monitored throughout the trial by medical staff, who will test the children for peanut allergy when they are 5 years old. The results are expected to provide valid scientific data that can form the basis for advice on how to minimise the risk of a child developing a peanut allergy.

1. Why do the researchers want to involve babies who have eczema and have tested positive for egg allergy in the trial?

2. Why are so many babies needed for this trial?

3. Suggest why the trial will last 7 years.

4. Why will half the babies in the trial not be given peanuts?

5. Give **two** reasons why parents of susceptible children may want their children involved in this trial and **two** reasons why they may not.

6. Scientists have found that food allergies and eczema are rare among babies who are only fed on breast-milk for the first 4 months of their lives. Some scientists think that this could be explained by natural selection. Explain how.

7. How could you investigate whether food allergies and eczema are more common among children who were fed on formula milk during the first 4 months of their lives?

AQA Examiner's tip

You need to know what an allergen is and about hay fever, food allergies, allergic asthma and hives as examples of allergic reactions. However, you do not need to know the specific examples given here, such as oilseed rape. In the exam you need to apply your knowledge to various examples.

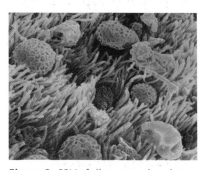

Figure 3 *SEM of allergens such as dust (pale blue) and pollen (pink) on the surface of the treachea. These allergens may cause asthma and hay fever*

Figure 4 *Oilseed rape crop*

■ **Summary questions**

1 What is an allergen?

2 What are the differences in the symptoms of hay fever and asthma?

3 Why do scientists think that changes in the environment are responsible for the increase in reported cases of allergy in Britain over the past 50 years?

There has been a huge increase in the consumption of processed foods in the UK in the past 50 years. Many of these foods contain peanut oil and sunflower oil. Oilseed rape is a crop that is now widely grown in Britain for the vegetable oil that is extracted from its seeds.

■ Application and How science works

Oilseed rape

Oilseed rape is a profitable crop to grow because it is used to make biodiesel and extra virgin rapeseed oil. As a result, is it extensively cultivated and covers about 600 000 hectares of British farmland. Cultivation of oilseed rape began in Britain about 30 years ago. As the crop became widely established in Britain, concern was raised about its potentially harmful effects. A scientific report published in the late 1990s concluded that there was no clear evidence that oilseed rape had adverse effects on human health. It stated that allergy to oilseed rape pollen was rare, even in areas of intensive cultivation, and individuals that were affected tended to be multiply-allergic to other substances with no indication that the allergic response was triggered by pollen from oilseed rape plants.

Most pollination of oilseed rape flowers is carried out by insects such as bees. The pollen is only transported about 50 metres away from the plant and so it is not considered to be a major cause of hay fever. It has been suggested that spores released from fungi found on the crop when the seed is ripening could trigger a reaction, but researchers have found no evidence for this. Oilseed rape does release organic compounds into the air that could irritate the mucus membranes of the eyes, mouth and throat. However, in other oilseed rape-growing countries such as France, Germany and Denmark, there is no public concern over the widespread cultivation of this crop.

1 Why is so much oilseed rape grown in Britain?

2 Oilseed rape is not considered to be a major cause of hay fever. Explain why.

Table 1 *Common UK allergies*

Type of allergy	Symptoms	Possible causes	Epidemiology (how often the disease occurs in people)
hay fever (seasonal allergic rhinitis)	sneezing a blocked or runny nose itchy eyes, nose and throat headaches In some people, pollen may also trigger asthma.	grass pollen – this is the case for 9 out of 10 people in the UK who have hay fever tree pollen – 1 in 4 people in the UK with hay fever are allergic to birch pollen, but different trees produce pollen at varying times of the year and it depends which one a person is allergic to as to when their symptoms are worst weeds and other plants, as well as spores from fungi and moulds	Mild winters and warmer springs mean that pollen production in the United Kingdom now starts earlier than it did 50 years ago. Therefore symptoms can be well established by the first week in May and peak around mid-June to early July. 3.3 million people in England have a recorded diagnosis of allergic rhinitis at some point in their life, and one person in every 135 of the population was diagnosed during 2005. Patients with allergic rhinitis may have symptoms for all or part of the year, so distinguishing between the causes in epidemiological surveys is difficult. A recent study found a very high occurrence of seasonal allergic rhinitis (hay fever) across Western Europe, but concluded that it is frequently undiagnosed.

Continued

Table 1 continued

Type of allergy	Symptoms	Possible causes	Epidemiology (how often the disease occurs in people)
food allergies	The first symptom is often itching and swelling in the mouth, tongue and throat. Some or all of the following symptoms may be experienced: ■ skin reactions, such as swelling and itching, ■ eczema and flushing ■ vomiting and/or diarrhoea ■ coughing, wheezing or a runny nose ■ swelling of the lips ■ sore, red and itchy eyes. Allergic reactions usually happen within a few minutes of eating food that triggers an allergy, but they can take several hours to develop.	In adults, the most common food allergies are to peanuts, tree nuts (such as walnuts and hazelnuts), fish and shellfish. Foods which commonly trigger allergies in children include cows' milk, eggs, peanuts, soya, wheat, tree nuts, fish and shellfish. An allergy to one food means that a person is more likely to have an allergic reaction to other foods (cross-reactivity). For instance, a person with an allergy to prawns may find that they are allergic to other shellfish. People with hay fever may also have a cross-reaction to certain foods.	Food allergies are most commonly found in children, with approximately 5–7% of infants experiencing an allergic reaction, although egg and milk allergies tend to get less severe with age. Some food allergies persist and around 1–2% of adults suffer from a food allergy. Although the persistence of childhood allergy is unusual, once a food allergy is established in an adult it is rarely cured. The increase in peanut allergy has been extraordinary. There was a 117.3% increase in the prevalence of peanut allergy from 2001 to 2005, and it is estimated that 25 700 people in England are affected. One in every 12 420 people was newly diagnosed during 2005. New food allergies are regularly being described, for example to fruits, vegetables, soya, sesame, mustard, chickpea and kiwi fruit (Chinese gooseberry), but the reasons for the rising trends of these new allergies are unclear. A recent study reported that approximately 30% of adults who are allergic to pollen also suffered from food allergies, particularly involving fruit or nuts.
allergic asthma	The symptoms of asthma may be mild, moderate or severe. They tend to be variable, may stop and start and they are usually worse at night. They may include: ■ coughing ■ wheezing ■ shortness of breath ■ tightness in the chest.	The exact cause of asthma isn't fully understood at present. Triggers can include: ■ viral infections such as colds and 'flu ■ irritants such as dust, cigarette smoke, fumes and chemicals ■ allergies to pollen, medicines, animals, house dust mite or certain foods ■ exercise – especially in cold, dry air ■ emotions – laughing or crying very hard can trigger symptoms, as can stress.	An estimated 5.7 million people in England are affected by asthma, and one person in 192 in the population was newly diagnosed during 2005. In 2006, it was reported that the incidence of asthma symptoms had risen in 6 to 7 year olds in the UK, from 18.4% to 20.9% over a period of approximately 5 years. The prevalence of asthma symptoms in 13 to 14 year olds was higher, at 24.7%, but the incidence within this age group had actually decreased from 31%. The occurrence of asthma has probably reached a plateau or may even possibly be falling. There has been a steady decline in child hospital admissions since 1990, and asthma-related deaths in childhood remain uncommon. The newly recorded incidence of asthma within primary care also decreased from 6.9 per 1000 person-years in 2001, to 5.22 per 1000 in 2005, possibly due to a greater awareness of the disease and the availability of more effective treatments.
hives (urticaria and angioedema)	itching and a swollen, red rash known as 'hives' or 'weals' on the surface of the skin (urticaria) or deeper in the skin, particularly around the mouth and eyes (angioedema) The rash occurs suddenly and usually disappears within 24–48 hours.	food allergy, especially to peanuts, tree nuts or shellfish Viral infection is more commonly the cause than food allergy.	Urticaria and angioedema are amongst the commonest problems referred to allergy specialists. Some studies estimate that one in five of the population have urticaria at some point in their lifetime, and hospital admission rates for urticaria more than doubled from 1990 to 2000. However, the rates for angioedema appeared unaltered.

10.2 Allergic responses produce illness

Learning objectives:

- ◼ What is the role of mast cells in an allergic reaction?

- ◼ What is anaphylaxis?

Specification reference: 3.5.3

◼ **Link**

Look back at Chapter 5 in *AS Human Biology* to remind yourself about antigens and how the immune system works.

In Topic 10.1 you were introduced to a variety of allergic reactions and the allergens that are thought to 'trigger' them. All true allergies are caused when a person's immune system produces an abnormal response to an allergen. Allergens can therefore be defined as antigens that produce an abnormal immune response.

◼ Hypersensitivity

For many people allergens do not produce allergic responses, but in those with allergies there is an over-reaction of the immune system. This is called **hypersensitivity**, and someone with an allergy is said to be **hypersensitive**. Hay fever, food allergies, allergic asthma and hives are all examples of hypersensitive allergic reactions. Figure 1 shows a summary of the sequence of events in an allergic asthma reaction.

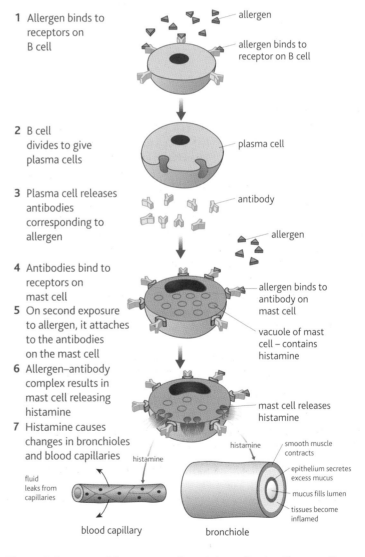

1 Allergen binds to receptors on B cell — allergen / allergen binds to receptor on B cell

2 B cell divides to give plasma cells — plasma cell

3 Plasma cell releases antibodies corresponding to allergen — antibody / allergen

4 Antibodies bind to receptors on mast cell — allergen binds to antibody on mast cell / vacuole of mast cell – contains histamine

5 On second exposure to allergen, it attaches to the antibodies on the mast cell

6 Allergen–antibody complex results in mast cell releasing histamine — mast cell releases histamine

7 Histamine causes changes in bronchioles and blood capillaries — histamine / smooth muscle contracts / epithelium secretes excess mucus / mucus fills lumen / tissues become inflamed

histamine / fluid leaks from capillaries / blood capillary / bronchiole

Figure 1 *Summary of the sequence of events in an allergic asthma reaction*

◼ **Hint**

The interactions between molecules involved in the allergic response depend on the specific shape of the allergen and antibody. The shape of the antibody is complementary to the shape of the allergen. IgE antibodies are involved in allergic responses. At AS level you learned how antigens lead to antibody production by B cells, and how these antibodies help to combat disease. The antibodies involved in combating disease are a different kind of antibody.

- When a person with an allergy is exposed to an allergen, their immune system recognises it as non-self.
- **B cells** then produce **IgE antibodies**. These are antibodies that belong to the immunoglobulin E group and are different from the antibodies normally produced in response to infectious disease.
- The IgE antibodies bind to **mast cells**. Mast cells are cells of the immune system that are found in all tissues.
- When the allergen binds to the IgE antibodies on the mast cell, a chemical called **histamine** is released.
- Histamine plays a part in producing the symptoms of the allergy.

The symptoms of allergies can be divided into two groups:

1 A localised response that affects a specific region of the body. For example, asthma and hay fever both affect the respiratory tract.

2 A severe, whole-body allergic reaction called **anaphylaxis** or **anaphylactic shock**. Anaphylaxis is a sudden, acute and potentially fatal reaction to an allergen. It can involve oedema (swelling) in the airways leading to the lungs, or a large and sudden fall in blood pressure. Sometimes a rash may accompany the onset of anaphylaxis, but in some cases anaphylactic shock following exposure to an allergen can occur without warning. The onset can be so rapid and severe that it can result in death by asphyxiation and/or lack of adequate blood circulation.

Application and How science works

Anaphylactic shock

Hospital admissions due to anaphylactic shock rose seven-fold from 1990/01 to 2003/04. During the 1990s, approximately 20 deaths each year were identified as having been caused by anaphylaxis. For the period 1992–1998, around half the number of anaphylaxis deaths were due to medical treatments such as drugs used in anaesthesia or injections for X-ray investigations, with the rest being caused by stings, foods or rare causes such as latex, hair dye or parasitic worms. The pattern of fatal anaphylaxis to food during this period was similar to that reported from 1999–2006, when 48 deaths occurred in people ranging from 5 months to 85 years old. These were caused by milk (6), peanuts (9), tree nuts (such as almonds, hazelnuts, walnuts and brazil nuts) (9), fish (1), shellfish (1), snail (1), sesame (1), egg (1), tomatoes (1) and 'uncertain' allergen deaths (18).

1 Give two possible causes for the seven-fold increase in hospital admissions due to anaphylactic shock from 1990/01 to 2003/04.

2 Convert the data for reported cases of fatal anaphylaxis to food from 1999–2006 into a pie chart.

3 Which of the known causes of food allergy seems to be responsible for most cases of fatal anaphylaxis during this period?

4 Suggest why 18 out of 48 deaths are classed as 'uncertain' allergen deaths.

Summary questions

1 What is an allergen?

2 Why is someone with an allergy said to be hypersensitive?

3 Describe how one of the polypeptide chains in an IgE molecule would be made in a B cell.

4 Produce a flowchart summarising the sequence of events that takes place in an allergic reaction. Use the terms 'specific' and 'complementary' in your answer.

5 Anaphylactic shock can be fatal. Explain why.

10.3 Detection and treatment of allergic response

Learning objectives:

- How do skin tests help to diagnose an allergy?

- Why are some types of allergic response treated with antihistamines, while others are treated with adrenaline?

Specification reference: 3.5.3

When an allergic reaction is experienced it is important that the allergen is identified. A GP will take a patient's full medical history and arrange for tests to be carried out to try to find the cause.

Skin tests for allergies

The first type of test to be carried out when an allergy is suspected is usually **skin-prick testing**. You can see this in Figure 1. It is a simple, quick and inexpensive form of testing that can give results within 10 or 20 minutes. It is suitable for most types of allergy and for most patients. However, the test is not appropriate for people who have suffered anaphylactic shock from the allergen. Skin-prick testing is usually carried out on the patient's inner forearm. Patients with severe eczema may have the test carried out on the back.

- Between 3 and 25 allergens can be tested.
- A pen is used to mark the arm with code numbers for the allergens to be tested. A drop of the allergen, or a solution containing an extract of it, is placed by each code number.
- The skin is then pricked through the drop using the tip of a sterile lancet or needle.
- If there is a positive reaction to an allergen, the skin at that spot becomes itchy, red and swollen within a few minutes. The 'weal' has a raised edge and expands to reach a maximum diameter of 3–5 mm in 15 to 20 minutes. The reaction usually lasts for about 1 hour.
- A negative control is used. This is saline solution and the patient is not expected to show a reaction to this.
- The patient is also tested with a positive control solution containing histamine. This is expected to give a positive reaction in all patients.
- A negative response to skin-prick testing usually indicates that the patient is not sensitive to that allergen. Negative reactions may occur if the patient is taking antihistamines or other medicines that block the effect of histamine.

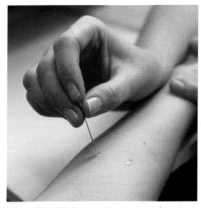

Figure 1 *Allergy testing – view of a child's arm during a skin-prick test for allergies*

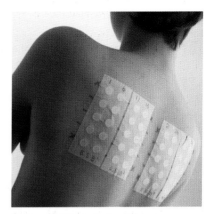

Figure 2 *Allergy test patches*

Patch testing is carried out if a patient suffers from a skin allergy known as eczema or contact dermatitis. You can see this in Figure 2.

- Allergens are applied to the skin on the back in petroleum jelly on hypoallergenic metal discs.
- The skin is coded and the discs are held in place with hypoallergenic tape.
- After 48 hours the discs are removed and the skin is observed for redness and swelling.
- The skin is examined again in a further 48 hours.

Antihistamines

The histamines that are released from mast cells during an allergic reaction bind to histamine receptors on the plasma membranes of many different types of cells in the body. These cells then respond to produce the symptoms of the allergy. For example, in the case of hay fever, histamines released from mast cells found in the nasal passages bind to histamine receptors on epithelial cells in the mucous membranes. This causes the membranes to become inflamed, which in turn produces symptoms such as an itchy, runny nose and sneezing. **Antihistamines** are drugs that prevent histamines from binding to the histamine receptors, so they prevent the symptoms of the allergy from developing.

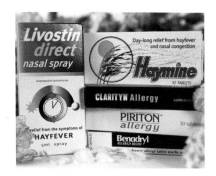

Figure 3 *Over-the-counter hay fever drugs*

Antihistamine drugs are available as tablets (oral), creams, eye drops, sprays and injections. The older antihistamine drugs also affect histamine receptors found in the brain and tend to cause drowsiness. The more recently developed, new antihistamines cause less drowsiness. The effects of 'new', non-sedating antihistamines also tend to last longer, so the medicine is taken less frequently.

Some antihistamines, such as creams for insect bites, and some tablets for hay fever, can be bought over-the-counter from a pharmacy. Others have to be prescribed by a doctor.

Antihistamine tablets can be used to reduce allergy symptoms such as itching of the nose and eyes, a runny nose and watering eyes, and sneezing. Antihistamines that are taken orally are also used for allergies to certain medicines such as aspirin. Topical antihistamines are antihistamine creams and ointments that are applied to the skin. Both antihistamine creams and tablets are effective in reducing urticaria. Oral antihistamines are also used to treat chronic (long-lasting) hives. Eye drops containing antihistamines can be applied to the eyes to treat allergic conjunctivitis (swelling and irritation of the conjunctiva of the eye).

In emergency treatment of serious allergic reactions such as anaphylactic shock, an adrenaline injection is given followed by an antihistamine injection.

Adrenaline

Anaphylactic shock is a severe, life-threatening, whole-body response to an allergen. Emergency treatment involves the injection of a drug called adrenaline (or epinephrine).

Adrenaline is a chemical that binds to adrenergic receptor molecules, which are found on cell membranes all over the body, especially in the heart, lungs and blood vessels. It is naturally produced by the adrenal glands, but when anaphylaxis occurs, an adrenaline injection must be given as soon as possible to alleviate the symptoms.

Hint

Look back at Topic 6.2 to remind yourself of the role of adrenaline in the fight or flight response.

When the adrenergic receptors are activated they cause the heart to beat faster and more strongly. The activation causes the width of the airways in the lungs to increase by causing the smooth muscle in the walls of the bronchioles to relax (bronchodilation). When adrenaline binds to its specific receptors it also causes smooth muscle in the walls of arterioles to contract (vasoconstriction), narrowing the lumen of the arteriole and increasing blood pressure. Under normal circumstances, all of these actions prepare the body to cope with a stressful situation and are commonly termed the 'fight or flight response'.

In the case of anaphylactic shock, one of the most important effects of adrenaline is to open up the airways to ease breathing. The narrowing of the blood vessels produced by the adrenaline also prevents or reverses falls in blood pressure. Adrenaline also relieves the itching, redness and swelling associated with anaphylactic reactions.

Anaphylactic reactions can develop within minutes of exposure to an allergen. Individuals who are at risk are sometimes advised to carry an automatic injection device that can be used to give a dose of adrenaline immediately.

Summary questions

1   Blood tests can also be used to detect allergies. The blood test measures the amount of specific IgE molecules in the blood following exposure to a suspected allergen, for example after a patient has suffered anaphylactic shock. Why might a blood test for an allergy be safer for some patients than a skin-prick test?

2   When a patient undergoes a skin-prick test, they are tested with a positive control solution containing histamine. Why is this expected to give a positive reaction in all patients?

3   The patient will be asked to stop taking some antihistamine medicines for about 6 weeks before a skin-prick test and some other medicines for about 5–6 days before the test begins. Explain why.

4   Why is adrenaline not used to treat hay fever?

5   Why do histamine receptors in plasma membranes bind only to histamines and not to other substances present in the body?

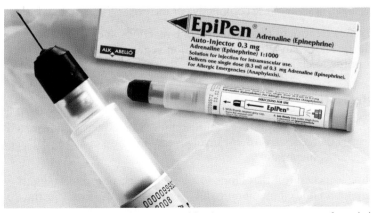

Figure 4 *EpiPen adrenaline syringe used for the emergency treatment of anaphylactic shock – a type of allergic reaction that can be fatal in minutes. The EpiPen is an auto-injector that can be self- administered.*

Allergy advice for schools

The Government recommends that unhealthy snacks such as cake and crisps should not be sold in schools. Instead, it recommends healthy snacks such as fruit, nuts or seeds. This of course poses a health risk for allergy sufferers. Traces of nuts and seeds can be easily transferred between children, and tiny traces could cause an allergic reaction.

The Anaphylaxis Campaign offers advice to parents, students and teachers regarding allergies in schools. It advises parents of non-allergic children not to give their child nuts or seeds for their lunch. Traces of nuts and seeds could be transferred to an allergy sufferer from the keys of a keyboard, or pages of a book.

For pupils who are not allergic, The Anaphylaxis Campaign recommends they familiarise themselves with what triggers their classmates' allergies. Pupils should make sure they wash their hands with soap after touching anything that could cause an allergic reaction in a fellow pupil.

For pupils who have a severe allergic reaction, The Anaphylaxis Campaign recommends that they should never share food, and only eat food that is clearly labelled. It also recommends that allergic pupils should wash their hands before eating, in case of contamination. Pupils with a severe allergy should inform their classmates of what to do in an emergency, and always carry with them emergency medication.

Head teachers are advised to ensure they have clear records of every student with a severe allergy. If possible, they should have a meeting with the pupils and parents to discuss the severity of the allergy and emergency procedures. They are also advised to arrange staff training by the school nurse or medical professional and carefully consider whether or not to encourage nuts and seeds in their school.

1. Do you think that parents of non-allergic children should be asked to avoid sending their child to school with foods containing nuts and seeds? Explain your answer.

2. Do you think a child of primary school age with a severe nut allergy should be taught to self-administer their EpiPen injection? Explain your answer.

10.4 Air pollution and respiratory illness

Learning objectives:

- Are respiratory illness and air pollution linked?

- Why do some people develop asthma and others do not?

Specification reference: 3.5.3

Figure 1 *The symptoms of asthma can be relieved by using an inhaler containing a chemical which works as either a 'reliever' or a 'preventer'. Relievers are chemicals that open up the airways; preventers are inhaled steroids that prevent inflammation and narrowing of the airways*

Asthma

In Topic 10.2 you saw that the symptoms of allergic asthma are produced when mast cells on the lining of the bronchi and bronchioles release histamine. The histamine then causes:

- the walls of the airways to become inflamed

- the epithelium lining the airways to secrete larger quantities of mucus than normal

- fluid to leave the blood capillaries and enter the bronchioles

- the smooth muscle surrounding the bronchioles to contract, causing narrowing of the airways.

These effects make it very difficult for air to move in and out of the lungs, producing the symptoms of wheezing, coughing and a feeling that the chest is being squeezed tight.

Asthma is very common in the UK – 5.2 million people are currently receiving treatment for the condition. Of these, 1.1 million are children. It is estimated that there is a person with asthma in one in five households in the UK. There are many triggers that can cause mast cells in the lungs to release histamine and produce the symptoms of asthma.

Scientific research suggests the following:

- You are more likely to develop asthma if you have a family history of asthma, eczema or allergies.

- Certain environmental factors combined with a family history of asthma, eczema or allergies influence whether or not someone develops asthma.

- Smoking during pregnancy significantly increases the risk of a child developing asthma.

- Children whose parents smoke are more likely to develop asthma.

- Atmospheric pollution may trigger an asthma attack, although infections and allergens are more likely to do so. Air pollution can make asthma symptoms worse but there is little evidence that air pollution itself causes asthma.

- Asthma may develop in adults after a viral infection.

- Changes in housing and diet and a more hygienic environment may have contributed to the rise in asthma over the last few decades. Can your suggest why this is?

Atmospheric pollution

Atmospheric pollution is associated with many health problems, especially those related to the respiratory system. The way in which a person responds to pollutants in the atmosphere depends upon:

- the age and health of the individual

- the type of pollutant

- the concentration and length of exposure to the pollutant

- the activity being undertaken when exposed to the pollutant.

AQA Examiner's tip

Never assume that a correlation proves a causal link!

People who live in cities are exposed to a lot more than just air pollution. Proving a causal link between one factor and an illness can be very difficult – especially if more than one factor contributes to the illness

If a person's health is good, the levels of air pollution usually experienced in the UK are unlikely to have any serious short-term effects. However, people with lung disease such as asthma or chronic bronchitis, or with heart conditions are at greater risk, especially if they are elderly. During serious air pollution episodes in the UK, admissions to hospital for treatment of respiratory disease may increase.

Table 1 shows the health effects that individuals with lung diseases might experience at very high concentrations of nitrogen dioxide, sulphur dioxide, ozone, particles and carbon monoxide.

Table 1 *Some common atmospheric pollutants and their effects at very high levels*

Pollutant	Source of pollutant	Health effects at very high levels
nitrogen dioxide	vehicle emissions, burning of fossil fuels and biomass in power stations	These gases cause irritation of the airways of the lungs, increasing the symptoms of those suffering from lung diseases such as asthma, chronic bronchitis and emphysema.
sulfur dioxide	burning of fossil fuels in power stations and oil refineries	
ground level ozone	Formed when nitrogen oxides and hydrocarbons react in the presence of sunlight. In polluted conditions, this can cause summertime 'smog'.	
particles / PM10 (particles with a diameter less than or equal to 10 μm that can be inhaled by humans)	combustion sources such as burning of coal and diesel combustion by road traffic; suspended soils; dusts produced by construction, mining and quarrying; sea salt; biological particles such as pollen and fungal spores Some PM10s are formed by chemical reactions in the atmosphere.	Fine particles can be carried into the lungs where they can cause inflammation and a worsening of heart and lung diseases.
carbon monoxide	incomplete combustion of fossil fuels, such as motor vehicles, power stations, waste incinerators, domestic gas boilers and cookers 50% of UK emissions of carbon monoxide come from road transport.	This gas prevents the normal transport of oxygen by the blood. This can lead to a significant reduction in the supply of oxygen to the heart, particularly in people suffering from heart disease.

Figure 2 *Aerial photograph of Canary Wharf, London, UK. The haze visible over the horizon is photochemical smog*

Scientists are still trying to understand the long-term health effects of exposure to air pollution. There is some evidence to suggest that cutting long-term exposure to PM10 by half could increase life expectancy by between 1 and 11 months on average. This is not as great as the effect of smoking on life expectancy (on average, non-smokers live 7 years longer than smokers).

How science works

Traffic pollution

Recent research into the effects of pollution have shown that exposure to traffic pollution can seriously effect the development of children's lungs. This means that children exposed to high levels of traffic pollution could have a lower lung capacity and weaker lungs than those exposed to less traffic pollution. The study tested the lung function of a group of children from the age of 10. They were tested every year until the age of 18. At 18, there was a much larger difference in lung function between those living close to traffic pollution and those living further away than there was at the age of 10. Factors such as socioeconomic status were controlled in the study.

There is great concern over the findings. Children exposed to high levels of traffic pollution could encounter serious health problems in later life. This reduced lung function in children is a risk factor for the development of asthma and chronic lung diseases.

1 Give an example of a chronic lung disease.

2 What is meant by socioeconomic status and why would this need to be controlled in this study?

3 Why do you think there is little difference in lung function between different groups of children at the age of 10, but a marked difference by the time they reach 18?

Summary questions

1 Emphysema is an example of an incurable chronic pulmonary disease. What is the name of one of the main substances used to treat it?

2 Why might a person suffering from chronic bronchitis find that their symptoms are worse when there is a summer 'smog' than on a cool and breezy summer's day?

3 Emerging evidence from the UK and other countries suggests a relationship between the number of people in the population with asthma at any given time (the prevalence) and local traffic density. Can it be concluded that vehicle emissions cause asthma? Explain your answer.

10.5 Water pollution and illness

Learning objectives:

- How is the quality of bathing water at Britain's beach resorts assessed?

- What is a blue flag beach?

Specification reference: 3.5.3

Figure 1 *Sewage pipeline depositing waste into the North Sea, Northumbria*

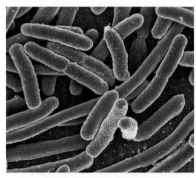

Figure 2 *Coloured SEM of* Escherichia coli *bacteria. E. coli is a coliform bacterium that lives in the intestine of humans and other endothermic animals. It is found in water that is contaminated with faeces*

◼ Monitoring bathing water quality

For centuries humans have used rivers and coastal waters as a convenient waste disposal system for untreated sewage, relying upon the action of the current or the tide to carry the waste out to sea.

Untreated sewage contains many pollutants, including faeces, and can pose a threat to the environment and to the health of people who may use rivers or stretches of coast for bathing. It releases large numbers of microorganisms into the water, including some **pathogenic** bacteria and viruses. If water that is contaminated with untreated sewage is accidentally swallowed or comes into contact with the body, the pathogenic organisms that it contains may cause gastroenteritis and ear and respiratory tract infections.

In order to protect public health and the environment in and around European bathing waters, the European Community has produced a bathing water directive, called the *1976 EC Bathing Water Directive (76/160/EEC)*. This defines bathing waters as all running or still fresh waters and sea water in which bathing is allowed and is traditionally practised by a large number of bathers. The Directive requires EC member states to identify popular bathing areas and to monitor water quality at these bathing waters throughout the bathing season (in England this runs from mid-May to September). The Directive sets a number of microbiological (biotic) and physicochemical (abiotic) standards that bathing waters must either comply with (these are called mandatory standards) or try to meet (theses are called guideline standards).

The two main standards used to assess the quality of bathing water are total coliforms and faecal coliforms. Coliforms are bacteria found in the guts of humans and other endothermic animals, and are indicators of faecal pollution. Figure 2 is a scanning electron micrograph of the faecal coliform bacterium *Escherichia coli*. High-quality bathing waters will have low levels of coliform bacteria. The EC Bathing Water Directive states that at least 20 bathing water samples must be taken at regular intervals throughout the bathing season. Table 1 shows the standards that are set by the EC Bathing Water Directive. If, for example 18 out of 20 tests for a particular parameter were below the guideline or the mandatory standard, then percentage compliance would be $18/20 \times 100 = 90\%$.

Table 1 *1976 EC Bathing Water Directive guideline standard for total coliform, faecal coliform and faecal streptococci parameters*

Parameter	Guideline standard – no more than/ per 100 cm³	Percentage compliance	Mandatory standard – no more than/ per 100 cm³	Percentage compliance
total coliform	500	80	10 000	95
faecal coliform	100	80	2 000	95
faecal streptococci	100	90	–	–

Beaches and Blue Flags

Coastal destinations may be awarded a Blue Flag on an annual basis if they have achieved very high standards of water quality, facilities, safety, environmental education and management. The Blue Flag is a prestigious, international award scheme that acts as a guarantee to tourists that a beach or marina they are visiting is one of the best in the world. It has acted as an incentive to many beach managers to improve the quality of the coastline. The number of beaches and marinas around the world gaining Blue Flag status increases every year and in 2007 more than 3300 Blue Flag Awards were made.

Application and How science works

Beach award schemes

There are a number of voluntary beach award schemes in operation in the UK which have bathing water quality as an essential requirement for the award.

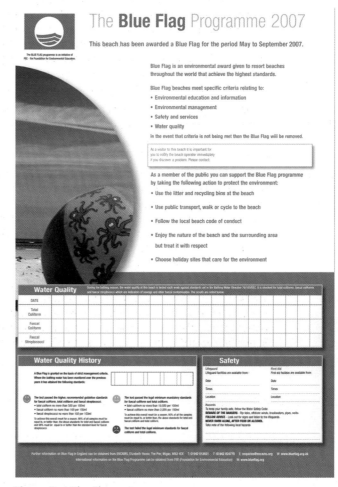

Figure 3 *A Blue Flag poster*

Name of campaign	Key features
Blue Flag Award	Annual International award owned by the Foundation for Environmental Education (FEE). Run in England by ENCAMS (Keep Britain Tidy Group), in Northern Ireland by Tidy Northern Ireland, in Scotland by Keep Scotland Beautiful, in Wales by Keep Wales Tidy, and in the Republic of Ireland by An Taisce (Irish National Trust). Awarded to well-developed resort beaches with good facilities, employing environmentally sensitive management practices and promoting public awareness of the coast. Bathing water quality must achieve the EC Guideline standard (best European standard).
MCS Recommended Beach	Marine Conservation Society (MCS) Recommended Beaches have the highest standards of any UK award for bathing water quality. The award only addresses water quality and does not grade anything else. MCS recommend a beach on the basis that it is likely to pose a minimum relative risk of sewage contamination and related diseases. 100% of water-quality samples pass the EC Mandatory Standard, 80% or more of the samples pass the EC Guideline Total & Faecal Coliform Standard, 90% or more of the samples pass the EC Guideline Faecal Streptococci Standard. The beach is unaffected by sewage from continuous unsatisfactory discharges (any sewage affecting the beach is treated to a minimum standard). Abnormal weather waivers are ignored.
ENCAMS Quality Coast Award	Launched in April 2007, the Quality Coast Award recognises different types of beaches throughout England. The Quality Coast Award website allows beaches to be identified on the basis of region, facilities and type of beach.
Green Coast Award	Annual award run by Keep Wales Tidy, and given to rural, unspoilt beaches in Wales and the Republic of Ireland for high water quality and best practice in environmental management. The Green Coast Award is intended to promote and protect the environment of rural beaches and places a strong emphasis on environmental and community activities. Bathing water quality must achieve the EC Guideline standard (best European standard).
Seaside 'Resort' and 'Rural' Awards	Two annual award schemes for well managed and maintained resorts and rural beaches. Run in Northern Ireland by Tidy Northern Ireland, in Scotland by Keep Scotland Beautiful, and in Wales by Keep Wales Tidy. Bathing water quality must achieve the EC Mandatory standard (basic European standard). Some of these beaches are affected by inadequately treated sewage.

1. What are the EC Bathing Water Guideline standards for total coliform, faecal coliform and faecal streptococci parameters?

2. After very heavy rain, untreated sewage may be discharged onto a beach. Under some bathing water quality schemes, samples taken at such times are not included in calculations of percentage compliance for the bathing season. This is called an 'abnormal weather waiver'. Which award scheme includes all water samples taken during the bathing season, including those taken during abnormal weather?

3. Why do you think there are so many different schemes in operation in the UK?

4. Explain why many stretches of the UK coastline with high-quality bathing waters do not have any beach awards.

Summary questions

1. What illnesses may be experienced if water that is contaminated with untreated sewage is accidentally swallowed or comes into contact with the body?

2. Why are coliform bacteria used as indicators of bathing water quality?

3. Explain why a Blue Flag is awarded on an annual basis.

10.6 *Cryptosporidium* and water pollution

Cryptosporidium is a single-celled parasite that infects the epithelial cells of the small intestine of a variety of mammals, including humans, leading to the illness cryptosporidiosis. Figure 1 shows an electron micrograph of *Cryptosporidium* parasites attached to the surface of the small intestine.

The symptoms of cryptosporidiosis are diarrhoea, vomiting, stomach cramps and fever. In healthy individuals the symptoms last for up to 2 weeks, but in immuno-suppressed patients, such as those with AIDS and the elderly, the disease is likely to be more serious. As yet there is no effective treatment for cryptosporidiosis.

Figure 1 *Coloured scanning electron micrograph (SEM) of the surface of the small intestine infected with* Cryptosporidium parvum *parasites (red), the cause of cryptosporidiosis. The parasite develops in the microvilli of epithelial cells that line the intestinal wall. Magnification: × 8000*

Cells of *Cryptosporidium* leave the gut of infected animals and humans as **oocysts** that are present in the faeces. Oocysts are very resistant to changes in their environment and can survive in a dormant state for many months in cool, damp conditions. If oocysts are ingested, they can infect a new host. Slurry from infected farm animals and sewage containing faeces from infected humans can carry oocysts into rivers used for drinking water abstraction. Outbreaks of cryptosporidiosis occur when drinking water remains contaminated with oocysts after it has passed from the water treatment plant into the drinking water supply.

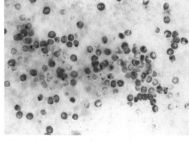

Figure 2 *Light micrograph of* Cryptosporidium parvum *oocysts (pink) in a sample of human faeces. Magnification: × 250*

The majority of water treatment plants cannot completely remove all *Cryptosporidium* oocysts from water. This is because the oocysts are very small and resistant to the levels of chlorine used to disinfect the water in these plants. However, levels of contamination of water with *Cryptosporidium* oocysts can be kept below the guideline 'formal notification level' of one or more *Cryptosporidium* oocysts per 10 litres and an 'alert level' of an average 0.1 oocysts in 10 litres by:

- limiting the contamination of river water by animals, manure or human sewage
- the use of ultra-filtration and ozone to purify the water
- careful maintenance of water-treatment systems.

How science works

Rabbit threatens the health of 250 000 people

Two hundred and fifty thousand people were exposed to the risk of developing cryptosporidiosis when routine sampling of the treated water leaving an East Midlands' reservoir detected *Cryptosporidium* oocysts above the accepted level. All affected customers were issued with a notice to boil their tap water before use, and an extensive cleaning and monitoring programme was carried out throughout the whole distribution system. Local health centres, hospitals and GPs were informed of the risk of an increase in the number of cases of cryptosporidiosis and were advised to test anyone with diarrhoea symptoms for the presence of *Cryptosporidium* oocysts. The public were asked to consult their GP or local health centre if they developed any of the symptoms of cryptosporidiosis. Twenty-four hours after *Cryptosporidium* was detected, routine sampling found the treated water from the reservoir to be free of *Cryptosporidium*, and 8 days later, the 'boil water' notice was lifted. The source of contamination was found to be a rabbit that had found its way into the water distribution system at the water treatment works. Genetic analysis confirmed that the strain of *Cryptosporidium* found in the water was the same as that carried by the rabbit. Genetic analysis also found that 13 of the 29 cases of cryptosporidiosis that were reported in the area around the time of the contamination incident were from strains of *Cryptosporidium* with the same genotype as that taken from the rabbit.

1. Why is the treated water that is leaving the reservoir tested for *Cryptosporidium* oocysts?
2. Why were customers issued with a 'boil water' notice?
3. Why wasn't the 'boil water' notice lifted as soon as routine sampling found treated water from the reservoir to be free of *Cryptosporidium*?
4. Why is it important to identify the genotype of the *Cryptosporidium* found in the water, the rabbit and the patients with cryptosporidiosis?
5. This incident has cost the local authority and the water utility company a great deal of money. List as many of the sources of expense as you can.

Link

Look back at Topic 10.7 of *AS Human Biology* to revise the adaptations shown by parasites. *Cryptosporidium* completes its life cycle within one host, unlike parasites such as *Plasmodium* (the cause of malaria) and *Toxocara*, which complete one part of their life cycle in a secondary host before infecting the host organism.

Summary questions

1. Which type of cells does *Cryptosporidium* infect?
2. Why are the symptoms of cryptosporidiosis more likely to be severe in immuno-suppressed patients?
3. How do the symptoms of cryptosporidiosis help the parasite to find a new host?
4. Why is it difficult to remove *Cryptosporidium* oocysts from water?
5. How may water from catchment areas in upland areas of the UK become contaminated with *Cryptosporidium* oocysts?
6. Explain why healthy people usually recover from cryptosporidiosis after about 2 weeks.

1 Scientists have developed a new hay fever vaccine. Conventional treatments consist of a series of injections containing pollen that gradually increase in strength. This eventually desensitises the immune system. It is a long, drawn-out process, requiring at least 100 injections spread over five years. The new vaccine requires just four injections over four weeks.

The scientists gave over 1000 volunteers either the new vaccine or a placebo just before the start of the pollen season. During the four peak weeks of pollen production, when hay fever is usually at its worst, symptoms were on average 13 per cent less severe in those vaccinated.

(a) Describe how pollen leads to the symptoms of hay fever in people with a pollen allergy. *(4 marks)*

(b) Explain why the vaccine needs to be given four times. *(2 marks)*

(c) (i) Some of the volunteers were given a placebo. What would the placebo contain?

    (ii) Neither the scientists nor the volunteers knew which volunteers were given the placebo. Explain why this is important.

    (iii) A large number of volunteers took part in this investigation. Explain why this was important. *(5 marks)*

2 Read the following passage.

A seven-year study is currently being conducted to try to find out whether exposing infants to peanuts lessens or increases their risk of an allergy developing.

Babies aged between four and 11 eleven months old will be taking part and will be offered a diet including peanuts. Each child already has either eczema or an allergy to eggs. Previous research indicates that 25% of those infants suffering with these disorders will go on to   5
develop a peanut allergy, which in some cases will be so severe it could cause anaphylaxis.

Scientists studying babies from different countries have noticed that there is a negative correlation between exposure to peanuts in infancy and peanut allergy. Current government advice in the UK is to avoid peanuts during pregnancy, breast feeding and the first three years of childhood.   10

(a) Suggest a suitable control for this investigation. *(2 marks)*

(b) (i) What is anaphylaxis? (line 6)

    (ii) Describe how anaphylaxis can be treated. *(4 marks)*

(c) (i) Explain what is meant by a negative correlation (line 7).

    (ii) A correlation does not show that one factor causes another. Explain why. *(2 marks)*

3 The EC Bathing Water Directive states that water from rivers, lakes and sea where people bathe or swim has to be tested regularly. A sample of water can be tested to determine the amount of coliform bacteria (e.g. *E.Coli*) present. A known volume of a sample of water is added to a sterile Petri dish containing McConkey agar. This is a special kind of agar that only coliform bacteria can grow on. The dish is incubated at 37°C for 24 hours.

(a) Drinking water is also tested for the presence of coliform bacteria. Explain why. *(3 marks)*

(b) Sterile technique must be used in performing the test for coliform bacteria. Give two examples of sterile technique which could be used in this test. *(2 marks)*

(c) (i) The dish containing the water sample should not be opened again after it has been incubated. Explain why.

    (ii) Suggest how you could use this test to estimate the number of coliform bacteria present in a sample of water. *(5 marks)*

4 Toxocariasis (*Toxocara canis*) is an infection caused by worms found in the intestines of dogs and cats. Humans can become infected by accidentally ingesting worm eggs, or by eating food that is contaminated with soil containing the eggs. *Cryptosporidium* is a microscopic parasite, which, if swallowed, can cause gastroenteritis in humans.

   (a) Describe how *Cryptosporidium* may be passed from one infected human to another. *(3 marks)*

   (b) Explain why *Cryptosporidium* is described as a parasite. *(2 marks)*

   (c) Give two ways in which *Cryptosporidium* is different from Toxocara. *(2 marks)*

5 (a) People suffering from an asthma attack have difficulty breathing. Explain why. *(2 marks)*

   Figure 1 shows the incidence of asthma and the concentration of the four major pollutants in traffic exhaust. Figures are given as a percentage of their value in 1976.

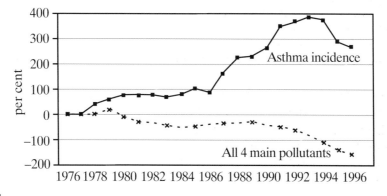

Figure 1

   (b) (i) Describe how the incidence of asthma changed from 1976 to 1996.

      (ii) Explain why the figures are given as a percentage of their value in 1976. *(2 marks)*

   (c) A motoring organisation claimed that these data provide evidence that traffic pollution does not cause asthma. Is this a valid conclusion? Give reasons for your answer. *(4 marks)*

Succession

Ecological **succession** is the gradual process by which ecosystems change and develop over time.

There are two types of succession: primary and secondary.

■ Primary succession is the series of community changes that occur on an entirely new habitat that has never been colonised before. For example, a newly quarried rock face.

■ Secondary succession is the series of community changes that take place in a habitat that has been previously colonised, but has been disturbed. For example, a freshly dug flowerbed or an area cleared of trees by fire or felling.

During succession, an area is initially colonised by certain organisms that are replaced over time by other organisms. Succession is an orderly sequence of events. At each stage, certain species can be recognised that change the environment so that it becomes more suitable for other species. The biomass (living material) that accumulates throughout the process of succession can increase the numbers of niches available for plants and animals and can increase the depth and nutrient content of the soil. This process continues until a relatively stable community develops that is in equilibrium with the existing environment. This community is called a **climax community**.

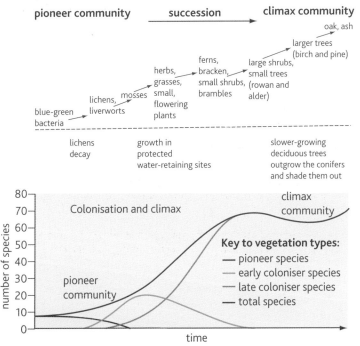

Figure 1 *Succession on bare rock, such as a disused quarry floor*

Succession is usually described in terms of the changes in the vegetation found in an area, although each stage in succession also has its own characteristic communities of animals and microorganisms. The communities change with time because of the interaction between species and their environment.

Figure 2 *Doulton's Claypit, Saltwell, West Midlands. Succession has produced gradual changes in the plant and animal communities living in this disused quarry since it closed in the 1940s*

Summary questions

1 What name is given to the first community to colonise an area that has never been inhabited before?

2 How do these organisms alter the environment so that it becomes more suitable for other organisms?

3 Describe and explain what happens to the total number of species within an ecosystem as succession proceeds from colonisation through to the climax community.

4 In the UK, the climax community for many ecosystems is deciduous woodland. Why is deciduous woodland scarce in the UK today?

11.2 Wasteland and brown-field sites

Figure 1 *Bee orchids, Ophrys apifera, growing beside the M1 motorway*

■ Wastelands are good for wildlife

Human activities often produce areas of bare land and open water that can become colonised by plants and animals. This wasteland is unmanaged and the plant and animal communities that live there are usually those associated with the early stages of succession. In urban areas, wasteland provides a mosaic of different, often unusual habitats, helping to increase biodiversity. Wasteland sites can become colonised by rare or exotic species that are not usually found in the geographic area and because of this, many areas of former industrial, mining and urban land have been designated as official Sites of Special Scientific Interest (SSSIs).

Wasteland includes corridor habitats, such as cuttings and embankments associated with railway tracks and roadsides. These are common in the built environment and allow plants and animals to move between habitats. Bee orchids are plants usually found in open chalk grasslands. Figure 1 shows bee orchids growing in an artificial habitat beside a very busy motorway. This species has colonised many dry, open wastelands in the south of England, including corridor habitats.

■ How science works

Butterflies use the railway

The Large Blue butterfly became extinct in the UK in 1979, but has since been successfully reintroduced to Somerset. It is vary rare in Britain, and when a colony of Large Blue butterflies spread from the initial reintroduction site to the embankment of the nearby Taunton to Paddington railway line, scientists were keen to ensure that the population survived there. A team effort by scientists and the operators of the railway line has ensured that the railway embankment is managed to provide the habitat needed for the survival of the Large Blue butterfly.

Railway sidings are increasingly important for butterflies and other species of wildlife because they link habitats together. These wildlife corridors can allow wildlife to reach new sites and are therefore valuable for their long-term survival.

1 Why are wildlife corridors important for the long-term survival of wild plants and animals?

2 Why may wildlife need to reach new sites?

3 Large Blue butterflies lay their eggs on wild thyme plants. Wild thyme grows in open habitats with thin soils. Why will the sections of railway embankment that have been colonised by the Large Blue require careful management if they are to provide suitable habitats for the long-term survival of this species?

Brown-field sites

Brown-field sites are areas that have previously been developed for human use. These sites are frequently overlooked for their wildlife value and they are often targeted for development or landscaping.

Brown-field sites are important to wildlife because:

- they offer a wide variety of habitats, important to the survival of many different species
- the habitats they contain may be rare or threatened by urbanisation or intensive agriculture.

Brown-field sites can be reclaimed to provide habitats for species that have lost their usual habitat. For example, intensive agriculture has caused the loss of flower-rich grasslands from the countryside, and many of the species of **flora** (plants) and **fauna** (animals) that depend on this type of habitat for survival have found refuge in brown-field sites. In Figure 2 you can see volunteers helping to manage a flower-rich brown-field site in an urban area in the West Midlands.

Figure 2 *Volunteers collecting wildflower seeds on a brown-field site during a habitat restoration project in the UK*

How science works

Brown-field sites and development

To reduce urban sprawl, Government policy adopts a 'brown-field first' approach, targeting new developments onto available sites within the urban area. Brown-field sites are often seen as 'useless' and people do not take into account their biodiversity. Public opinion of brown-field sites is generally low, as they can be associated with antisocial behaviour and drug abuse. Green spaces that surround urban areas have strong protection from development even though they may not be very valuable for wildlife. Some local authorities choose to 'tidy-up' brown-field sites but this can greatly affect the biodiversity of the area.

1. Why does Government policy target new developments onto urban brown-field sites?

2. Why is the development of brown-field sites generally welcomed by the public?

3. Why are some brown-field sites more valuable for wildlife than green spaces in the countryside?

4. Why may the Government be reluctant to adopt a 'biodiversity-first' approach to planning?

Summary questions

1. Why are many areas of wasteland designated as Sites of Special Scientific Interest?

2. What is a brown-field site? Give an example.

3. Why are brown-field sites important to wildlife?

4. Describe **one** example of the habitats found on a brown-field site and explain its importance for wildlife.

5. A supermarket wishes to build an outlet on a brown-field site that contains large, open areas of flower-rich grassland. Local residents are welcoming the proposal because they feel the derelict site is an eyesore where a great deal of antisocial behaviour takes place, and they would like to have the facilities that the supermarket can offer. Opponents of the scheme are concerned that if this brown-field site is developed, the survival of many species of birds and insects in the area will be threatened because they will lose their habitat. A public enquiry decides that the development can go ahead, as long as the developers work with scientists to ensure the long-term survival of the populations of plants and animals that have made this brown-field site their home. Suggest **three** ways in which the developers can ensure that a suitable habitat remains or is provided in the area.

11.3 Biodiversity and the urban environment

Learning objectives:

- What is biodiversity?

- Should small ecosystems be linked together to form larger ones?

Specification reference: 3.5.4

Link

Refer back to Topic 9.1 to make sure that you understand what an ecosystem is.

Enhancing biodiversity

Ecosystems range in size from very small, such as a pond, to very large.

Farming, road-building and industrial and housing developments break up large ecosystems into small, isolated fragments. This can cause the numbers of species present in the remaining fragments of the ecosystem to decrease for the following reasons:

- There may be long distances between some ecosystem fragments so it is difficult for species to disperse from one ecosystem fragment and to colonise new areas.

- Some of the plant species that remain may be cut off from their pollinators so they cannot reproduce and may become extinct.

- The food chains supported by these plants may collapse, causing a reduction in the numbers of species present.

- The small population supported by an ecosystem fragment may be too small to recover from a catastrophe such as a fire or a flood.

- The fragment of the ecosystem that remains may not provide the access to food, water, shelter and mates needed for the survival of the species.

Scientists have found that if the area of an ecosystem is increased by a factor of 10, the number of species present approximately doubles. This means that larger sites are important in enhancing the range and variety of living organisms (the **biodiversity**) in both rural and urban environments.

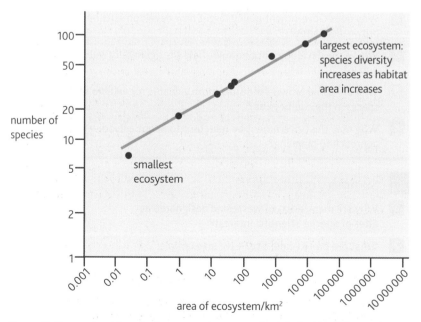

Figure 1 *The number of species in an ecosystem approximately doubles if the area is increased by a factor of ten. The scales used on the axes of this graph are logarithmic*

Living Landscapes – the way forward for Britain's wildlife?

The wildlife of the UK will need to move along wildlife corridors if it is to adapt to climate change. Britain's wildlife has done this before when native species migrated north from southern Europe at the end of the last glaciation, 10 000 years ago. However, this time the migrating wildlife will find their way blocked by artificial obstacles such as towns and cities, motorways and agricultural land, and the wildlife will only survive if humans give it a helping hand. Britain has many well-managed nature reserves, but these small wildlife oases will need to be linked together by landscapes that can support the migrating wildlife. This means that the surrounding, human-dominated landscapes must be managed sympathetically if wildlife is to stand a chance of finding safe passage through them. The UK wildlife conservation charity, The Wildlife Trust, thinks that this can be achieved through teamwork on a large scale. Their hope is that community groups, businesses, land managers and local authorities can work with them to manage a series of Living Landscapes, linked together at local, regional and country level. Living Landscapes are large landscape features, such as river catchment areas, providing shelter and migration routes for wildlife and access and enjoyment for people. Living Landscapes consist of:

Figure 2 *Derelict land next to a canal in an industrial area in Walsall, West Midlands. The habitats found in this area could be managed as part of a network of green corridors through the region*

- core areas of high-quality habitat – protected areas, SSSIs and Nature Reserves
- inter-linkages between core areas that will turn a landscape of isolated patches into one large unit
- land between core areas and its connections that allow wildlife to move through it. It may not be the perfect habitat, but it is good enough to allow wildlife to migrate through it safely.

One of the Wildlife Trust's Living Landscape projects focuses on the Black Country, a major urban area in the UK comprising Wolverhampton, Dudley, Walsall and Sandwell. This region has a remarkable industrial heritage, and, for its size, the most diverse geology in the world. Its industrial past has contributed to an image of low environmental quality, yet Birmingham and the Black Country together are one of the UK's most diverse areas for wildlife, with more rivers and canals than Venice! Two million people live alongside otters, water voles, peregrine falcons, great crested newts, threatened crayfish and huge numbers of unusual plants. The Wildlife Trust plans to create multi-use green corridors through the region that are rich in wildlife and that will link together the key population centres, key nature reserves and other natural heritage features.

1 Why will Britain's wildlife need to be able to move along wildlife corridors if it is to survive in the future?

2 What **three** key features should a Living Landscape contain?

3 What are the benefits of a Living Landscape project for **a** the people living in the West Midlands and **b** the wildlife in the region?

Summary questions

1 Using Figure 1, describe and explain what happens to the number of species present in an ecosystem when its area is increased by a factor of 10.

2 In the past, some nature reserves have consisted of small areas of land where a rare species or an unusual plant or animal community survives. What advice would you give to the manager of a small nature reserve to help ensure the long-term survival of the wildlife living there? What scientific evidence would you use to support your argument?

11.4 Measuring biotic and abiotic factors in an ecosystem

Learning objectives:

- How would you measure changes in the plant community in a local nature reserve with time?

- How would you estimate the number of organisms in an area?

Specification reference: 3.5.4

Link

Look back at the How science works box in Topic 11.2. The scientists monitoring the success of the project to reintroduce the Large Blue butterfly to the UK need to measure the size of the populations of Large Blue butterflies on a regular basis.

When a nature reserve is managed for the benefit of its wildlife, scientists need to be able to measure the types and numbers of the organisms found there. They also need to make measurements of the physical and chemical factors that may have an effect on the plant and animal communities in the nature reserve. This allows the reserve to be managed to meet the habitat requirements of the organisms living there. Changes in the size of populations of animals and plants can be used to find out how successful a particular management technique has been and whether extra effort needs to be made to conserve a species or habitat.

It is impossible to identify and count every organism present within a particular habitat, so instead, samples are taken which are thought to be representative of the whole habitat.

Sampling techniques

By sampling a small area of the ecosystem or habitat we can estimate the numbers of organisms living in a much larger area. Sampling can be carried out randomly or systematically.

Random sampling

- The study area is divided up into a grid system.
- Random numbers are used to generate sampling coordinates within the grid.
- Each number must have an equal chance of being chosen.
- A sample is taken at the intersection of each pair of coordinates.
- The technique assumes that the samples are representative of the whole population.

Random sampling ensures that every organism in the sampling grid has an equal chance of being sampled, helping to remove sampling bias. Sampling bias occurs when the sampling sites are not chosen at random.

Table 1 *Some abiotic factors that may need to be measured when carrying out a habitat survey*

Abiotic factor	How is it measured?
light intensity	light meter
temperature	thermometer or thermistor
pH	soil pH test or electronic probe and meter
oxygen	oxygen probe and meter
relative humidity	hygrometer
levels of ions such as nitrates, phosphates or chloride ions	soil test or electronic probe and meter
soil moisture or soil humus	soil tests

Systematic sampling

■ The area is divided into a grid system.

■ Sampling points are located at regular intervals.

To increase the reliability of the results obtained, many samples should be taken from each grid. A variety of measurements will be taken at each sampling point. Some of these will be of **abiotic** factors – physical and chemical factors that can influence the numbers and types of organisms present. Table 1 shows some of the abiotic factors that may need to be measured when carrying out a habitat survey. In Figure 1 you can see some research scientists measuring abiotic factors in a freshwater wildlife habitat at a reclaimed waste-disposal site.

Biotic factors make up the living part of an organism's environment. The biotic factors measured at the sampling points may include the numbers of a particular species present, the numbers of different species present, the percentage cover of plant species present or the relative abundance of a particular species. Table 2 shows some techniques that can be used to gather data about the organisms present at the sampling points.

Figure 1 *Researchers monitoring the water quality of an area reclaimed from a waste-disposal site. The female researcher is using a dissolved oxygen meter to assess the overall health of the water, while her colleague is checking on emissions of methane gas*

Table 2 *Some sampling techniques*

Technique	Method	Organisms that can be sampled
quadrat sampling	use a metal frame to mark out a small area within the grid	animals and plants within the small area
transect sampling	mark out a line or a narrow belt and study along it	organisms living in an area where the abiotic conditions change with distance along the transect
sweep netting	sweep a sampling net through water or plants such as grasses	freshwater organisms or flying insects
kick sampling	kick stones or gravel on a river bed and catch the disturbed organisms with a net placed downstream	freshwater invertebrates
trapping	pitfall traps	crawling insects
	longworth traps	small mammals
	light traps	moths and other night-flying insects
indirect methods	count droppings, dead animals, nest sites, burrow entrances	larger organisms that are hard to sample by any of the direct methods above

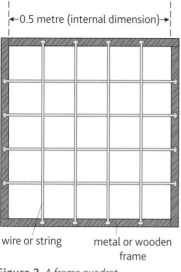

wire or string   metal or wooden frame

Figure 2 *A frame quadrat*

Frame quadrats

A **frame quadrat** is a square frame, sometimes divided by string or wire into subdivisions. You can see a common design in Figure 2.

A quadrat is placed on the ground at a specific sampling point; the species present are identified and their abundance recorded using a number of methods:

■ The **density** of a species is calculated by counting the number of times a species occurs within the quadrats used and then calculating the mean number of individuals per unit area.

$$\text{mean density} = \frac{\text{total number of individuals counted}}{\text{number of quadrats} \times \text{area of each quadrat}}$$

This method can be time-consuming and can be difficult with plants that form clumps, such as grasses.

■ **Percentage cover** is an estimate of the area within the quadrat that a particular species covers. This is useful for clump-forming species, where it may be difficult to count the number of individual plants. Percentage cover may be estimated by eye, i.e. the proportion of the ground area, viewed from directly above, covered by above-ground parts of the species. This can be difficult with plant species that overlap.

■ **Abundance scales** give the relative abundance of a particular species but can be subjective.

Point quadrats

A **point quadrat** consists of vertical legs, across which is fixed a horizontal bar with 10 small holes along it. Long pins are placed in each of the holes. Each time a pin touches a species, it is recorded as a 'hit'. The hit values are converted to percentage cover. Point quadrats are useful where the vegetation is dense as they can sample at many different levels.

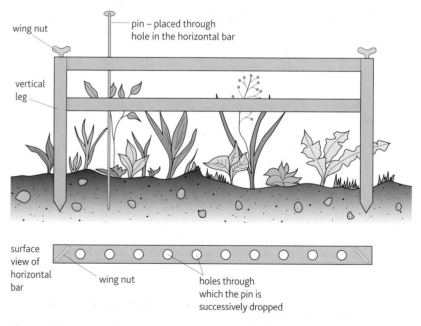

Figure 3 *A point quadrat*

Transects

Transects are used where conditions and organisms change over distance. A tape is stretched across a habitat and quadrat samples are taken along a straight line at regular intervals. This is a systematic sampling method. If three or more parallel transects are carried out across the same habitat, the reliability of the data can be improved.

▦ A **line transect** records the organisms that lie underneath a string or tape that is stretched out across the ground.

▦ A **belt transect** is a strip across a study area, usually 0.5 to 1 metre wide. The species occurring within the belt are recorded. Alternatively, a quadrat can be placed alongside a line transect and the species within it recorded. If the transect is short, sampling can be continuous along its entire length. If large distances are involved then samples are taken at intervals along the belt.

Figure 4 *Students carrying out a transect survey of the plants in a grassy area*

Summary questions

1 Many urban nature reserves are on areas of open land, in the early stages of succession. Using an example to illustrate your answer, suggest why is it important that these sites are managed, rather than leaving them to grow completely wild.

2 Why is the size of a population in an area estimated using random samples?

3 a Describe how you would measure the size of the population of bee orchids growing in grassland that had become established on an area of wasteland.

b How would you measure the changes in the size of the bee orchid population over a 5-year period?

c Why would abiotic factors be measured at the same time as the population estimate was carried out?

11.5 Dealing with data

Learning objectives:

■ How can ecological data be presented?

■ What is a statistical test for?

Specification reference: 3.5.4

If theories about the best way to encourage local wildlife are to be tested, then field data must be carefully collected and analysed.

■ Data collection and analysis

There are four stages involved in the collection and analysis of fieldwork data:

1 The data need to be accurately recorded in a well-organised results table.

2 The data about abiotic and biotic factors that are collected during fieldwork are manipulated to find values such as the mean, median, mode, the standard deviation and standard error.

3 Graphs or charts are drawn to provide information about trends or correlations within the data.

4 The data undergo a statistical test, which allows a comparison to be made between two or more sets of data. The statistical test allows any differences between the sets of data to be tested for their significance. If the statistical test finds that the differences between the data sets are significant, then it can be concluded that they are the result of something other than chance, and the possible causes of these differences need to be investigated.

■ Link

In Topic 2.13 you were introduced to the terms mean, median, mode and standard deviation. These are examples of descriptive statistics. Look back at Topic 2.13 to remind yourself of how and why these calculations are made.

Suitable graphs

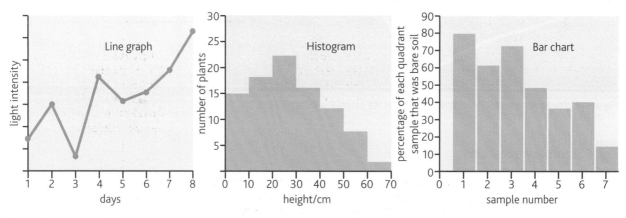

Figure 1 *Examples of the types of graphs that can be used to represent ecological data*

Line graphs, scatter graphs, histograms, bar graphs, pie charts and pyramids are all suitable ways of representing ecological data.

A kite diagram is a special type of bar chart. It provides a visual display of the changes in population numbers over distance or time. The width of each band represents the relative abundance of each organism. Kite diagrams are often used to display transect data. Figure 2 shows an example of a kite diagram.

■ Statistical tests

The results of any investigation could have a genuine scientific explanation but they could be due to chance. Chance is essentially the same as luck. If

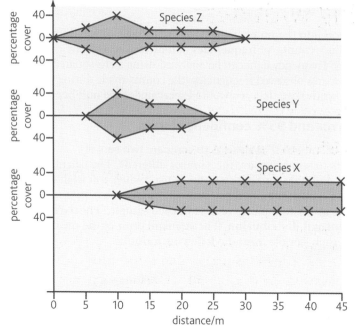

Figure 2 *A kite diagram*

Link

Look back at Topic 2.10 to remind yourself of the meaning of the term **null hypothesis** (H_0).

a coin is tossed in the air, whether it comes down heads or tails is purely due to chance. Scientists carry out statistical tests to assess the probability of the results of an investigation being due to chance. Probability is the likelihood of an event occurring. It differs from chance in that it can be expressed mathematically. In statistical tests, probabilities are usually expressed as a decimal fraction of one. Thus a probability of 0.05 means that an event is likely to occur 5 times in every 100.

It can be difficult to decide which statistical test should be carried out. Figure 3 shows a flowchart that can help you to make this decision.

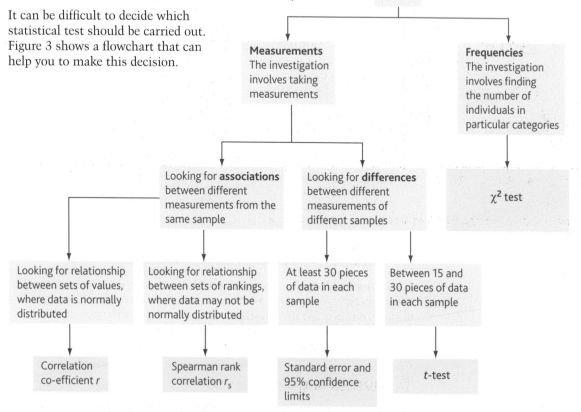

Figure 3 *Deciding which statistical test to use*

■ Link

Remind yourself of how carry out a χ^2 test by looking back at Topic 2.10.

Chi-squared χ^2 test

An investigation into the numbers of plants or animals in two different areas will produce counts, or **frequencies** of individuals, and not measurements. Frequency data can be analysed using a chi-squared test (χ^2). This compares observed frequencies (the counts made during the investigation) with those that you would expect under the null hypothesis.

Standard error and 95% confidence limits

Another type of statistical analysis is to compare two sets of **measurements** to see if they are the same or different. The calculation of standard error and 95% confidence limits is used when you wish to find out if there is a significant difference between the means of two samples when there are at least 30 pieces of data in each sample. The data must also show a **normal distribution**. The standard error of the mean, SE, is calculated for each sample from the following formula:

$$SE = \frac{SD}{\sqrt{n-1}}$$

where SD = the standard deviation and
n = sample size.

Worked example:

Scientists involved in bluebell conservation are concerned that Britain's native bluebell populations may decline as populations of hybrid bluebells increase. Spanish bluebells are often grown in gardens and they will pollinate native bluebells, producing hybrid seeds which grow to become hybrid plants. Researchers used random sampling in a $4\,m \times 4\,m$ grid to measure the leaf diameter of 32 native bluebells. In the same woodland, they also measured the leaf diameter of 32 bluebells randomly sampled from a $4\,m \times 4\,m$ grid containing many bluebells that were hybrids between native and Spanish bluebells. They were testing the hypothesis that native bluebells have narrower leaves than Spanish bluebells, and this can be used as a reliable method for identifying bluebell type.

Using this data, the mean and standard deviation of leaf diameter at each site were calculated.

	Native bluebells	Hybrid bluebells
Mean leaf diameter/mm	6.57	7.73
SD	0.81	1.45

Null hypothesis: There is no significant difference between the mean leaf diameter of the native and Spanish bluebells.

Calculation of the standard error: $\quad SE = \frac{SD}{\sqrt{n-1}}$

SE native bluebells $= \dfrac{0.81}{\sqrt{n-1}} = 0.15$

SE Spanish bluebells $= \dfrac{1.45}{\sqrt{n-1}} = 0.26$

	Native bluebells	Hybrid bluebells
Mean leaf diameter/mm	6.57	7.73
SD	0.81	1.45
SE	0.15	0.26

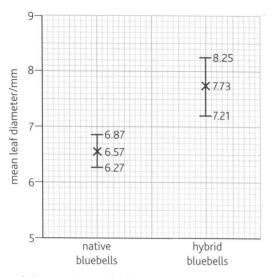

Figure 4 *Interpreting standard error*

Interpreting the values:

The 95% confidence limits are 1.96 standard errors above the mean and 1.96 standard errors below the mean (1.96 can be rounded up to 2.0). If the 95% confidence limits do not overlap, there is a 95% chance that the two means are different. In other words, there is a significant difference between the means at the 5% level of probability. A graph is plotted showing the mean values for each sample as crosses, and bars are drawn to represent 2 standard errors on either side of each of these mean values. In this case, there is no overlap between the bars so the null hypothesis is rejected and it is concluded that there is a significant difference between the mean leaf diameters of the two bluebell samples.

t-test

The Student's *t*-test is used to determine if there is a significant difference between the means of two samples where the data consists of measurements. The samples should be from a normally distributed population, and there should be at least 15 values in each sample

1 The number of degrees of freedom is given by $n = (n_1 + n_2 - 2)$.

2 The value is calculated from:

$$t = \frac{\bar{x}_1 - \bar{x}_2}{\sqrt{\dfrac{s^2_1}{n_1} + \dfrac{s^2_2}{n_2}}}$$

Where: $\bar{x}_1$ = mean of sample set 1

$\bar{x}_2$ = mean of sample set 2

s^2_1 = variance of set 1 = $\dfrac{\Sigma(x_1 - \bar{x}_1)^2}{n-1}$

s^2_2 = variance of set 2 = $\dfrac{\Sigma(x_2 - \bar{x}_2)^2}{n-1}$

n_1 = number of samples in set 1

n_2 = number of samples in set 2.

NB: The variance is simply the standard deviation squared.

Worked example:

The effect of soil contamination on the size of woodlice was investigated. Two similar areas of scrub were selected – one on the site of a disused metal smelting factory with contaminated soil, the other on nearby wasteland that was not contaminated. The head-width of 15 woodlice was measured in each area.

Null hypothesis: There is no significant difference in mean head diameter of the woodlice from the site with contaminated soil and those from the site with uncontaminated soil.

Carrying out the t-test:

Sample	Contaminated soil			Uncontaminated soil		
	Head diameter /mm x_1	$x_1 - \bar{x}_1$	$(x_1 - \bar{x}_1)^2$	Head diameter /mm x_2	$x_2 - \bar{x}_2$	$(x_2 - \bar{x}_2)^2$
1	1.65	0.050	0.0025	1.98	−0.07	0.0049
2	1.78	0.18	0.032	2.31	0.26	0.068
3	1.44	−0.16	0.026	2.45	0.40	0.16
4	1.98	0.38	0.14	1.79	−0.26	0.068
5	2.01	0.41	0.17	1.68	−0.37	0.14
6	1.63	0.03	0.00090	2.03	−0.020	0.00040
7	1.55	−0.05	0.0025	2.64	0.59	0.35
8	1.67	0.07	0.0049	1.78	−0.27	0.073
9	1.59	−0.01	0.00010	1.55	−0.50	0.25
10	1.47	−0.13	0.017	2.57	0.52	0.27
11	1.02	−0.4	0.16	2.31	0.26	0.068
12	1.95	0.35	0.12	2.16	0.11	0.012
13	1.53	−0.07	0.0049	1.84	−0.21	0.044
14	1.38	−0.22	0.048	1.92	−0.13	0.017
15	1.29	−0.31	0.096	1.70	−0.35	0.12
Σ	23.94		0.83	30.71		1.65
Mean	1.60			2.05		
Variance	0.059			0.12		

$$t = \frac{\bar{x}_1 - \bar{x}_2}{\sqrt{\dfrac{s_1^2}{n_1} + \dfrac{s_2^2}{n_2}}}$$

$$t = \frac{1.60 - 2.05}{\sqrt{\dfrac{0.059}{15} + \dfrac{0.12}{15}}} = \frac{-0.45}{\sqrt{0.0039 + 0.008}} = \frac{-0.45}{0.11}$$

$$t = 4.09$$

Interpreting the value of *t*:

For the *t*-test, the number of degrees of freedom is calculated as follows:

Degrees of freedom = $n_1 + n_2 - 2 = 28$

With 28 degrees of freedom, the calculated value of '*t*' (4.09) is larger than the critical value of 2.048 given in Table 1. Therefore if the calculated value of t is greater than the critical value found in the '*t*' tables with 28 degrees of freedom, then the probability of obtaining such a value of '*t*' under the null hypothesis is less than 5%, so the null hypothesis is rejected. We can now conclude that there is a statistically significant difference between the two samples and that this difference is produced by something other than chance. We can conclude that there is a significant difference between the mean head diameter of the woodlice from the site with contaminated soil and the woodlice from the site with uncontaminated soil.

Pearson's correlation coefficient, *r*

Correlation is a technique for investigating the relationship between two quantitative, continuous variables, for example, body length and mass. Pearson's correlation coefficient (*r*) is a measure of the strength of the association between the two variables that are normally distributed.

The first step in studying the relationship between two continuous variables is to draw a scatter plot of the variables to check for linearity. The correlation coefficient should not be calculated if the relationship is not linear.

Pearson's correlation coefficient, *r*, is calculated from the following equation:

$$r = \frac{\Sigma d_x d_y}{\sqrt{(\Sigma d_x^2 \Sigma d_y^2)}}$$

Worked example:

The body mass and nose to tail length of seven animals was measured. The following results were obtained.

Null hypothesis: There is no association between the mass of the animals and their nose-to-tail length.

Animal	Mass/ arbitrary units	Length/ arbitrary units
1	1	2
2	4	5
3	3	8
4	4	12
5	8	14
6	9	19
7	8	22

Carry out the test:

Plot the results on graph paper. This is the essential first step because only then can we see what the relationship might be.

The relationship seems to be linear.

Set out a table as follows and calculate Σx, Σy, Σx^2, Σy^2, Σxy, mean of *x* and mean of *y*.

Animal	Mass/x	Length/y	x^2	y^2	xy
1	1	2	1	4	2
2	4	5	16	25	20
3	3	8	9	64	24
4	4	12	16	144	48
5	8	14	64	196	112
6	9	19	81	361	152
7	8	22	64	484	176
Total	$\Sigma x = 37$	$\Sigma y = 82$	$\Sigma x^2 = 251$	$\Sigma y^2 = 1278$	$\Sigma xy = 553$
Mean	5.286	11.714			

Table 1 *A t distribution table*

Degrees of freedom	Critical value
1	12.71
2	4.30
3	3.18
4	2.78
5	2.57
6	2.45
7	2.36
8	2.31
9	2.26
10	2.23
11	2.20
12	2.18
13	2.16
14	2.15
15	2.13
16	2.12
17	2.11
18	2.10
19	2.09
20	2.09
21	2.08
22	2.08
23	2.07
24	2.06
25	2.06
26	2.05
27	2.05
28	2.05
29	2.04
30	2.04
40	2.02
60	2.00
120	1.98

AQA Examiner's tip

You will not need to learn the formula for the statistical tests but you will need to know how to use them.

Table 2 A table showing the critical values of r for different numbers of paired values

Degrees of freedom	Critical value P=0.05
5	0.754
6	0.707
7	0.666
8	0.632
9	0.602
10	0.576
11	0.553
12	0.532
13	0.514
14	0.497
15	0.482
16	0.468
17	0.456
18	0.444
19	0.433
20	0.423
25	0.381
30	0.349
70	0.232
80	0.217

Calculate $\Sigma d_x^2 = \Sigma x^2 - \dfrac{(\Sigma x)^2}{n} = 55.429$

Calculate $\Sigma d_y^2 = \Sigma y^2 - \dfrac{(\Sigma y)^2}{n} = 317.429$

Calculate $\Sigma d_x d_y = \Sigma xy - \dfrac{\Sigma x \Sigma y}{n} = 119.571$

(this can be positive or negative)

Calculate r (correlation coefficient):

$$r = \frac{\Sigma d_x d_y}{\sqrt{(\Sigma d_x^2 \Sigma d_y^2)}} = 0.901$$

Look up and interpret the value of r:

Table 2 is a table of values of r. Look up r in the table (ignoring + or – sign). The number of degrees of freedom is two less than the number of pairs of data (five in this example because we have seven pairs). If our calculated r-value exceeds the tabulated value at $p = 0.05$ then the correlation is significant. The probability of obtaining a value of 0.901 under the null hypothesis is less than 5%, i.e. we would have obtained a correlation coefficient as high as this in less than 5 in 100 times by chance. So we can conclude that weight and length are positively correlated in our sample of animals.

Spearman rank correlation test

This statistical test is used to find out if there is a significant association between two sets of measurements, consisting of between 7 and 30 pairs.

It is usually chosen when the data is not normally distributed.

Spearman's rank correlation, r_s, is calculated from the following equation:

$$r_s = 1 - \frac{6\Sigma d^2}{n(n^2 - 1)}$$

where n is the number of pairs of items in the sample and d is the difference in rank.

Worked example:

The date on which chaffinches lay their first eggs has been recorded for many years throughout the UK. Scientists decided to look at this data to find out if there was any correlation between the date of first egg-laying and the temperature at that time.

Mean of March and April temperature/°C	5.1	5.8	6.1	6.9	7.4	7.8	8.2	8.7	9.1
Annual median laying date/days after 1st January	131	135	125	130	125	123	120	115	112

Null hypothesis: There is no association between the ranks of the March and April temperature and annual median laying date.

Calculate the correlation coefficient:

Start by ranking the temperature and the egg-laying date. Note that when two or more values are of equal rank, each of the values is given the average of the ranks which would otherwise have been allocated. Calculate the difference between the rank values and square this difference. Find the sum of the squares of the differences (Σd^2).

Sample	Mean of March and April temperature / °C	Rank	Annual median laying date/110 = 20th April, 121 is 1st May	Rank	Difference in rank / d	d^2
1	5.1	9	131	2	7	49
2	5.8	8	135	1	7	49
3	6.1	7	125	4.5	2.5	6.25
4	6.9	6	130	3	3	9
5	7.4	5	125	4.5	0.5	0.25
6	7.8	4	123	6	−2	2
7	8.2	3	120	7	−4	16
8	8.7	2	115	8	−6	36
9	9.1	1	112	9	−8	64
Σ						231.5

Table 3 *A table showing the critical values of r_s for different numbers of paired ranks*

Number of pairs of measurements	Critical value P=0.05
5	1.00
6	0.89
7	0.79
8	0.74
9	0.68
10	0.65
12	0.59
14	0.54
16	0.51
18	0.48

$$r_s = 1 - \frac{6\Sigma d^2}{n(n^2 - 1)}$$

$$r_s = 1 - \frac{6 \times 231.5}{9(81 - 1)} = 1 - \frac{1389}{9(80)} = 1 - \frac{1389}{720} = 1 - 1.93$$

$$r_s = -0.93$$

Look up and interpret the value of r_s:

The value of r_s will always be between 0 and either +1 or −1.

A positive value indicates a positive association between the variables concerned. A negative value shows a negative association. A number near 0 shows little or no association.

Table 3 is a table of values of r_s. Look in the table under the number of pairs of measurements used in the test.

With nine pairs of ranked values, the calculated value of r_s is larger than 0.68 (the value in the table for nine pairs of values).

The calculated value of r_s is greater than the critical value so we reject the null hypothesis and say that there is a significant negative correlation between the ranked values of the mean March and April temperatures and the annual median laying date.

AQA Examiner's tip

It is important to remember that a correlation shows that there is a relationship between two variables, however, it might not be a causal one.

Summary questions

1 Describe how the scientists looking at head diameter in woodlice would have sampled the two sites scientifically to obtain representative samples of woodlice.

2 What should the scientists investigating bluebell conservation do to increase the reliability of their data?

3 Light traps were set up in four different locations. The table shows the total numbers of moths trapped at each site:

Moth trap location	A	B	C	D
Observed numbers	32	26	21	9

4 Use Figure 3 to help you decide which statistical test should be used to test the significance of the difference between the moths caught in the four traps. When you have decided, carry the test out. Is there a significant difference between the moths caught at the four locations, or could these results have been obtained by chance?

11.6 Waste disposal should be environmentally sustainable

Learning objectives:

■ What is the waste hierarchy?

■ Why should we recycle?

Specification reference: 3.5.4

Almost everything we do produces waste or rubbish that we throw away because we do not want or need it. Worldwide studies have shown that waste creation is closely linked to prosperity, with people who live in more economically developed countries producing the greatest amount of waste.

Waste management is a serious issue because when something is thrown away:

■ the raw materials, time and energy used to make it are lost

■ more raw materials, time and energy must be used to make a replacement and the environment must cope with the waste

■ it is not being valued as a resource.

Using up the Earth's natural resources in this way cannot continue indefinitely. It is not **sustainable**.

■ The waste hierarchy

The **waste hierarchy** sets out the main options for the management of waste. It is the primary tool for assessing the 'Best Practical Environmental Option' (BPEO) for waste management. The BPEO is the option which provides the most benefit or least damage to the environment as a whole, at an acceptable cost in both the long and short term.

The options within the waste hierarchy are presented on a sliding scale with the most sustainable option first (reduction) and the least sustainable option last (disposal).

1 Waste should be prevented or reduced at source. (Don't produce waste unless you have to.)

2 Waste materials should be reused.

3 Waste materials should be recycled and used as a raw material. (Use it to make other things.)

4 Waste that cannot be reused should be used as a substitute for non-renewable energy sources. (Use it as fuel.)

5 Only waste which cannot be treated in any of the above ways should go to landfill. (If you can't use it for any of these things, then bury it!)

The waste hierarchy has played a very important part in helping the UK to reduce the amount of waste that is sent to landfill. It has provided a national framework in the UK for the:

■ introduction of a landfill tax

■ setting of national recycling and recovery targets

■ encouragement of energy recovery schemes.

Local Authorities are set recycling targets by the Government. In order that these targets are met or exceeded, Local Authority waste management plans now incorporate recycling and recovery options for all waste.

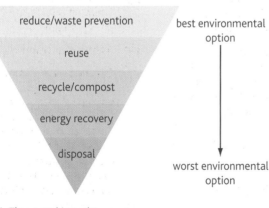

Figure 1 *The waste hierarchy*

How science works

Recycling in the UK

Households in England produced on average 23 kg of waste per week in 2005/06. Within the UK, Northern Ireland produced the most waste at 26 kg and London the least with 20 kg per household each week.

Landfill is the most common method for Local Authorities to dispose of waste in Great Britain. It accounted for 64% of the 34 million tonnes of waste disposed of in 2005/06. Wales and Scotland disposed of the highest proportion in landfill, with 73%. West Midlands disposed of only 44% of waste by this method. However, the use of landfill decreased by an average of 10 percentage points in all regions between 2003/04 and 2005/06.

Figure 2 *Recycling bins in an English park. Recycling is now encouraged by schemes operated by local authorities in the UK*

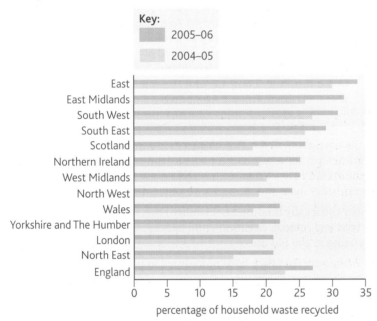

Key:
- 2005–06
- 2004–05

percentage of household waste recycled

Figure 3 *Proportion of household waste recycled, 2004/05 and 2005/06*
Source: UK Statistics Authority
Crown Copyright material is reproduced with the permission of the Controller, Office of Public Sector Information (OPSI)

Across Great Britain 27% of waste was recycled or composted and 9% was incinerated with energy recovery. However, 31% of West Midlands' waste was disposed of by incineration with energy recovery. This method was not used at all in Wales and the South West.

In 2005/06, 36% of recycled household waste in England was composted. This ranged from 25% in London to over 40% in the East Midlands, West Midlands and the East. Other significant proportions of materials recycled across England were paper and card (22%), other materials (20%), glass (11%) and scrap metal/white goods (8%).

1 Why did the use of landfill decrease across Great Britain between 2003/04 and 2005/06?

2 Which region recycles the greatest proportion of household waste?

3 Which region(s) recycle(s) the least?

4 Suggest a reason for the variation shown in the amount of recycling that goes on in different parts of the country.

Summary questions

1 Why is waste creation closely linked to prosperity?

2 What does 'Best Practical Environmental Option' (BPEO) mean?

3 What is the waste hierarchy?

4 England currently recycles approximately 33% of its household waste. Recycling targets of 50% for household waste look set to become law across Europe. How could local authorities encourage more household waste to be recycled to meet this target?

5 Find out how the landfill tax operates and produce an information leaflet about it.

11.7 The polluter pays

Learning objectives:

- What is the Polluter Pays Principle?

- How does the average householder already pay some of the costs of pollution?

Specification reference: 3.5.4

The Polluter Pays Principle states that the polluter pays for the direct and indirect environmental consequences of their actions.

Application and How science works

Clean-up costs

A small UK company faced a £600 000 clean-up bill following a fire at its premises, even though it had insurance cover in place. The chemicals needed to fight the fire had seeped into nearby streams and the Environment Agency charged the company for the emergency clean-up needed as a result. The cost of this emergency clean-up operation was not covered by the company's insurance policy. In the court case that followed, the court found in favour of the insurance company and the small company had to pay the £600 000 Environment Agency bill.

In the past, the Polluter Pays Principle has focused on the costs of cleaning up *after* a pollution incident. However, in recent years, the focus of the Polluter Pays Principle has shifted towards avoiding pollution, minimising waste and managing resources sustainably.

Under the legally binding EU Liability Directive, the polluter pays to prevent and remedy environmental damage. It means that all businesses operating in the EU must:

- make sure that they have procedures in place to prevent environmental damage

- make sure that their company is able to cover the cost of cleaning up and returning a polluted environment to its original state if they do cause damage to it.

Landfill tax

Large landfill sites have numerous environmental impacts. They use large areas of land, represent a source of visual pollution and may cause contamination of groundwater, residual soil contamination after landfill closure and the release of methane gas from decaying organic waste. The organisations responsible for running the sites have to put measures in place to control the amount of contamination, but pollution of the local environment may still occur. Faced with a rapidly growing waste disposal problem, the UK Government introduced a landfill tax whose aims are:

- to promote the 'polluter pays' principle, by increasing the price of landfill to better reflect its environmental costs

- to promote a more sustainable approach to waste management in which less waste is produced and more is recovered or recycled.

Landfill tax is an environmental tax paid on top of normal landfill rates by any company, local authority or other organisation that wishes to dispose of waste in landfill. The UK Government see the landfill tax as the key to reducing the UK's dependence on landfill as a method of dealing with waste. They also hope that it will help the UK to meet the European waste targets set in the EU Landfill Directive.

Figure 1 *Landfill site – bulldozer moving rubbish around at a landfill site, where domestic waste is buried; photographed near Southend, Essex, England*

Application and How science works

Taxing waste disposal

Every household in the UK recieves an annual Council Tax bill from their Local Authority. The money collected by the Local Authority from this tax is used to provide local services, including the collection and disposal of household waste (which includes the payment of landfill tax). The greater the amount of household waste sent to landfill, the higher the landfill tax bill paid by the Local Authority. If households are encouraged to recycle more, the volume and cost of waste sent to landfill decreases. However, with only about 31% of all household rubbish thrown out in Britain being recycled – one of the worst rates in Europe – how can households be encouraged to recycle? One idea is that local authorities should charge households for the amount of waste they send to landfill. So-called 'pay-as you-throw' schemes are already operated throughout Europe. An alternative to this is that households or local communities should receive financial incentives for the amount of waste they recycle. This could include a reduction in the Council Tax bill, or the provision of improved facilities for the local community.

Figure 2 *The Arpeley Landfill site in Warrington, Cheshire, is the largest landfill gas power generation scheme in the UK. Covering an area of 2 km², the landfill site currently receives over one million tonnes of waste each year. At full capacity the plant extracts some 9000 cubic metres of landfill gas per hour containing 50% methane, via a network of wells and extraction pipes*

1. Give **two** advantages and **two** disadvantages of a scheme that charges households for the amount of waste thay send to landfill.

2. Which of the two methods do you think would be most effective at encouraging people living in the UK to recycle more of their waste? Give **two** reasons to support your answer.

▪ Methane

Landfill gas is produced when microorganisms in a landfill site break down biodegradable waste, such as food waste and paper. In the absence of oxygen, anaerobic bacteria produce methane gas. In the past, the methane produced at landfill sites was simply 'flared off' (burnt) to produce carbon dioxide. This not only reduced the risk of explosions but also prevented the release of a gas that is about 21 times more potent than carbon dioxide as a greenhouse gas. Allowing methane to escape into the atmosphere would have significant climate change implications. However, landfill gas is now recognised as a valuable fuel source and increasingly landfill gas is being used to generate electricity or heat. It is collected through a series of wells and pipes constructed within the waste, and is used to produce electricity, combined heat and power, or heat only. Over the last decade, electricity generation from landfill gas has kept pace with the overall increase in renewable electricity generation.

The EU Landfill Directive requires biodegradable municipal waste in England to be reduced to 5.2 million tonnes by 2020. This figure represents 35% of the amount of biodegradable waste sent to landfill in the UK in 2003–2004. The smaller the quantity of organic waste that is sent to landfill, the less landfill gas is produced.

Summary questions

1. What is the Polluter Pays Principle?

2. What is landfill tax?

3. Why will the amount of electricity generated from landfill gas in the UK reduce?

4. Under the EU Landfill Directive, much of the biodegradable waste that would previously have been sent to landfill has to be dealt with by other means. Alternatives include incineration, anaerobic digestion in a controlled environment and composting. For each alternative, find out how it is carried out and give **one** advantage and **one** disadvantage of using it to deal with biodegradable waste.

AQA Examination-style questions

1   (a)   What is the *waste hierarchy*?   *(5 marks)*

       (b)   Give **two** reasons why it is important to reduce the amount of waste that goes into landfill.   *(2 marks)*

2   **Figure 1** shows the UK emissions of methane from 1990 to 2006.

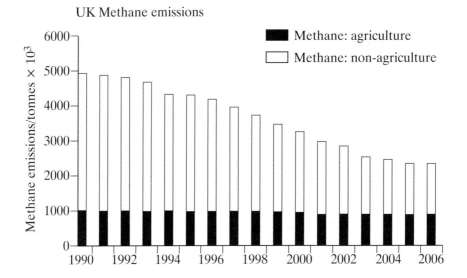

Figure 1

       (a)   Describe **one** way in which agriculture contributes to methane emissions.   *(2 marks)*

       (b)   Total UK methane production in 1990 was 4900 tonnes x 10^3, and in 2006 total UK methane production was 2400 tonnes x 10^3. Calculate the percentage reduction in UK methane emissions between 1990 and 2006. Show your working.   *(2 marks)*

       (c)   The main source of non-agricultural methane production from 1990 to 2006 was from landfill sites. Suggest **two** reasons why non-agricultural methane production has fallen significantly over this time.   *(2 marks)*

3   Scientists studying dormice in a woodland nature reserve in the south of England found that a major road was causing problems. Dormice living in the woods on one side of the road were isolated from those on the other side of the road. Scientists feared this would lead to lack of genetic diversity.

       To solve this problem, scientists placed a rope from trees on one side of the road to those on the other side. This formed a high-level 'bridge' for dormice to move between the two areas of woodland.

       (a)   (i)   Explain how isolating these two populations of dormice could lead to lack of genetic diversity.   *(2 marks)*

             (ii)   Explain why genetic diversity is important in a population.   *(4 marks)*

       (b)   After the rope had been put in place, scientists wanted to evaluate whether the rope improved movement of dormice from the woodland on one side of the road to the other. Suggest an investigation they could carry out.   *(4 marks)*

4   Scientists investigated the distribution of bumblebee nests in six countryside habitats. Volunteers were each asked to observe a small area of land with a known area for 20 minutes on the same day. They looked for bees moving into and out of nest entrances in the area.

The total number of nests found in each habitat surveyed by the volunteers was divided by the total area of the habitat to calculate the nest density. The table shows some of the results.

Habitat	Nest density /nests hectare^{-1}
Short grass	11.4
Long grass	14.6
Woodland	10.8
Fenceline	37.2
Hedgerow	29.5

(a) Suggest why the scientists counted bumblebee nests, rather than numbers of bumblebees. *(1 mark)*

(b) Explain why the scientists calculated the nest density rather than just counting the total number of nests in each habitat. *(2 marks)*

(c) A greater density of nests was found in hedgerows and fencelines than in the other three habitats listed. A scientist suggested that this might be because these are corridor habitats. Explain why this might explain the higher nest density in these habitats. *(2 marks)*

(d) Suggest two ways in which modern farming practices could reduce the population of bumblebees. *(2 marks)*

5   In a sand dune succession the pioneer community (A) colonises bare sand. This community is replaced over time by other communities (B and C) until a climax community of woodland (D) is formed.

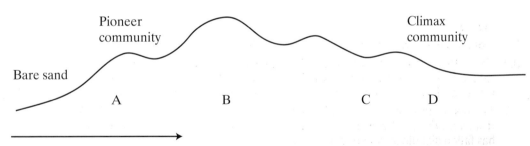

Direction of succession

(a) The communities **A** to **D** are composed of different species. Explain how the change in species composition occurs in a succession. *(3 marks)*

(b) Which community, **A** to **D**, is the most stable? Explain what makes this the most stable community. *(2 marks)*

(c) Explain why it would be more appropriate to use a transect rather than random quadrats when investigating this succession. *(1 mark)*

AQA, 2006

12.1 Your carbon footprint

The Earth's climate has changed throughout its history. However, the term **'climate change'** generally refers to changes that have been observed since the early part of the 1900s. The release of extra **greenhouse gases** such as carbon dioxide (CO_2) from the burning of fossil fuels is thought to be responsible.

The **carbon footprint** is a measure of the impact that human activities have on the amount of greenhouse gases produced, measured in terms of kilograms of carbon dioxide produced per year. Carbon footprints can be calculated for individuals, organisations, events or products.

Calculating your carbon footprint

A carbon footprint has two parts:

■ The **primary** footprint is a measure of how much CO_2 is emitted directly by energy consumption. This part of the carbon footprint measures the impact that your demands for heating, lighting and transportation have on your overall footprint. Individuals have direct control of their primary carbon footprint.

■ The **secondary** footprint is a measure of the indirect CO_2 emissions that we are responsible for when we buy products, including their manufacture, distribution and eventual breakdown. The more products bought, the more emissions will be caused and the greater the secondary footprint.

There are many on-line carbon footprint calculators. They ask questions about how much energy a person consumes in the home and when travelling. Table 1 shows some examples of the questions that are asked when a primary carbon footprint is being calculated. The answers to these questions are converted into equivalent masses of carbon dioxide and are added together to give a carbon footprint. This can be compared with national averages for primary carbon footprints. Calculating a carbon footprint represents a first step towards identifying ways in which carbon emissions can be reduced.

In total, it is estimated that the UK produces 648 million tonnes of carbon emissions every year. This means that each person in the UK can be linked to an average of 11 tonnes of carbon emissions annually.

Reducing the size of a household's carbon footprint

Heating, lighting and running our homes accounts for more than a quarter of the UK's carbon emissions. Table 2 shows some of the ways in which an average household can reduce their emissions and hence the size of their carbon footprint.

Figure 1 *Thermograph showing the distribution of heat over the external surface of a domestic house. The colour-coding ranges from white and orange for the warmest areas through to green and blue for the coolest. The picture shows that the greatest heat loss is through the single-glazed windows. The roof is relatively well insulated, but there is heat loss indicated by the red areas*

Table 1 *Some of the questions that are asked when calculating a carbon footprint*

At home	When travelling
What kind of property do you live in?	What type of vehicle do you own?
When was your home built?	Which type of fuel does your vehicle use?
How many bedrooms do you have?	What size is your vehicle's engine?
How many people live in your home?	What is your annual mileage?
Which of the following do you have? – cavity wall insulation – loft insulation – double glazing – condensing boiler	How often do you have your vehicle serviced?
What is the main heating source for your home?	How often do you check your tyre pressure?
How much is your annual heating bill?	How many hours a week do you spend in cars or on motorbikes for personal use?
How much is your annual electricity bill?	How many personal return flights (not for business) do you take in a year? – flights up to 1 hour long – flights between 1 and 4 hours – flights longer than 4 hours
How many of your light bulbs are energy efficient?	
Which of the following appliances do you have? – dishwasher – washing machine – fridge – tumble dryer	
How many of your appliances are energy efficient?	
Do you regularly turn off lights and not leave your appliances on standby?	

The UK Government hopes that by 2016, all new homes will be built to a zero carbon standard. This means that the buildings will be energy efficient and will be approaching energy self-sufficiency. A zero carbon building will pay back the carbon invested in its construction through exporting zero carbon energy back into the National Grid. This is just one of the ways in which the UK hopes to achieve an 80% reduction in carbon emissions by 2050.

How science works

Zero carbon homes

The first zero carbon apartments in the UK will be built in Manchester. The development of 61 one- and two-bed apartments will help to meet targets for affordable housing and carbon emissions. The properties will be completely carbon neutral, or will use less energy than they generate. This will be achieved through the use of specially designed features, such as energy-efficient walls and windows, solar panels and biomass generators. The development's roof garden will support native wildflowers, increasing local biodiversity, and 122 secure bicycle spaces will encourage residents to cut carbon emissions by cycling.

Hint

Greenhouse gases include carbon dioxide from burning fossil fuels and methane from landfill sites and cattle. Almost anything you buy adds to your carbon footprint – for example, transport involves energy use, you getting to a shop, getting goods to the shop, getting raw materials to manufacturers, getting workers to both places.

Table 2 *Ways in which the size of a household's carbon footprint can be reduced*

Primary footprint		Secondary footprint
Methods that will not cost any money to set up	**Methods that will pay for themselves in three to four years through savings on energy bills**	
Sign up to a green energy supplier, who will supply electricity from renewable sources.	Fit energy-saving light bulbs.	Try to avoid highly processed foods.
Turn off appliances such as lights, television, DVD player and computer when not in use.	Install thermostatic valves on all radiators.	Reduce meat consumption.
Turn down the central heating by 1 to 2 °C and the water heating setting by 2 °C.	Insulate the hot water tank, loft and walls.	Do not buy bottled water if the tap water is safe to drink.
Check the central heating timer setting.	Replace an old fridge/freezer (if it is over 15 years old) with a new one with an energy efficiency rating of 'A'.	Do not buy fresh fruit and vegetables which are out of season. They may have been flown in.
Only use dishwasher and washing machine when full.	Replace an old boiler with a new energy-efficient condensing boiler.	Try to buy products made closer to home.
Fill the kettle with only as much water as is needed.		Buy organic produce.
Unplug a mobile phone as soon as it has finished charging.		Do not buy over-packaged products.
Defrost the fridge/freezer regularly.		Recycle as much as possible.
Do the weekly shopping in a single trip.		Think about the carbon emissions caused by leisure activities and hobbies.
Hang out the washing to dry rather than use a tumble dryer.		
Go for a run rather than driving to the gym.		
Car-share, use public transport or walk or cycle.		
Avoid using domestic flights – use trains or buses.		

Summary questions

1. What is a carbon footprint?

2. Give an example of a source of carbon emissions that may be difficult to reduce.

3. Why do you think the UK Government wishes all new homes to be built to a zero carbon standard by 2016?

4. What does carbon-offsetting mean?

5. Produce a list of all the contributions that you make to your carbon footprint when you buy a DVD. Remember to sort your contributions into primary and secondary footprints.

Carbon offsetting

Everybody has a responsibility to reduce the size of their carbon footprint. However, it may not be practical or possible to reduce some emissions, and these emissions can be compensated for by paying someone to make an equivalent greenhouse gas saving. This is called '**carbon offsetting**'.

The projects that can generate carbon offsets include energy efficiency projects (e.g. installing energy-saving technologies in housing developments), renewable energy schemes (e.g. wind farms) or tree-planting schemes. Renewable energy and energy efficiency projects can have immediate benefits to the environment. However, it can take many years for the environmental benefits of tree-planting to be felt, and it is difficult to measure how much carbon is actually saved.

12.2   Our climate is changing

Learning objectives:

- Is climate change affecting the distribution of plants and animals in the UK?

Specification reference: 3.5.5

Climate refers to the average weather experienced in a region over a long period, typically at least 30 years. This includes temperature, wind and rainfall patterns. The climate of the Earth has changed many times in the past in response to a variety of natural causes, such as interactions between the ocean and the atmosphere, changes in the Earth's orbit, fluctuations in energy received from the Sun and volcanic eruptions. Recent changes in global climate are likely to be due to a combination of both natural and human causes. The main human cause of global climate change is likely to be emissions of **greenhouse gases** such as carbon dioxide and methane.

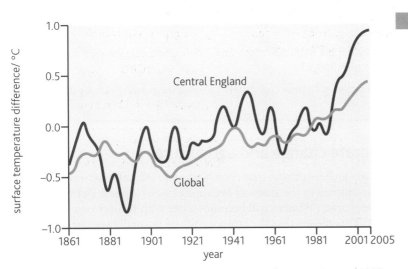

Figure 1 *Average UK and global surface air temperatures between 1861 and 2005, compared to the average between 1961 and 1990*
Source: UK Statistics Authority
Crown Copyright material is reproduced with the permission of the Controller, Office of Public Sector Information (OPSI)

The climate of the United Kingdom is getting warmer

Climate change in the UK can be measured in records that have been kept for over 350 years. The Central England Temperature (CET) series is one of the longest continuous temperature records in the world and this shows that since 1659, temperatures in the UK have increased by 0.7 °C, with an increase of around 0.5 °C occurring in the 20th century. Figure 1 shows the average UK and global surface air temperatures between 1861 and 2005, compared to the average between 1961 and 1990.

Climate change and wildlife

All species are adapted to their environment. Changes to temperature, rainfall and wind patterns (abiotic factors) will therefore affect the species that can survive in an area. Existing species may need to migrate into new habitats to find favourable conditions and new species that migrate into an area may fill **niches** and out-compete existing populations.

Application and How science works

Climate change may produce succession on a very large scale. The concepts involving communities and how they can change over time apply here.

Animals, birds and plants will need to move their range north and westwards across the British Isles in search of suitable homes and food as weather patterns change. Some species are already shifting, such as the comma butterfly which has been seen further north, while European species are adapting to life in the UK. Birds like the little egret are now seen frequently in southern Britain, since their arrival from the Mediterranean around 20 years ago. However, development and loss of habitat are blocking movement for other species. One of

Link

Look back at the application in Topic 9.2. This deals with changes in the migration patterns in the chiffchaff, a small bird.

the species threatened is the hazel dormouse, which can only survive in hazel woodland. Many of the UK hazel woods only exist as small, scattered fragments, and as dormice will not cross open ground they will need linked woodland if they are to move between habitats.

Warmer temperatures will push northern and mountain species to the edge of their range. The wildlife living in the arctic-like conditions on the Cairngorm mountains in Scotland is likely to be affected considerably. Animals such as the mountain hare may become isolated at the mountain tops, while a lack of snow could threaten birds such as the ptarmigan whose white winter plumage would no longer make them difficult for predators to see. Ptarmigans and other birds such as the dotterel, won't be able to survive in warmer temperatures and so they will have to migrate further north, away from the UK.

If climate change occurs very quickly, some species may not be able to adapt and move quickly enough and so may not survive.

1. Why will the hazel dormouse find it difficult to migrate in response to climate change?

2. Explain why climate change will threaten the survival of the ptarmigan in the UK.

Figure 2 *Male ptarmigan (Lagopus mutus) in winter plumage. Photographed in the Cairngorms National Park, Aberdeenshire, Scotland, UK, in January*

Climate change and agriculture

Climate researchers predict that the UK climate will become warmer, with high temperatures in the summer becoming more frequent and very cold winters more rare. Winters will become wetter with heavier rain more common. This means that the crops we would normally see growing in the south of the country will be able to be grown further north. Increased dryness in the summer could affect the quality and yields of the crop due to an increase in drought and heat waves. This will mean that the type of crops grown in the UK will have to change. Increasing demands are likely to be made on water, and in parts of the UK, irrigation systems may need to be implemented so that winter rain can be stored for agricultural use in the summer.

Populations of agricultural pests, such as aphids, are reproducing rapidly earlier in the year. Aphids can cause a great deal of damage to crops and if their populations increase rapidly early in the growing season, they can cause serious damage to the young, developing crop.

Climate change and disease

If average global temperatures increase, it may be possible for pathogens and their carriers (**vectors**) to move to areas that were previously too cold for them to survive. The pathogens that cause diseases such as lyme disease, dengue fever, cholera and yellow fever could spread into areas that were previously unaffected, greatly increasing the incidence of these diseases. It is even thought possible that a mild strain of malaria could become established in localised parts of the UK for up to 4 months of the year. Bacteria are more likely to survive through the milder winters and this could mean that diseases caused by bacterial pathogens could become more widespread.

Summary questions

1. Explain why the effects of climate change can be thought of as an example of succession on a very large scale.

2. The changes observed in the egg-laying dates of British birds, recorded since the mid 1960s are believed to be in response to increasing temperatures. In 1966, the average date for chaffinches laying their first egg in the UK was May 11, but by 2006 this date had moved forward to May 2. For the robin, the average dates have moved from April 28 in 1966 to April 22, in 2006. Use the information in this topic and Topic 9.2 to give examples of the effects of climate change on:

   a the natural range of species

   b the availability of food for some species at key times.

12.3 Photosynthesis

- What happens during photosynthesis?

- Where does photosynthesis happen?

Specification reference: 3.5.5

You will remember from GCSE that plants take in carbon dioxide during the process of photosynthesis and produce organic compounds such as sugars. They use the energy from sunlight to do this. Before looking at this in more detail, you need to understand a little about the structure of a chloroplast. This is the organelle inside plant cells where photosynthesis occurs. The structure of a chloroplast is shown in Figure 1.

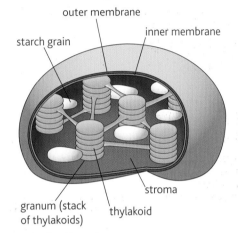

Figure 1 *The structure of a chloroplast*

The chloroplast has an inner and an outer membrane. Inside are **thylakoid membranes** arranged in stacks which contain pigments, including chlorophyll. These stacks of membranes are called **grana**. They are surrounded by a watery fluid that contains enzymes. This is called the **stroma**.

Photosynthesis is a two-stage process. The first stage is the **light-dependent stage** and this takes place in the grana of the chloroplast. The second stage of photosynthesis is the **light-independent stage** and this takes place in the stroma. Figure 2 shows a summary of what happens during photosynthesis.

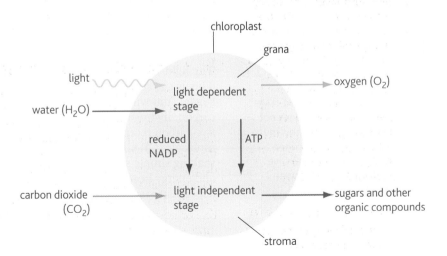

Figure 2 *A summary of photosynthesis*

■ The light-dependent stage

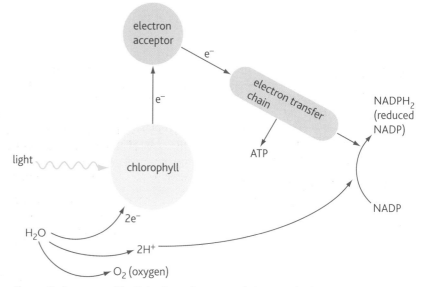

Figure 3 *Summary of the light-dependent stage of photosynthesis*

The light-dependent stage of photosynthesis is summarised in Figure 3. You will see that light is absorbed by a chlorophyll molecule. The energy in the light is used to 'excite' an electron in the chlorophyll, which gains energy and passes to a protein in the thylakoid membrane, called an **electron acceptor**. From here, the electron is passed through a series of membrane proteins in the thylakoid, collectively called an **electron transfer chain**. As the electron passes along the electron transfer chain it loses energy. This energy is used to make ATP. You will remember that ATP can be used to provide energy for many biological processes in the cell.

The electrons that chlorophyll has lost need to be replaced. The chlorophyll regains electrons from water. Light energy splits water into electrons, hydrogen ions and oxygen. Oxygen is given off as a waste product and the electrons replace those that chlorophyll has lost. The hydrogen ions join up with the electrons that have passed through the electron transfer chain to form hydrogen atoms again. These hydrogen atoms join to a coenzyme called NADP to form **reduced NADP**. The reduced NADP and ATP produced in the light-dependent stage are used in the light-independent stage.

■ The light-independent stage

The light-independent stage takes place in the stroma of the chloroplast. This is the stage in which carbohydrates are produced, in the form of sugars. You will remember from AS level that carbohydrates contain carbon, hydrogen and oxygen. In this stage, carbon dioxide binds to a five-carbon receptor molecule to form two molecules of a three-carbon molecule, GP. GP is then used to make simple sugars. The process of making simple sugars from GP requires hydrogen from reduced NADP, and energy from ATP. Reduced NADP and ATP were made in the light-dependent stage.

■ Hint

A coenzyme is an organic molecule (but *not* a protein) that usually contains a vitamin or mineral. It works in conjunction with enzymes. Coenzymes 'pick up' hydrogen atoms and become reduced. When they release their hydrogen, they are said to be oxidised again.

■ Summary questions

1. Give one similarity between a chloroplast and a mitochondrion.

2. ATP is also made in respiration. Explain why plants could not use ATP from respiration in the light-independent stage of photosynthesis.

3. The light-independent stage of photosynthesis does not require light. Does this mean that it could occur when no light is available? Give reasons for your answer.

12.4 Food chains

Learning objectives:

- How does energy enter an ecosystem?

- How is energy transferred through an ecosystem?

- How can planting trees help off-set carbon emissions?

Specification reference: 3.5.5

Photosynthesis is the major route by which energy enters an ecosystem. You will remember that green plants are called **primary producers** because they are able to produce organic materials from the inorganic molecules carbon dioxide and water, in photosynthesis. Organisms that eat plants and obtain their energy from respiring molecules obtained from plants, are called **primary consumers**. Organisms that feed on primary consumers are called **secondary consumers**. These different levels in a food chain are called **trophic levels**. This is shown in Figure 1.

When organisms die, their dead bodies are broken down by **decomposers** such as bacteria and fungi. Decomposers also break down animal faeces. Dead organic matter is also broken down by **detritivores**. These are animals such as earthworms, millipedes and woodlice.

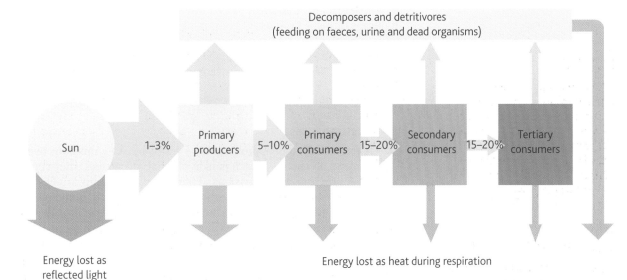

Figure 1 *Energy flow through a food chain*

Efficiency of energy transfer

Only about 1% of the energy from sunlight that falls on a plant is actually used in photosynthesis. There are several reasons for this, including the following:

- Some of the light is the wrong wavelength. Only red and blue light can be used in photosynthesis.

- Some of the light does not land on a chloroplast and gets transmitted through the leaf.

- Some of the energy is lost as heat.

Only about 5–10% of the energy in plant tissues is passed on to primary consumers for the following reasons:

- A large amount of the plant tissue is composed of lignin and cellulose which cannot be digested by enzymes in animals' guts.

- Parts of the plant are not available to be eaten, for example the roots of trees, or grass that is very close to the ground.

About 10–20% of the energy in primary consumers is passed on to secondary consumers. This is higher than the transfer at the previous stage because:

▪ animal tissue is made mainly of proteins and fats which have a higher energy value than plant tissue
▪ animal tissue is more digestible than plant tissue.

However, some parts of animals, e.g. bones and hooves, are not readily available so the transfer of energy is still not very efficient.

Furthermore, at every trophic level, organic molecules are used in respiration to produce ATP. This is used for processes such as muscle contraction, synthesis of large molecules and active transport. Ultimately, the energy in the organic molecules used in respiration is lost as heat.

▪ Off-setting carbon dioxide emissions

One way to reduce carbon dioxide emissions is to plant more trees. Trees take in carbon dioxide in photosynthesis and use this to produce organic molecules that form plant tissue. We call the plant tissue made in photosynthesis **biomass**. This carbon dioxide is released again if the biomass in the tree is burned as fuel, or decomposed by decomposers. However, because trees take many years to grow, the carbon dioxide is 'locked up' or **sequestered** in the biomass of trees for many years.

How science works

Planting trees

A tree will take in approximately 1 tonne of CO_2 emissions over its lifetime. In theory, if each person planted seven trees, they could off-set their carbon emissions. There have been many studies into how trees can help off-set carbon dioxide emissions. Some research suggests that as trees are surrounded by more carbon dioxide, they grow more tissue but they will not necessarily absorb more carbon. Trees that received plenty of water and nutrients were most efficient at storing carbon dioxide. So, for planting more trees to off-set carbon emissions effectively, sufficient water and nutrients must be available.

Summary questions

1 Some farmers keep animals indoors where their movement is restricted. Suggest why this is a more efficient method of rearing them.

2 It has been estimated that people eating a meat-based diet need seven times more land to supply their food than if they eat a vegetarian diet. Explain why.

3 Food chains rarely have more than five trophic levels. Explain why.

12.5 Biofuels instead of fossil fuels?

Learning objectives:

- What are the environmental advantages of using biofuels?
- What are the environmental disadvantages of using biofuels?

Specification reference: 3.5.5

Biofuels are **renewable energy sources**. They include:

- **biomass** from fast-growing plants such as miscanthus grass and willow, used as fuel for burning
- vegetable oils, such as those obtained from oilseed rape and the fruit of the oil palm. These can be used as diesel substitutes (**biodiesel**)
- ethanol from the fermentation of plant material such as sugar-cane waste, used as a petrol substitute or as a fuel additive (**bioethanol**). When used as an additive in petrol, ethanol improves the efficiency of the combustion of the fuel.

Figure 1 *Willow growing in Nottinghamshire*

Figure 1 is a photograph of willow. It is grown as a biofuel. The wood is periodically harvested by cutting the trees near to ground level. This stimulates regrowth.

The use of biodiesel and bioethanol as transport fuels could reduce emissions of greenhouse gases when compared to the emissions from conventional transport fossil fuels. Burning the biofuels releases carbon dioxide, but the growing biofuel crop plants absorb a comparable amount of the gas from the atmosphere during their lifetime. However, energy is used in growing and processing the crops and this can make greenhouse gas emissions from biofuels as great as those from petroleum-based fuels. To produce significant reductions in the use of fossil fuels, the plants used have to be grown on a very large scale. This will have impacts on the environment, reducing the available habitats for wild animals and plants and affecting the availability of food for human consumption.

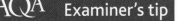

AQA Examiner's tip

You need to develop the skill of looking for the pros and cons of apparently attractive solutions to problems.

Figure 2 *African oil palm trees* (Elaeis guineensis) *growing in a plantation in Sabah, Malaysian Borneo. The fruit of this plant contains oils that are used in many food products and non-food household products. It is also potentially useful for the production of biodiesel*

How science works

Just how sustainable is the production of biofuels?

Biofuels make up just over 2% of the UK's vehicle fuel. A recent report commissioned by the UK Government found that manufacturers were unable to say where almost half the biofuels supplied at British petrol pumps had been grown. The report warned that the figures suggesting that biofuels reduce greenhouse gas emissions (when compared to fossil-fuel emissions) may be over-optimistic because they do not take into consideration indirect climatic effects, such as clearing rainforest and draining tropical marshland in South East Asia for the cultivation of oil palm as a biofuel. It may take centuries for these areas to grow enough palm oil to offset the greenhouse gases released when they are drained. Biofuel manufacturers will need to know where their supplies come from if these indirect costs are to be included in any calculations of overall greenhouse gas savings. The report also suggests that the benefits of using biofuels, measured as the reduction in greenhouse gas emissions compared to fossil fuels, is least for soya oil from Brazil (10%), while, by the same comparison, bioethanol produced by fermentation of Brazilian sugar cane reduced greenhouse emissions by about 70%. Cooking oil and tallow (animal fat) from within the UK, the US and elsewhere reduced greenhouse gas emissions by more than 80%, palm oil from south-east Asia by about 40% and oil-seed rape grown in Europe and North America by about 30%. The UK does not use large amounts of ethanol made from corn, which has been shown to use more fossil fuel in its manufacture than is saved when it is used as a biofuel.

1 Why must manufacturers of biofuels know where their supplies come from?

2 Suggest why greenhouse gases are released when tropical marshland is drained for the cultivation of oil palm.

3 Give four examples of indirect climate effects that may reduce the overall benefits of using biofuels to reduce greenhouse gas emissions.

Summary questions

1 What is a renewable energy source?

2 Some scientists predict food riots in various parts of the world as a result of switches to biofuels. Explain why.

3 Give **two** environmental advantages and **two** environmental disadvantages of using biofuels.

12.6 ATP and aerobic respiration

Learning objectives:

- What is ATP?

- How is ATP made in aerobic respiration?

Specification reference: 3.5.5

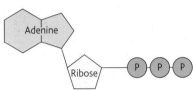

Figure 1 *The structure of ATP*

ATP

ATP stands for adenosine triphosphate. It is the immediate source of energy for biological processes. In other words, it used for almost all energy-requiring processes in a cell.

Figure 1 shows the structure of ATP. You will see that there are three phosphate groups attached to the adenosine. The second two phosphate groups release quite a lot of energy when they are broken. This energy can be used for energy-requiring processes in the cell. Usually it is only the third phosphate group that is removed.

$$ATP \longrightarrow ADP + P_i + energy$$

adenosine   adenosine   inorganic
triphosphate   diphosphate   phosphate

You learned in Topic 12.3 that ATP is made during the light-dependent stage of photosynthesis. However, this ATP is used to produce carbohydrate in the light-independent stage of photosynthesis. The ATP that is needed for energy-requiring processes in the cell is produced in cellular respiration.

An overview of respiration

There are several stages in respiration. These are shown in Figure 2.

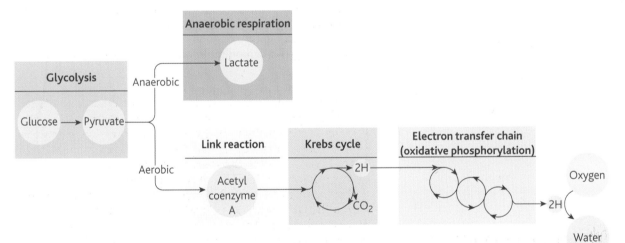

Figure 2 *Overview of respiration*

The first stage of respiration, glycolysis, takes place in the cytoplasm of the cell. This stage does not require oxygen, so we can describe this stage as **anaerobic**. However, the link reaction, Krebs cycle and oxidative phosphorylation all take place in the mitochondrion. These stages only take place when oxygen is available, so we say that they are **aerobic**.

Glycolysis

Glycolysis means 'sugar-splitting'. During this stage, glucose (which is a 6-carbon monosaccharide) is broken down in stages to two molecules of a 3-carbon compound, pyruvate. During glycolysis, energy is released

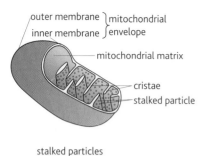

outer membrane ⎤ mitochondrial
inner membrane ⎦ envelope
mitochondrial matrix
cristae
stalked particle

stalked particles

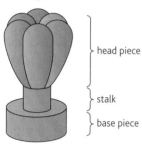

head piece

stalk

base piece

Figure 3 *The structure of the mitochondrion*

by breaking the chemical bonds in glucose, and this is used to produce four molecules of ATP per glucose (therefore a net gain of two molecules of ATP per glucose). As glucose is broken down in stages, hydrogen atoms are released. These are picked up by the coenzyme, NAD, to form reduced coenzyme (NADH).

Glycolysis can occur in both aerobic and anaerobic conditions. However, if oxygen is not available, pyruvate is converted to lactate. This means that the glucose is not fully broken down, and only two ATP per glucose are formed.

The link reaction

If oxygen is available, the pyruvates formed in glycolysis enter the mitochondrion. The structure of a mitochondrion is shown in Figure 3.

The link reaction takes place in the matrix of the mitochondrion. During this stage, each 3-carbon pyruvate molecule is broken down into a 2-carbon compound, acetylcoenzyme A, and carbon dioxide is given off. More reduced coenzyme is also formed (NADH).

Krebs cycle

The Krebs cycle also takes place in the matrix of the mitochondrion. Acetyl coenzyme A joins with a 4-carbon compound to make a 6-carbon compound. This is broken down into a 5-carbon compound, with the release of carbon dioxide. The 5-carbon compound breaks down to re-form the 4-carbon compound, releasing carbon dioxide. This is shown in Figure 4. Reduced coenzymes are also produced. Some of these are NADH, but some are another reduced coenzyme, FADH. As the reactions of the Krebs cycle occur, enough energy is released to produce one ATP molecule for each acetyl coenzyme A that enters.

In Krebs cycle, the remains of the glucose have been completely broken down. You will notice that two carbon molecules enter the cycle as acetyl coenzyme A. These two carbon molecules are both given off in the form of carbon dioxide.

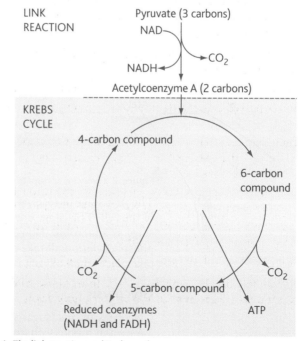

Figure 4 *The link reaction and Krebs cycle*

Oxidative phosphorylation

In this stage, the reduced coenzymes that have been produced in the earlier stages of respiration are used to make ATP. This happens in the inner membranes of the mitochondria. You can see what happens in Figure 5.

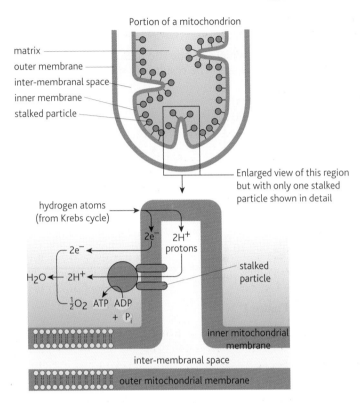

Figure 5 *Oxidative phosphorylation*

In the inner membrane of the mitochondrion are special proteins that form an **electron transfer chain**. There are also large complexes which are visible on electron micrographs of the mitochondrion as stalked particles. These stalked particles are complex structures also known as **ATPases**.

- Reduced coenzymes bring their hydrogen atoms to the first protein in the electron transfer chain.
- The hydrogen atoms split into a hydrogen ion and an electron.
- The electron is passed through the electron transfer chain in the inner membrane.
- The hydrogen ions pass across the membrane into the inter-membranal space – the space between the inner and outer mitochondrial membranes.
- The hydrogen ions diffuse back across the membrane through the ATPases. This releases enough energy to make ATP from ADP and P_i.
- The hydrogen ions recombine with the electrons from the electron transfer chain, re-forming hydrogen atoms.
- The hydrogen atoms combine with oxygen, forming water.

In this way, a great deal of ATP is made. In fact, most of the ATP produced in respiration comes from oxidative phosphorylation.

Summary questions

1 Explain why the link reaction and Krebs cycle cannot take place unless oxygen is present.

2 List **three** uses of ATP in a cell.

1   Grouse are birds that live in moorland. Ticks are large parasitic mites that attach to the bird's skin and feed on its blood.

Scientists investigated the percentage of young grouse chicks at one site in Scotland with at least one tick. They recorded the data over 20 years. The results are shown in **Figure 1**.

(a) Describe the trend shown by these data. *(2 marks)*

(b) Explain why a tick is described as a parasite. *(2 marks)*

(c) One explanation for these results could be that the weather has been progressively milder over the last twenty years, increasing the length of the time when the ticks are able to feed. Do these data support this hypothesis? Explain your answer. *(3 marks)*

(d) Suggest two factors, other than factors to do with the weather, that might also affect tick populations. *(1 mark)*

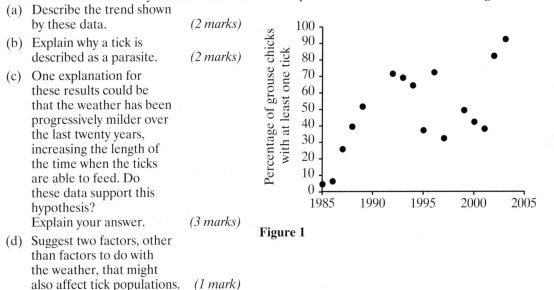

Figure 1

2   Scientists have developed a genetically modified strain of corn that contains cellulase enzymes that break down its own cellulose cell walls. They hope that the corn might be used to make ethanol to use as a biofuel.

The gene that codes for the cellulase enzyme was removed from a bacterium found in a cow's gut, and transferred to the corn. This enzyme breaks down the cellulose into sugars. These sugars can be readily fermented into ethanol.

(a) Name the enzyme that would be used:

   (i)   to remove the gene for cellulase from the bacterium;

   (ii) to join the gene for cellulase to the DNA of corn. *(2 marks)*

(b) Name:

   (i)   the type of reaction involved in breaking down cellulose into sugars;

   (ii) the sugar that is released when cellulose is completely digested. *(2 marks)*

(c) Corn naturally contains the enzyme amylase, which digests starch into sugars. Explain why this enzyme cannot digest cellulose. *(2 marks)*

(d) (i)   Use your knowledge of photosynthesis to explain why ethanol made from corn could help to reduce carbon dioxide emissions. *(2 marks)*

   (ii) Explain one reason why some people are opposed to using corn to produce biofuels. *(2 marks)*

3   (a) What is a carbon footprint? *(2 marks)*

Figure 2 shows the carbon footprint for a typical person in the UK.

(b) (i)   What percentage of this person's carbon footprint is made up of secondary contributions? *(1 mark)*

   (ii) Describe and explain one way in which this person could reduce the contribution made by food and drink to their carbon footprint. *(2 marks)*

(c) What is meant by carbon off-setting? *(2 marks)*

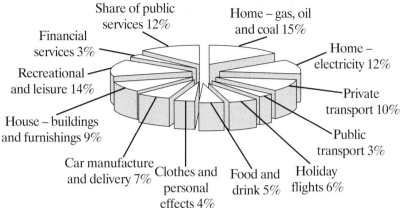

Figure 2

4   **Figure 3** shows the energy transfer through the trophic levels in an ecosystem.
The numbers in the boxes show the amounts of energy in the biomass at each trophic level.

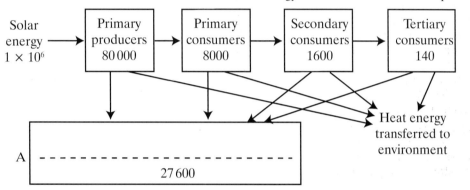

Figure 3

(a) Complete box A in the diagram with the name of a group of organisms.   *(1 mark)*

(b) Suggest suitable units for energy transferred between trophic levels.   *(2 marks)*

(c) Give three explanations for the difference between the amount of solar energy
reaching the primary producers and the energy in the biomass of the primary
producers.   *(3 marks)*

AQA, 2006

5   Lemmings are small mammals which live in the Arctic. Their main predator is the
stoat, a small carnivorous mammal, which feeds almost entirely on lemmings. The
Figure 4 shows the changes in the numbers of lemmings and stoats from 1988 to 2000.

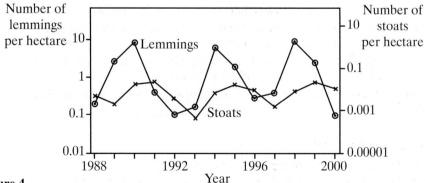

Figure 4

Describe and explain the changes which occur in the lemming and stoat populations.   *(6 marks)*

AQA, 2006

13.1 The human ecosystem

Learning objectives:

- What does the human ecosystem consist of?

Specification reference: 3.5.6

Ecosystems may be very large, such as a lake or forest, or small, such as a pond. The human body can be thought of as a small ecosystem, supporting characteristic **populations** of bacteria and some fungi. The microorganisms that share the body's environment live in **communities** and occupy **niches** in which they feed. They show adaptations to the **abiotic** and **biotic** factors in their environment and **compete** with each other for nutrients.

Living with your microorganisms

Inside the uterus, the fetus develops in a sterile environment. Microorganisms begin to colonise the surfaces of the human body during or shortly after birth, and populations of these microorganisms remain throughout life, forming the body's **normal flora**. The normal flora in humans usually develops in an orderly sequence, or **succession**, after birth, leading to the stable populations of bacteria that make up the normal adult flora.

Link

Topic 1.4 of *AS Human Biology* looks at how a human gut flora is established in a baby. Look back at this topic to remind yourself.

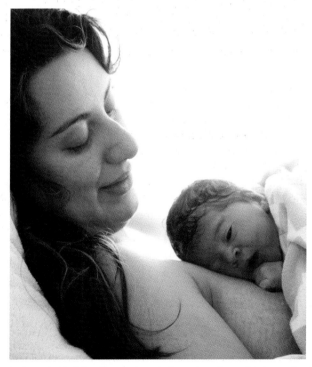

Figure 1 *A mother holding her newborn baby on her chest after breastfeeding*

Link

You learned about the structure of prokaryotic cells in Topic 4.1 of *AS Human Biology*.

Normal flora can be found in many parts of the human body including the skin, respiratory tract (particularly the nose), urinary tract and the gut. The normal flora of humans consists of a few **eukaryotic** fungi and protoctists, but the most numerous components are bacteria. **Prokaryotic** (bacterial) cells associated with the human body outnumber the body's own eukaryotic cells ten to one! Bacteria live on every external surface (including the epithelial cells that line the gut) of the human body. It is estimated that a human adult may have about 10^{12} bacteria on the skin, 10^{10} in the mouth and 10^{14} in the gut.

In a healthy animal, the internal tissues, e.g. blood, brain, muscle, etc., are normally free of microorganisms, with only the surface tissues, such as the skin and mucous membranes of the gut, being readily colonised by microbes.

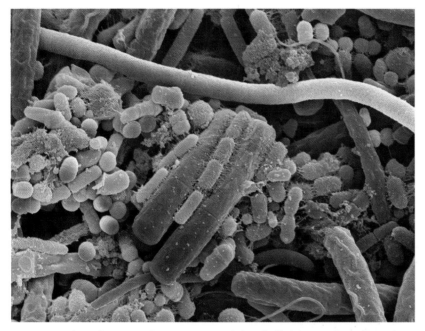

Figure 2 *Coloured scanning electron micrograph (SEM) of bacteria on the surface of a human tongue*

A person's normal flora consists of communities of microorganisms that function as microbial ecosystems. The microorganisms in these communities obtain the nutrients they need from their environment by carrying out **extracellular digestion** of biological molecules. This means that they secrete enzymes to bring about the hydrolysis of larger molecules found on the surface of the tissues that make up the microbial habitat. The soluble products of this digestion are then absorbed by the bacteria and can be used for their own metabolism. In this way, communities of microorganisms, such as those living on the skin, can recycle chemical elements from human cells, or, in the case of the microbes living in the large intestine, make use of material that has not been digested higher up the gut. The surfaces of the human body provide habitats where the microorganisms can occupy niches. Here, the populations of microorganisms will compete with each other for nutrients, helping to maintain a dynamic population balance and preventing the growth of populations of potential **pathogens**.

The use of antibiotics, medical procedures such as surgery, tissue damage, changes in diet, and the introduction of new species of microorganisms, can affect a person's normal flora. Such disturbances to the microbial ecosystems of the body can affect a person's health.

Summary questions

1 Produce a table of differences between prokaryotic and eukaryotic cells.

2 Why is the environment inside the uterus free from bacteria?

3 Explain why the first microorganisms to colonise the gut of a baby delivered vaginally will differ from those that first colonise the gut of a baby born by Caesarean section.

4 Suggest why so many different species of bacteria can live in the human ecosystem.

13.2 Microorganisms and the human skin

Learning objectives:

- What are spots?

- Do spot treatments really work?

Specification reference: 3.5.6

Link

MRSA and antibiotic resistance were introduced at AS Level and are covered in Topic 4.5 of *AS Human Biology*.

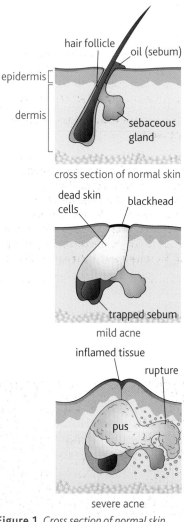

Figure 1 *Cross section of normal skin, skin with mild acne and skin with severe acne*

The adult human is covered with approximately 2 square metres of skin. The skin surface is relatively dry and slightly acidic, and the main food sources for any microorganisms living on the skin are dead cells and sebum, an oily secretion from the sebaceous glands. In places where the moisture content of the skin is high, such as the armpit, groin and in the areas between the toes, relatively high population densities of bacterial cells can be found. Elsewhere the population densities of bacteria are low, generally in 100s or 1000s per square cm.

Human skin supports a community of many microorganisms, including:

- *Staphylococci*
- *Micrococci*
- *Corynebacterium*
- Fungi, such as yeast.

These microorganisms are called **commensals** because they live harmlessly on the skin, helping to prevent infections by out-competing potential **pathogenic** microorganisms. A recent study into human skin bacteria suggests that the range of bacterial species present on the skin is unique to each individual. The scientists who carried out this work also found that there were differences between men and women in terms of the bacteria that are found on the skin and that the variety of skin bacteria may change over several months. The study also found that for each individual tested, some bacterial populations were always present, while the populations of other species seemed to change.

Spots

A number of skin conditions are caused when populations of commensal skin bacteria begin to grow rapidly in response to a change in their environment. An example of such a condition is **Acne vulgaris**, which affects about 80% of teenagers between the ages of 13 and 17. Acne is caused by over-activity of the sebaceous glands. Figure 1 shows their position, just below the surface of the skin. The sebaceous glands release sebum into the hair follicles. From here it can reach the surface of the skin, helping to keep the skin supple and waterproof. The sebaceous glands of people with acne are especially sensitive to normal blood levels of the hormone, testosterone, found naturally in both men and women. The levels of testosterone in the blood begin to increase during puberty and this stimulates the sebaceous glands on the face, chest and back to produce excess sebum. If this can't leave the hair follicles because they are blocked by dead skin cells, the sebum builds up inside the follicles, causing blackheads and whiteheads (spots) to form.

Sometimes, the build-up of sebum inside the hair follicles provides an ideal environment for a normally harmless skin bacterium called *Propionibacterium acnes* to multiply rapidly in and near sebaceous glands in the skin. The increased bacterial population produces an immune response that inflames the skin and creates the redness associated with spots. Sometimes this inflammatory acne can become severe, and cysts (cavities filled with dead skin cells, sebum and pus that form when the sebaceous glands rupture) can develop beneath the skin's surface that produce scarring of the skin, as seen in Figure 2.

Treating Acne vulgaris

There are a number of treatments that can be used to control the populations of the bacteria that can cause Acne vulgaris.

How science works

Table 1 *Treatments available for Acne vulgaris*

Type of treatment		Examples	How the treatment works
soaps and cleansers		mild soap, mild antiseptic cleanser	Remove some of the surface oil and reduce the size of populations of some bacteria living on the skin. The **antiseptics** may kill the bacteria (**bactericidal**) or they may simply inhibit further growth of the bacterial population (**bacteriostatic**).
over-the-counter remedies		benzoyl peroxide, e.g. Clearasil	Antibacterial cream or lotion. Dries out the skin and encourages it to shed the surface layer of dead skin, making it harder for follicles to become blocked and for infection to develop. Can take weeks, sometimes months, for significant effects to be noticeable.
prescriptions	topical (creams and lotions)	retinoids, e.g. Adapalene	Medicines based on vitamin A, which you can rub into your skin daily. They work by encouraging the outer layer of skin to flake off.
		an antibiotic lotion, such as clindamycin or erythromycin	Applied to the skin. Can be used to control the *P. acnes* bacteria. Need to continue this treatment for at least 6 months.
	systemic (taken as tablets)	antibiotics, e.g. tetracycline	Prescribed for inflammatory acne. Taken daily for around 3 months, although it might take 4 to 6 months for noticeable improvement. The success of this treatment can be limited because the strains of bacteria are often resistant to the common antibiotics. Benzoyl peroxide is often also prescribed at the same time.
		oral contraceptive (in women)	Suppresses testosterone activity. This drug has been shown to reduce sebum production so is often used in women with acne.
		isotretinoin, e.g. Roaccutane (an oral retinoid)	Inhibits secretion of sebum. It tends to be prescribed to people with severe forms of acne that have proved resistant to other treatments. There are a number of serious side-effects of this drug, such as liver disorders and depression. It must not be taken by pregnant women as it can cause the death of the fetus. For safety reasons, isotretinoin is only prescribed under the supervision of a dermatologist (a doctor specialising in skin conditions). Many thousands of people have benefited from treatment with isotretinoin without serious side-effects.

Can over-the-counter acne treatments be as effective as antibiotics?

Scientists investigating the effectiveness of acne treatments given over an 18-week period have found that an over-the-counter remedy is 12 times as cost effective as an antibiotic prescribed to treat the condition.

The effectiveness of an acne treatment containing 5% benzoyl peroxide, used twice a day, was compared with the effectiveness of a daily 100 mg oral dose of the antibiotic minocycline. Effectiveness of the treatment was measured by comparing the appearance and frequency of spots on the skin before and after the trial. Despite the minocycline treatment being 12 times more expensive than benzoyl peroxide, there was little difference in the effectiveness of the two treatments over the 18 weeks of the trial. No antibiotics tested in the trial worked significantly better than benzoyl peroxide and tetracyclines sometimes worked less well because of antibiotic resistance in the populations of bacteria causing the skin infection.

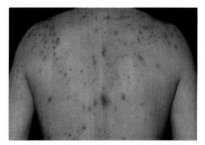

Figure 2 *Acne vulgaris on a 24-year-old man's back*

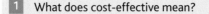

1 What does cost-effective mean?

2 Explain why the antibiotic tetracycline sometimes worked less well than benzoyl peroxide in treating acne.

3 Why may antibiotics still be prescribed to treat Acne vulgaris?

How science works

Comparing the effectiveness of face washes

Manufacturers of antiseptic face washes make all kinds of claims for their products. An investigation was carried out to compare the ability of two different antibacterial face washes to prevent the growth of bacteria, and to find out which concentration was most effective.

Using **aseptic technique**, 30 bacterial lawns were prepared, using fixed volumes of a pure bacterial culture of *P. acnes*. The plates were divided into two groups of 15, and the base of each plate was marked as shown in Figure 3. Fifteen plates were marked face-wash 1, and 15 were marked face-wash 2.

Using aseptic (sterile) technique, a 5 mm disc of sterile filter paper that had been impregnated with a fixed volume of face-wash solution at a known concentration was placed in the centre of each quarter of the bacterial lawns in each dish. Fifteen bacterial lawns were prepared in this way for each face-wash, and the dishes were then sealed with two tabs of tape. This is shown in Figure 4.

The dishes were incubated at 30°C for 24 hours. After incubation there was an even growth of bacteria across the surface of each plate, except around some of the discs where the bacteria had been killed; these are clear areas called inhibition zones. The diameter of each inhibition zone was carefully measured. The results are shown in Table 2.

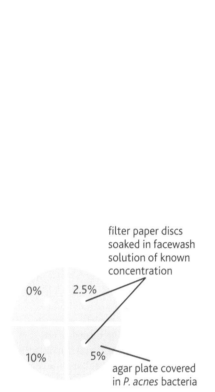

Figure 3 *Agar plate before incubation*

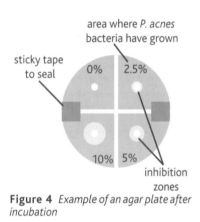

Figure 4 *Example of an agar plate after incubation*

Table 2 *Results*

Diameter of zone of inhibition/mm	Face wash 1 Concentration/%				Face wash 2 Concentration/%			
	0	2.5	5	10	0	2.5	5	10
	5	8	12	20	5	8	10	14
	5	9	16	24	5	8	9	15
	5	7	17	21	5	7	9	21
	5	9	14	18	5	9	12	16
	5	10	11	19	5	10	11	15
	5	10	17	18	5	8	10	18
	5	9	13	22	5	7	11	11
	5	10	12	20	5	7	8	15
	5	8	15	18	5	10	10	13
	5	8	16	21	5	8	10	14
	5	9	18	22	5	8	8	16
	5	10	12	25	5	8	11	16
	5	11	15	21	5	7	10	12
	5	9	11	19	5	7	10	13
	5	8	11	27	5	8	11	16

1. What is aseptic technique?

2. Why was it used in this experiment?

3. Calculate the mean diameter of the zone of inhibition at each concentration of face wash.

4. Why was a 0% face-wash disc tested on each plate? What do the results for the 0% discs tell you?

5. Which face wash appears to have been most effective against *P. acnes*?

6. Which statistical test would you use if you wanted to find out if the difference between the effectiveness of the two face washes is statistically significant?

7. State the null hypothesis that you would use to test these results against.

8. Carry out this statistical test using the results obtained for the 10% face-wash solutions. Decide whether you will accept or reject your null hypothesis.

9. Give **four** limitations of this experiment as a method for testing the effectiveness of face washes against the bacterium that causes Acne vulgaris.

Summary questions

1. Why are the populations of microorganisms that make up the skin flora called commensal?

2. Explain why many teenagers develop acne.

3. Describe how an antiseptic may affect a bacterial population.

4. Copy and complete the following table by placing a tick in the appropriate box to summarise the action of the spot treatments shown in Table 1.

Treatment	Bactericidal/ bacteriostatic effect on bacterial populations	Encourages shedding of surface layer of skin cells	Reduces production of sebum
antiseptic cleanser			
benzoyl peroxide			
retinoid cream			
antibiotic tablets, e.g tetracycline			
oral contraceptive			
isoretinoin tablets			

13.3 The bacterial community of the gut

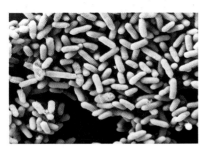

Figure 1 *Coloured scanning electron micrograph of Bifidobacteria infantis, which is the main bacterial species found in the large intestine of breastfed infants*

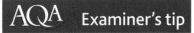

Link

Probiotic bacteria were studied in Topic 1.4 of *AS Human Biology*.

AQA Examiner's tip

Your gut is adapted to function with its usual community of bacteria. A change in this community is likely to adversely affect the functioning of the gut.

More than 500 different species of bacteria live in the human gut. The bacterial community of the gut represents 80% of the body's normal flora and adds at least one kilogramme to the weight of an adult human! The gut of a newborn baby will be colonised by the bacteria that are encountered during birth and feeding. This means that the gut flora of a baby that is delivered by caesarean section and is then bottle-fed will be very different from that of baby a that is delivered vaginally and is breastfed.

After just a few weeks of breastfeeding, over 90% of a breastfed baby's intestinal flora is made up from *Bifidobacteria*. These bacteria are known to make the baby's gut acidic, helping to prevent infection by pathogenic bacteria that may enter the gut. *Bifidobacteria* do not survive well on the proteins that are found in the formula milk given to bottle-fed babies, and this can mean that formula-fed babies may be more susceptible to infection by pathogenic bacteria than breastfed babies. Scientists have discovered that if susceptible infants are given formula milk that contains strains of *Lactobacillus* and *Bifidobacteria*, their risk of infection from pathogenic bacteria is reduced. Foods that contain these *Lactobacillus* and *Bifidobacteria* are called **probiotics**.

The baby's gut flora will continue to change as solid foods are introduced into its diet, allowing the growth of different types of bacteria. By the time a child has been completely weaned off milk, at about the age of two, its gut flora will resemble that of a normal adult.

Antibiotics and the gut flora

A course of antibiotics will hopefully help to destroy the populations of pathogenic bacteria that are making a person ill. However, the antibiotics will also wipe out members of the normal gut flora, and this can prevent the gut from functioning normally. For example, Vitamin K production by gut bacteria in the large intestine can be reduced. Losing the gut's 'good' bacteria can allow populations of potentially harmful gut bacteria to increase in size because they no longer have to compete with the 'good' gut bacteria for nutrients or space. These disease-causing bacteria can cause symptoms such as diarrhoea, and it can take many weeks for the populations of normal gut bacteria to become re-established.

How science works

Probiotics and health

Antibiotics are sometimes prescribed to patients who do not have infections but who are at risk of developing them. Patients who have undergone major surgery are considered to be at risk from infection by pathogenic bacteria. A team of surgeons discovered that between 50% and 80% of the patients receiving antibiotics after major surgery in their hospital unit developed hospital-caught infections. The team also found that such infections did not occur amongst the patients in the unit who did not receive antibiotics after their surgery. The surgeons wondered if the antibiotics given after surgery were actually harming the patients' immune systems. Research has shown that **probiotics** can boost the effectiveness of the body's immune system, so the surgeons decided that instead of giving antibiotics to their patients after surgery (which would wipe out their normal gut flora), they would give them a course of probiotic *Lactobacillus* bacteria (to allow a rich and varied bacterial community to survive in the gut). They supplied the probiotic bacteria as a drink, together with the prebiotic carbohydrate molecules that are essential for the growth of the probiotic bacteria. Their trial showed that only one out of 33 patients treated in this way developed an infection.

This is just one example of an increasing amount of evidence for the beneficial effects of probiotic use amongst seriously ill people. It is not known whether regular consumption of off-the-shelf probiotic products, such as live yoghurt drinks, by healthy people is beneficial. Not all the strains of bacteria found in the supermarket products are probiotic, and current research suggests that it is *specific* strains of *Lactobacillus* bacteria which grow on plants that are more likely to survive the acidity in the stomach or the bile content in the small intestine. A diet containing plenty of fresh fruit and vegetables and non-processed food may provide a reliable source of the probiotic bacteria that we need for a diverse bacterial community in a healthy gut.

Figure 2 *A diet containing plenty of fresh fruit and vegetables and non-processed food may provide a reliable source of the probiotic bacteria that we need for a diverse bacterial community in a healthy gut*

1. Why may the doctors have given the patients antibiotics straight after surgery?

2. Explain whether you think that the results of the surgeons' trial supports the hypothesis that the antibiotics given after surgery were harming the patients' immune systems?

3. Why may seriously ill people benefit from the use of probiotics?

4. Suggest why healthy people may not need to take probiotic supplements such as live yoghurt drinks.

▮ At home in the gut

The varied diet of our early human ancestors would have allowed them to develop much more diverse bacterial communities in the gut than modern humans. Scientific research over the past hundred years has focused our attention on the dangers of contamination of food by pathogenic bacteria such as *Salmonella*. Modern humans in many parts of the world are obsessed with eradicating bacteria – both the 'bad' and the 'good' bacteria, because they are afraid that they will spread disease. As a result large amounts of antibacterial agents have been introduced into the environment of bacteria to destroy them. We now realise that the frequent use of antibacterial agents such as antibiotics can select for populations of bacteria that are resistant to them and humans must now face the threats posed by strains of pathogenic bacteria that are untreatable with antibiotics. A diet rich in fresh fruit and vegetables and non-processed food will introduce a huge variety of commensal gut flora to the many ecological niches available in the human gut. This competition can prevent the growth of populations of potentially pathogenic bacteria that may also pass through this diverse ecosystem.

▮ Link

The evolution of resistance to antibiotics has been covered in Topic 4.5 of *AS Human Biology* and in Topic 8.2 of this book.

▮ Summary questions

1. What is a probiotic?

2. What is 'good' about the 'good bacteria' that manufacturers claim to have in probiotic foods?

3. Give **three** reasons why breastfeeding a baby may be considered more beneficial to the baby's health than feeding the baby on formula milk.

4. Why may it take many weeks for the populations of normal gut bacteria to become re-established after a course of antibiotics?

1 The genome of the acne bacterium, *Propionibacterium acnes*, has recently been sequenced. This shows that the bacterium produces proteins that actively cause acne.

Some of its genes code for enzymes that break down human skin. It also produces proteins that trigger the immune response.

Scientists are worried that *P. acnes* could contaminate blood. They have found the bacterium in blood donated for transfusion. Scientists have shown that *Propionibacterium acnes* is sometimes presented in blood bank samples. By sequencing the genome of this bacterium, scientists hope to discover new cures for acne. Severe acne used to be treated with antibiotics, but the bacteria are developing resistance.

   (a) Suggest how having enzymes to break down human skin can help a bacterium to cause an infection. *(2 marks)*

   (b) Describe how a gene in *Propionibacterium acnes* can lead to the production of an enzyme. *(6 marks)*

   (c) Suggest how scientists can test a sample of blood to find out whether *Propionibacterium acnes* is present. *(4 marks)*

   (d) (i) Use your knowledge of natural selection to explain how the bacteria that cause acne are developing resistance to antibiotics.

      (ii) Some scientists suggested that they might be able to study the genome of *Propionibacterium acnes* to help them to develop a vaccine against acne. Suggest how they might do this. *(8 marks)*

2 Some species of deer have sweat glands near their hooves. Scientists have discovered that sweat produced from these glands contains a chemical that might be useful as a treatment for acne.

   (a) Describe how scientists could use agar plates and a culture of the bacterium that causes acne to test whether this chemical kills the bacterium. *(4 marks)*

   (b) Use your knowledge of natural selection to explain how the ability to produce this chemical in sweat may have developed in deer. *(4 marks)*

3 Acne is one health problem associated with eating processed bread and cereal products. Some scientists believe that it is the sugars, released from these foods during digestion, that stimulate the production of male hormones. It is known that the hormones cause the release of excess sebum and that this can lead to the growth of bacteria that cause acne. To test their idea, scientists will study the effect of placing sixty teenage boys with acne on a low-carbohydrate diet for three months to see if it makes a difference.

However, acne is not the only health problem that might be caused by eating highly processed bread and cereal products.

   (a) (i) Describe a suitable control group for this study.

      (ii) Explain why a control group is necessary. *(4 marks)*

   (b) It is important that the boys in the experimental group should be as similar as possible to the boys in the control group. Explain how the scientists could ensure that this happens. *(2 marks)*

4   A study was carried out to find out whether probiotic yoghurt drinks were effective in reducing symptoms of hay fever. People were divided into two groups.

Group **A** drank a probiotic yoghurt drink every day for 5 months.

Group **B** drank a similar drink that did not contain bacteria every day for 5 months.

Blood samples were taken from each person before the grass pollen season, when the grass pollen season was as its peak, and again four weeks after the end of the season. The blood samples were tested for the presence of IgE antibodies. There were no significant differences in levels of IgE in the blood between the two groups at the start of the study, but IgE levels were lower in the probiotic group both during the peak season and afterwards.

(a) Explain why the blood was tested for IgE antibodies. *(2 marks)*

(b) The scientists identified no significant differences in the levels of IgE in the blood between the two groups at the start of the study. Explain what this means. *(1 mark)*

(c) Another group of scientists that reviewed this study suggested that more data was required before it was possible to conclude that probiotic yoghurt drinks reduce the symptoms of hay fever. Explain why. *(3 marks)*

5   The human appendix is a small dead-end tube in the gut which is connected to the large intestine. Scientists believe that it must have a useful function since few other mammals have an appendix. Where an appendix is present, it is very different from the human appendix.

One theory is that the appendix is a region separated from the rest of the gut where commensal bacteria can survive, away from the gut contents. During an infection of the gut, it is thought that the bacteria could 'hide' in the appendix, ready to re-populate the gut when the infection is over.

(a) Explain the evidence in the passage to explain how the appendix might have developed through natural selection. *(4 marks)*

(b) Give two functions of commensal gut bacteria. *(2 marks)*

(c) It would be very difficult to carry out an investigation to test the scientists' theory. Explain why. *(4 marks)*

Unit 5 questions: The air we breathe, the water we drink, the food we eat

1 Scientists used a line transect to find the distribution of three species of *Ranunculus* (buttercup) in a field. The field consisted of a series of ridges and furrows. **Figure 1** shows the distribution of the species of *Ranunculus* along the line transect.

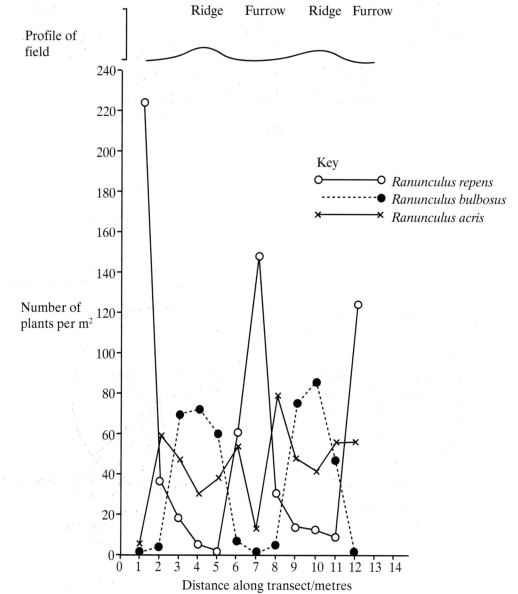

Figure 1

(a) Describe how you would use a line transect to obtain data on the distribution of the *Ranunculus* species as shown in **Figure 1**. *(2 marks)*

(b) Other than soil moisture, give one abiotic factor and explain how it could lead to the abundance of *Ranunculus repens* in the furrows of this field. *(2 marks)*

(c) The scientists then investigated the effect of soil moisture on seed germination of the three species of *Ranunculus*. They planted seeds of species A in three sets of pots. The soil in one set of pots was maintained at 25% water content, the soil in the second set was maintained at 50% water content and the soil in the third set was maintained at 100% water content. They repeated this with seeds of species B and species C. Four weeks later, the scientists recorded the mean number of seedlings in the pots in each set. Their results are shown in **Figure 2**.

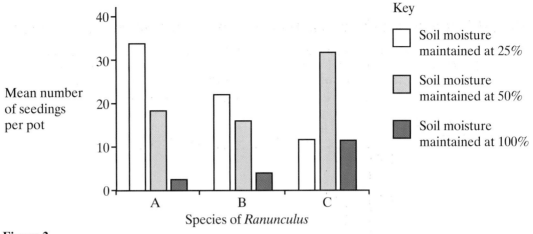

Key

☐ Soil moisture maintained at 25%

▨ Soil moisture maintained at 50%

■ Soil moisture maintained at 100%

Figure 2

(i)   Suggest two factors which should be controlled during this investigation.

(ii) Use information from **Figure 1** and **Figure 2** to identify each species of buttercup, **A – C**.

(iii) Describe how you could determine whether two *Ranunculus* plants belong to the same species.

(5 marks)

2   The pie charts in **Figure 3** show the percentage biomass of different mammals in Great Britain. Chart **A** shows all mammals. Chart **B** shows different groups of wild mammals.

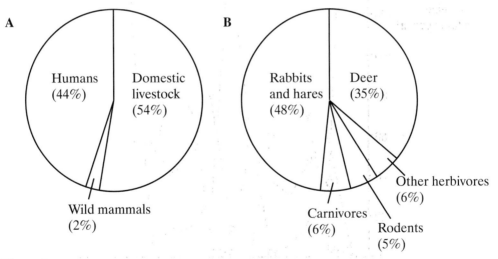

Figure 3

(a) The total biomass of the carnivores is less than the total biomass of all the other groups shown in Chart B. Explain this difference.   *(2 marks)*

(b) The biomass of deer in Great Britain is 45 000 tonnes. Calculate the biomass of the domestic livestock in Great Britain. Show your working.   *(2 marks)*

AQA, 2008

3 (a) Scientists investigated the changes in the different types of plant species growing on a disused football pitch. Eventually the grass became replaced by a woodland community. Explain how this happened. *(3 marks)*

(b) Bluebells are small herbaceous plants that grow in woodland. Scientists investigated the abiotic factors which might affect the distribution of bluebells.

Give three abiotic factors the scientists would measure and explain how each factor might affect the distribution of bluebells. *(3 marks)*

AQA, 2008

4 (a) A farmer changed the use of his land from raising animals to growing cereal crops. This is more efficient is terms of total food production. Explain why. *(2 marks)*

(b) The farmer was paid to stop using one of his fields for farming. Over the next 20 years the species in the field changed. Describe a method that could be used to measure these changes. *(4 marks)*

AQA, 2008

5 The following is a list of ecological terms.

A Food chain	**F** Population	**J** Trophic level
B Community	**G** Biome	**K** Biomass
C Abiotic	**H** Ecosystem	**L** Habitat
D Quadrat	**I** Transect	**M** Biotic
E Niche		

Copy and complete the table to give the letter to match the term to the appropriate definition. The first one has been done as an example.

Definition	Letter
The role of an organism in the community	E
Non-living factors, such as temperature, that affect the distribution of living organisms	
All the living organisms in a defined area	
The total number of a species living in a defined area	
The place where an organism lives	
A technique for measuring the influence of an environmental gradient on organisms	

(5 marks)

AQA, 2008

6 (a) What is the **waste hierarchy**? *(4 marks)*

(b) A magazine article advised readers on ways to reduce their carbon footprint. Explain the reason for the following pieces of advice:

(i) Avoid highly processed foods

(ii) Reduce your meat consumption. *(2 marks)*

7 Recently, the European Union developed a policy that 10% of road fuels used in Europe should come from plants.

(a) Explain **one** advantage and **one** disadvantage of this policy. *(2 marks)*

(b) What is meant by **carbon offsetting**? *(2 marks)*

Glossary

abiotic: an ecological factor that makes up part of the non-biological environment of an organism, e.g. temperature, pH, rainfall and humidity.

accommodation: the way the *ciliary muscles* cause the lens to change shape.

Acne vulgaris: a skin condition caused by over-activity of the sebaceous glands.

acrosome reaction: a reaction that occurs in the acrosome of the *sperm*, that enables the sperm to digest the zona pellucida.

action potential: change that occurs in the electrical charge across the membrane of a *neurone* when it is stimulated, causing a nerve impulse.

aerobic respiration: respiration when oxygen is available. This includes *glycolysis*, the link reaction, *Krebs cycle* and oxidative phosphorylation.

allele: an alternative form of a *gene*.

allele frequencies: the number of times an *allele* occurs within the *gene pool*.

allergen: a normally harmless substance that causes the immune system to produce an immune response in someone who is allergic.

allopatric speciation: occurs when populations are prevented from interbreeding because they become geographically isolated.

amnion: the membrane sac that protects the *fetus*.

amniotic fluid: the fluid that is secreted by the *amnion*.

anaerobic respiration: respiration in the absence of oxygen. This consists only of *glycolysis*.

anaphylactic shock: a sudden, acute and potentially fatal reaction to an *allergen*.

antagonistic pair: a pair of muscles that work in opposite ways to each other.

anticodon: a sequence of three adjacent nucleotides on a molecule of transfer RNA that are complementary to a particular *codon* on a messenger RNA molecule.

antihistamines: drugs that prevent *histamines* from binding to the histamine receptors, so preventing the symptoms of an allergy from developing.

antiseptic: a substance that inhibits or prevents the growth of microorganisms that cause infection.

apoptosis: this is programmed cell death, for example, when tumour suppressor genes cause cells with badly damaged DNA to 'commit suicide'.

aqueous humour: clear, watery fluid of the eye found in front of the lens.

aseptic technique: a series of procedures used to transfer microorganisms without contaminating the microbial culture with environmental microorganisms, or allowing microorganisms from the culture to escape into the environment.

ATP (adenosine triphosphate): the immediate source of energy for biological processes. When ATP is hydrolysed to ADP, energy is released.

autonomic nervous system: the part of the nervous system that carries nerve impulses to glands and muscles that are not under voluntary control.

axon: a single, long projection from the nerve cell body that carries nerve impulses away from the nerve cell body.

B cell (B lymphocyte): type of white blood cell that is produced and matures within the bone marrow. B cells produce antibodies as part of their role in immunity.

basal metabolic rate: the rate at which we use energy to keep ourselves alive while we are at rest, but not asleep.

binocular vision: vision resulting from the visual fields of two eyes overlapping.

biodiversity: the variety of different organisms in a particular area.

biofuels: renewable energy sources made from *biomass*.

biomass: the total mass of living organisms in a community.

biotic: an ecological factor that makes up part of the living environment of an organism. Examples include food availability, competition and predation.

bivalent: the pair of homologous chromosomes that pair and form *chiasmata* in the first division of meiosis.

blastocyst: an early stage in embryo development consisting of a hollow ball of cells. This is the stage at which the embryo implants into the uterus.

cap: a barrier method of contraception that consists of a latex or silicone cap designed to fit over the cervix.

capacitation: changes that occur in the *sperm* while they are in the female reproductive tract to make them capable of fertilising the *oocyte*.

carbon footprint: a measure of the impact that human activities have on the amount of greenhouse gases produced, measured in terms of kilograms of carbon dioxide produced per year.

carbon offsetting: compensating for emissions by making an equivalent greenhouse gas saving.

cardiac output: the volume of blood pumped by the heart in one minute.

carrying capacity: the population size that can be sustained by an ecosystem.

central nervous system: the part of the human nervous system that consists of the brain and the spinal cord.

cerebral cortex: the outer part of the brain concerned with conscious thought, interpretation of stimuli and memory.

chiasmata: the place at which the chromatids of homologous chromosomes wrap around each other and exchange pieces.

chorionic villi: the villi that make up the surface of the placenta.

choroid layer: the middle layer of the eye that contains blood vessels and melanin pigment.

ciliary muscle: the smooth muscle of the eye that is responsible for *accommodation*.

climate: average weather experienced in a region over a long period.

climate change: changes in climate that have been observed since the early part of the 1900s as a result of human activity.

climax community: a stable community that is in equilibrium with the existing environment (as a result of ecological succession).

codominant: condition in which both *alleles* for one gene in a *heterozygous* organism contribute to the *phenotype*.

codon: a sequence of three adjacent nucleotides in mRNA that codes for one amino acid.

colostrum: the first-formed breast milk. It is high in antibodies.

commensals: organisms that live symbiotically, e.g. microorganisms that live on the skin and out-compete potential pathogens.

community: all the living organisms in an ecosystem at any given time.

concordant: twins are said to be concordant for a characteristic if they are both similar for that characteristic.

condom: widely used barrier method of contraception that consists of a thin latex sheath that is placed over the erect penis.

copulation: sexual intercourse.

cornea: the front, transparent part of the sclera that refracts light rays entering the eye.

corpus luteum: a hormone producing structure, formed from the ruptured follicle.

cortical reaction: a reaction that alters the zona pellucida in order to prevent other *sperm* entering the *oocyte*.

counter current flow: the flow of the mother's blood and the fetal blood in opposite directions, that ensures a concentration gradient all the way along the capillary in the placenta.

crossing over: when the chromatids break, and equivalent pieces of chromatid are exchanged at the *chiasmata* during meiosis.

D

decomposers: organisms such as bacteria and fungi that break down faeces and the bodies of dead organisms.

degenerate: a genetic code in which some amino acids may be encoded by more than one *codon* each.

dendrite: an extension of a nerve cell body that carries nerve impulses towards the cell body.

detritivores: animals such as earthworms that break down dead organic matter.

discordant: twins are said to be discordant for a characteristic if they are different for that characteristic.

DNA ligase: enzyme used in *genetic engineering* to join pieces of DNA, e.g. to insert a *gene* into a cut *plasmid*.

DNA methylation: when methyl groups $(-CH_3)$ are added to cytosine bases on the DNA, which causes the repression of transcription of the DNA.

dominant: a term applied to an *allele* that is always expressed in the *phenotype* of an organism.

E

ecological niche: the role, activities and location of a population within a habitat.

effectors: an organ that responds to stimulation by a nerve impulse resulting in a change or response.

electron acceptor: a protein that accepts electrons, for example, in the *thylakoid membrane* that accepts excited electrons in the *light-dependent stage* of photosynthesis.

electron transfer chain: a series of *electron acceptors* such as

found in the inner mitochondrial membrane.

endocrine system: system of glands that secrete hormones.

endometrium: the lining of the uterus.

environmental factors: factors of an organism's surroundings and experiences, such as diet and food availability.

environmental resistance: the collective term for the limiting factors that keep a population at carrying capacity.

epigenetic imprinting: when during gametogenesis, certain genes in both sperms and oocytes are modified by the addition of methyl $(-CH_3)$ groups.

eukaryotic: a type of cell that contains a membrane-bound nucleus and membrane-bound organelles.

extracellular digestion: the digestion of matter that occurs outside the cell.

F

fauna: all the animal life in an ecosystem.

fetus: the term for an unborn baby from 10 weeks after fertilisation, when all the main organs have developed.

fight or flight response: refers to the body's reaction when faced with a threat. It results from the activity of the sympathetic nervous system.

flora: all the plant life in an ecosystem.

follicle stimulating hormone (FSH): a hormone secreted by the *pituitary gland* that causes follicle development.

frame quadrat: a square frame, sometimes divided by string or wire into subdivisions, that is used in sampling.

G

gamete: a sex cell that contains the haploid number of chromosomes.

gametogenesis: the formation of *gametes*.

ganglion: collection of nerve cell bodies.

gene: a length of DNA that occurs at a specific *locus* on a chromosome and codes for a particular protein or polypeptide.

gene pool: all the genetic information (genes) present within a population at a given time.

generator potential: depolarisation of the membrane of a receptor cell as a result of a stimulus.

genetic engineering: general term that covers the processes by which *genes* are manipulated, altered or transferred from organism to organism.

genetic factors: factors such as the presence of certain *alleles*.

genetic marker: the resistance of a gene to a specific antibiotic.

genetically modified organism: an organism that has had its DNA altered as a result of recombinant DNA technology.

genotype: the genetic composition of an organism.

geographically isolated: the isolation of populations of a species by a geographical barrier, such as a river, so they can no longer interbreed.

germinal epithelium: the epithelium that divides to produce *gametes*, for example, the outer layer of the *seminiferous tubule*.

glycolysis: first part of cellular respiration in which glucose is broken down anaerobically in the cytoplasm to two molecules of pyruvate.

gonadotrophin releasing factor (GnRF): a hormone released by the *hypothalamus* that stimulates the release of FSH and LH.

Graafian follicle: a ball of cells surrounding an *oocyte*.

grana: stacks of *thylakoid membranes* in the chloroplast.

greenhouse gases: gases present in the atmosphere which trap out-going long wave radiation, causing a rise in temperature, e.g. carbon dioxide and methane.

H

habitat: the place where an organism normally lives, which is characterised by physical conditions and the types of other organisms present.

heterozygous: condition in which the *alleles* of a particular *gene* are different.

histamine: a chemical that is released when *IgE antibodies* bind to *mast cells* in an allergic reaction. Histamine plays a part in producing the symptoms of the allergy.

homozygous: condition in which both *alleles* of a particular *gene* carried by one individual are identical.

human chorionic gonadotrophin (hCG): a hormone secreted by the *blastocyst* and the developing placenta that forms the basis of pregnancy tests.

hyperglycaemic: a person is hyperglycaemic if their blood glucose level rises significantly above the normal level.

hypersensitive: overreaction of the immune system in reaction to *allergen* exposure.

hypoglycaemic: a person is hypoglycaemic if their blood glucose level drops significantly below the normal level.

hypothalamus: the area of the brain that regulates many homeostatic processes such as body temperature and blood water potential, and stimulates the release of hormones by the *pituitary gland*.

hypothermia: when body temperature falls significantly, and the normal heat conservation and heat generation responses of the body are unable to bring the body temperature back up.

I

IgE antibodies: antibodies that belong to the immunoglobulin E group that are produced during an immune response.

interspecific competition: competition between organisms of different species.

interstitial cells: cells inbetween the *seminiferous tubules* that secrete *testosterone*.

intraspecific competition: competition between organisms of the same species.

intrauterine device: a non-barrier birth control device that is placed inside the uterus. It works by preventing the implantation of an embryo.

introns: portions of DNA within a *gene* that do not code for a polypeptide. The introns are removed from messenger RNA after transcription.

***in vitro* fertilisation (IVF):** a method of assisted conception in which fertilisation takes place in a glass dish.

islets of Langerhans: groups of cells in the pancreas comprising large α cells, which produce the hormone glucagon, and small β cells, which produce the hormone insulin.

K

karyotype: the arrangement of chromosomes into their homologous pairs, starting with the largest pair and ending with the smallest.

Krebs cycle: series of aerobic biochemical reactions in the matrix of the mitochondria of *eukaryotic* cells. ATP is produced through the oxidation of acetyl coenzyme A produced from the breakdown of glucose.

L

light-dependent stage: the stage of photosynthesis that is dependent on light and takes place in the *grana*.

light-independent stage: the stage of photosynthesis that is not dependent on light and takes place in the *stroma*.

limiting factors: factors such as food supply that keep a population at a maximum size that can be sustained by an ecosystem.

locus: the position of a *gene* on a chromosome.

luteinising hormone (LH): a hormone secreted by the *pituitary gland* that stimulates the formation of the *corpus luteum*.

M

mammography: taking a breast radiograph designed to screen for breast cancer.

mast cells: cells of the immune system that are found in all tissues. They release *histamine* during an immune response.

mean: the average of a set of values.

median: the middle value of a set of values.

menarche: the time of a female's first menstrual cycle.

Mendelian ratio: the expected ratio of *phenotypes* resulting from a genetic cross.

menopause: the time when a female permanently stops having menstrual cycles.

menstruation: the shedding of the outer layer of the *endometrium*.

mode: the most frequently occurring value in a set of values.

morula: an early stage in embryo development consisting of a ball of cells that forms 3–4 days after fertilisation. The morula develops into the *blastocyst*.

motor neurones: *neurone* that transmits action potentials from the *central nervous system* to an *effector*, e.g. a muscle or gland.

mutation: a permanent change in the amount or arrangement of a cell's DNA.

myelin sheath: a sheath consisting of a fatty substance that surrounds *axons* and *dendrites* in certain *neurones*. It is formed from *Schwann cells* that wrap themselves around the neurone.

myofibrils: smaller fibres that make up muscle tissue.

N

natural selection: the process by which the best adapted organisms in a population survive, reproduce and pass on their *alleles* to their offspring.

negative feedback: a series of changes, important in homeostasis, that result in the body's internal environment being restored to its normal level.

neuromuscular junction: a *synapse* that occurs between a nerve cell and a muscle.

neurone: a nerve cell, comprising a cell body, *axon* and *dendrites*, which is adapted to conduct *action potentials*.

neurotransmitter: one of a number of chemicals that are involved in communication between adjacent nerve cells or between nerve cells and muscles.

niche: *see ecological niche.*

nodes of Ranvier: gaps in the *myelin sheath* that surrounds the *axon* of a *neurone*.

normal distribution: a bell-shaped curve produced when a certain distribution is plotted on a graph. The *mode*, *median* and *mean* are equal.

null hypothesis: the hypothesis that there is 'no significant difference between observed and expected data'.

O

oestrogen: the female sex hormone that stimulates the thickening of the *endometrium*.

oocyte: an immature ovum.

ova: female *gametes*.

ovaries: where *ova* are produced.

oviduct: tube that carries *oocytes* to the uterus.

oxytocin: a hormone secreted by the *pituitary gland* that stimulates the contraction of the myometrium and the release of breast milk.

P

parasympathetic nervous system: part of the autonomic nervous system that is active under normal conditions.

passive immunity: when a person receives ready-made antibodies.

patch testing: a type of allergy test that is carried out if a person already suffers from a skin allergy such as eczema or contact dermatitis.

pathogen: a microorganism that causes a disease.

peptide bond: the chemical bond formed between two amino acids during condensation.

peripheral nervous system: the nerves and the receptors of the nervous system.

phenotype: the characteristics of an organism resulting from its *alleles*.

phenotypic variation: the total variation in the characteristics of an organism, usually visible, resulting from both its genotype and the effects of the environment.

pituitary gland: the endocrine gland at the base of the brain.

plasmid: small circular piece of DNA found in bacterial cells and used as a *vector* in *genetic engineering*.

point quadrat: a sampling tool that consists of vertical legs, across which is fixed a horizontal bar with ten small holes along it.

polar bodies: haploid nuclei produced in oogenesis when cells divide unequally.

polygenic inheritance: where more than one *gene* contributes to a *phenotype*.

polymerase chain reaction: process of making many copies of a specific sequence of DNA or part of a *gene*. It is used extensively in gene technology and genetic fingerprinting.

positive feedback: process which results in a factor in the body's internal environment, such as temperature, that departs from its normal level becoming further from its norm.

postzygotic mechanisms: reproductive isolating mechanism that takes effect after fertilisation.

prezygotic mechanisms: reproductive isolating mechanism that takes effect before fertilisation.

primary consumers: organisms that eat plants, and obtain their energy by respiring molecules obtained from plants.

primary producers: organisms that are able to produce organic materials from the inorganic molecules, carbon dioxide and water, in photosynthesis.

primers: short single-stranded nucleic acid sequences used in the *polymerase chain reaction*.

probiotics: substances that contain beneficial bacteria such as *Lactobacillus* and *Bifidobacteria*.

progesterone: a hormone that inhibits FSH and LH, and stimulates the growth of blood vessels in the *endometrium*.

prokaryotic: a type of cell that does not have a membrane-bound nucleus or membrane-bound organelles. These are bacteria cells.

prolactin: a hormone secreted by the *pituitary gland* that stimulates the production of milk.

R

receptors: a cell adapted to detect changes in the environment.

recombinant DNA: modified DNA that results from *genetic engineering*.

refractory period: period during which the membrane of a *neurone* cannot be depolarised and no new *action potential* can be initiated.

relaxin: a hormone secreted late in pregnancy that softens the cervix and makes the pelvis more flexible.

renewable energy resources: energy sources that are sustainably regenerated by natural processes.

reproductive isolation: when groups within the population become isolated from one another and cannot interbreed.

resting potential: the difference in charge between the negatively charged inside of *neurone* and the outside of the neurone, when the neurone is at rest.

restriction enzyme: an enzyme that cuts DNA molecules at a specific sequence of bases called a recognition sequence.

retina: the inner layer of the eye that contains light sensitive rod and cone cells.

reverse transcriptase: an enzyme found in the HIV virus that is used in genetic engineering to make a single-stranded DNA copy of RNA.

RNA polymerase: enzyme that joins together nucleotides to form messenger RNA during transcription.

S

saltatory conduction: propagation of a nerve impulse along a myelinated dendron or *axon* in which the *action potential* jumps from one *node of Ranvier* to another.

sarcomere: a section of *myofibril* between two Z-lines that forms the basic structural unit of skeletal muscle.

Schwann cell: cell around a *neurone* whose cell surface membrane wraps around the dendron or *axon* to form the *myelin sheath*.

sclera: the tough and fibrous outer layer of the eye.

scrotum: a sac of skin that contains the testis.

secondary consumers: organisms that feed on *primary consumers*.

selection pressure: the environmental force altering the frequency of *alleles* in a population.

selective advantage: a variation that gives one organism an advantage over another organism, making it more likely that it will survive and reproduce.

seminiferous tubules: long tubules in the testes where sperm are produced.

sensory neurones: a *neurone* that transmits an *action potential* from a sensory receptor to the *central nervous system*.

Sertoli cells: cells that provide developing sperm with nutrients.

skewed distribution: a non bell-shaped distribution in which the *mean, median* and *mode* are all different.

skin prick testing: the first type of test carried out when a person is suspected to be an allergy sufferer.

somatic nervous system: the part of the nervous system that carries impulses to skeletal muscles that are under conscious control.

speciation: the process by which new species develop.

species: a group of similar organisms that can interbreed and produce fertile offspring.

sperm: the male *gamete*.

stroma: watery, enzyme containing fluid that surrounds the *grana* in the chloroplast.

succession: the natural, progressive sequence of events where one community replaces the previous community over time, and which ends in the development of a *climax community*.

sympathetic nervous system: part of the *autonomic nervous system* responsible for reacting to stress, emergencies and danger.

sympatric speciation: occurs when populations living together become reproductively isolated.

synapse: the point at which two *neurones* communicate with each other.

synaptic cleft: the gap between two *neurones*.

T

tamoxifen: a drug that blocks the action of oestrogen, and is used to treat oestrogen positive breast cancer.

testes: where sperm are produced.

testosterone: the male sex hormone that promotes the development of male sex characteristics.

thermoreceptors: sensory receptor cells found in the *hypothalamus* that monitor blood temperature.

thylakoid membranes: pigment containing membranes arranged in stacks inside chloroplasts.

transects: a sampling tool used where conditions and organism may change. A tape is stretched across a habitat and quadrat samples are taken along a straight line at regular intervals. This is a systematic sampling method.

transgenic: an organism is transgenic when its genetic composition has been altered by the addition of foreign DNA.

trichromatic theory: the theory that refers to the perception of colours dependent on the proportions of the three different types of cone that are stimulated.

trophic levels: the different levels of the food chain.

trophoblast: the outer layer of cells of the *blastocyst*.

U

umbilical cord: the cord that connects the *fetus* to the placenta and carries blood between the fetus and the placenta.

V

vector: a carrier such as a *plasmid*, which transfers DNA into a cell.

vitreous humour: clear-jelly like fluid found behind the lens in the eye. It holds the eyeball in shape.

W

waste hierarchy: sets out the main options for the management of waste and is the primary tool for assessing the 'Best Practical Environmental Option' (BPEO) for waste management.

Z

zygote: fertilised ovum.

Answers

1.1

1 Helps to propel the sperm and secretions towards the urethra.

2 Lining of the respiratory tract.

3 Helps to move the ovum (and possibly zygote) towards the uterus.

1.2

1 If FSH not inhibited, too many follicles would develop and several oocytes would be released. Could result in several embryos developing, which would not survive.

2 **a** The male would have helped to provide food for the female and their children, in return for the female being sexually receptive to him most of the time.

   b This would have been an advantage to a female needing a good food supply to sustain her during pregnancy and in breast-feeding a big-brained infant who has high energy demands.

3 Only certain cells have a protein receptor in their cell membranes with a receptor site of the correct shape for the hormone to fit into it.

1.3

1

Spermatogenesis	Oogenesis
Occurs all the time in a man after the age of puberty.	All the stages up to the first division of meiosis occur before a woman is born.
Every primary spermatocyte produces four sperm cells.	Every primary oocyte produces only one ovum – the other products are polar bodies which are not functional.

2 **a** 12 **b** 6 **c** 3

3 They are haploid, so have different alleles from each other and from the man's body. This means they will have different proteins (antigens) in their cell-surface membranes.

Older fathers run higher risk of fetal defects

1 May argue that the findings are invalid because they only studied 18 donors and the increase in frequency of diploidy was from 0.2% to 0.4%. May comment that a large sample of sperm was studied although the number of donors was small.

2 Down's Syndrome results from non-disjunction, i.e. a zygote contains three copies of chromosome 21. Either the ovum or the sperm contains two copies of chromosome 21.

3 Oogenesis occurs in a female before birth, so the oocyte ovulated by an older woman has spent much longer 'arrested' in meiosis than a younger woman's. Over this time, it may have become damaged.

4 Sertoli cells nourish the developing sperm and protect them from the immune system. A defective Sertoli cell may not provide sufficient nourishment or protection for the sperm.

1.4

1 $\text{magnification} = \dfrac{\text{measured size}}{\text{actual size}}$

sperm scale bar measures 9 mm = 9000 μm

$\text{magnification} = \dfrac{9000}{5}$

$= 1800$

oocyte scale bar measures 28 mm = 28 000 μm

$\text{magnification} = \dfrac{28\,000}{100}$

$= 280$

2 Oocyte: Good food supply (yolk droplets) to supply the embryo with nutrients until implantation has taken place. Sperm: small so they can move to the female gamete using a minimum supply of energy.

3 These carry out aerobic respiration and supply the sperm with ATP for movement of the flagellum.

4 Yes: it gains its nutrients from digestion of the mother's tissues. No: it is not a different species, and it does not cause the mother any harm.

1.5

1 Large surface area: contains many villi; villi have microvilli to further increase the surface area. Large concentration gradient: the maternal arteries constantly bring fresh oxygenated blood to the placenta and remove deoxygenated blood; the fetal arteries constantly bring deoxygenated blood and remove oxygenated blood; there is counter current flow between the two blood systems, maintaining a concentration gradient all the way along the capillaries. Short diffusion pathway: substances only have to diffuse across the thin flattened endothelial cells of the fetal capillary, the connective tissue layer and the thin cells on the surface of the villi.

2 The mother's immune system would mount an immune response to the fetal cells as these are genetically different from the mother; the fetus may have a different blood group from the mother; pathogens in the mother's blood would pass to the fetus.

3 The fetus does not produce the antibodies itself, so there are no memory cells present.

4 Small and lipid-soluble, passes very easily across cell membranes.

5 Oxygen would pass from the mother's blood to the fetus's blood until equilibrium is reached – at this point both blood vessels would each contain similar concentrations of oxygen. (With counter current flow, the mother's blood is able to give up almost all of its oxygen.)

6 Progesterone inhibits uterine contractions, that would otherwise expel the fetus; it inhibits menstruation; it inhibits FSH so no more oocytes can be fertilised.

Screening in pregnancy

1 Chromosomes can't be seen unless the cells are dividing.

2 Prophase/accept metaphase. Reason: Spindle development is inhibited so the cells cannot get past metaphase.

3 **a** To spread the chromosomes out so they can seen individually in the photograph.

   b Salt solution will have a higher water potential than the cell, so water will enter the cell by osmosis down a water potential gradient.

4 Whether the fetus has the correct diploid number of chromosomes. Triple 21 leads to Down's Syndrome. See if there are damaged chromosomes present, or whether the fetus is male or female.

Pregnancy testing

1. The shape of the antibody binding site is specific for hCG and not any other substance.
2. Antibodies are too small to be seen. However, a cluster of antibodies, each with a latex particle attached, can be seen on the dipstick

1.6

1. The fetus receives antibodies, but does not make memory cells itself. Immunity only lasts while the antibodies are present in the fetus's body.
2. As the baby's suckling stimulates her nipple, the pituitary gland is stimulated to release oxytocin. Causes contraction of the milk ducts and contraction of the muscles in the uterus wall.

Wet nursing

1. The more the nipple is stimulated by suckling, the more milk is produced. The wet nurse will produce more milk.
2. TB bacteria may be present in the mother's blood and pass into the milk she produces.
3. Better for a baby, nutritionally, than formula milk; if the wet nurse is healthy there is no risk of disease, baby benefits from some passive immunity; it comforts the baby. Against: the wet nurse may pass on diseases; reduces the special bond between a mother and her baby.

Breast milk

1. Apart from nutrients, it contains antibodies that help the baby to fight disease.
2. As the baby grows, it feeds more. This provides more stimulation to the nipple. More nerve impulses pass to the hypothalamus, causing the pituitary gland to secrete more prolactin.
3. Two babies feeding provides more stimulation to the nipple. As a result, more nerve impulses pass to the hypothalamus, causing the pituitary gland to secrete more prolactin.

Breastfeeding could slash breast cancer risk

1. Socio-economic class might indicate the income level of the woman, which might affect type of diet eaten. Women with a higher educational level might be more inclined to breast feed their children as they may be more aware of its benefits. Such women might have breast feed for a shorter time and/or have smaller families as they may wish to have a career.
2. It combines a large number of studies, so the sample size is very large and includes women from many countries. However, when studies are combined, this may mean that each study was carried out a little differently, so combining the findings may not be reliable. Although many factors were considered, it is not possible to look at each factor in isolation.
3. High prolactin or high progesterone levels reducing oestrogen, which can stimulate the growth of some kinds of breast cancer; pregnancy could bring about structural changes in breast tissue that protect against cancer; carcinogens that accumulate in breast tissue might somehow be broken down or expelled through breastfeeding.

1.7

1. After childbirth the woman's uterus and cervix may be larger. This means her old cap may not fit properly, and if it is not a good fit it will be ineffective as a contraceptive.

2.

Method of birth control	Advantages	Disadvantages
combined pill	Very reliable contraceptive May reduce menstrual cramps and reduce bleeding	Does not protect against STIs Slightly increased risk of thrombosis
male condom	Reliable and protects against STIs	Not quite as reliable as the pill, although still reliable
IUD	Once in place, does not need to be thought about Very reliable	Some people do not like the idea that it may prevent implantation of an embryo In a few women, it may cause menstrual cramps, heavy bleeding or may be expelled by the uterus Needs to be fitted and removed by a medical practitioner
cap		Needs to be fitted before sex and left in place for some hours afterwards Does not protect against STIs

There are other acceptable answers that could also be offered

3. After ovulation and during pregnancy, progesterone is already high.

Relative effectiveness of different birth control methods

1. Couples may be of different ages and levels of fertility; may have sexual intercourse more or less frequently than each other; may not all be as careful in using birth control as each other.
2. People who become pregnant while using the method are asked whether they used the method properly or whether they could think of a reason why the method failed. A failure of birth control due to human error like this would not count towards theoretical effectiveness figures but would be count towards the actual use figures.
3. The woman may not have had a new cap fitted, so the old one may not fit properly. The cervix may be larger after she has had a child. May be harder to fit the cap properly over the cervix when it enlarges following a baby's birth.
4. Condoms protect against sexually transmitted diseases, but the pill does not.
5. The pill is not effective unless taken at the same time each day. If the woman has a bout of sickness or diarrhoea, the hormones in the pill may not be absorbed properly.

1.8

1 FSH (follicle stimulating hormone)

2 causes ovulation

3　**a** Separates the sperm from other components in the semen, reduces the chance of passing on infections such as HIV.

　b This allows changes to take place in the membrane surrounding the sperm head. As a result, the acrosome reaction can occur and the sperm is able to penetrate the oocyte.

4　**a** Invasive process that takes place under sedation or even general anaesthetic. Doctors wish to collect as many oocytes as possible in one session, so that the woman should not need to go through this procedure more often than necessary.

　b If more than two embryos develop in the uterus, there is a lower chance of a successful pregnancy.

Youthful 'nurse' cells could restore male fertility

1 Sertoli cells nourish the developing spermatids, lack of them means that developing sperm do not gain enough nutrients to mature. A lack of Sertoli cells may mean that developing spermatids are destroyed by the immune system.

2 Mature cells, have differentiated and become specialised. This means that some genes have been 'switched off'. As a result, the cell cannot undergo mitosis.

3 Two large groups of men, of similar age, lifestyle, health, etc. Tissue samples from all the men should be examined to find the number of healthy Sertoli cells in a specific amount of tissue, sperm count could also be taken. One group would have their hormones suppressed for a fixed period of time, followed by FSH treatment. Other group would not have any such treatment. (Probably a placebo should be used.) After this time, the men's Sertoli cells should be examined and the number of healthy Sertoli cells in a specific amount of tissue should be counted. A sperm count might also be taken.

4 While Sertoli cells are dividing, they are not protecting developing spermatids from the immune system. It is possible that cells of the immune system might cause some damage to the seminiferous tubules or developing spermatids during this time.

1.9

1 There are various answers here: students may suggest destroying them, offering them to other infertile couples, or using them for research. They may also suggest that the parents have the right to decide the fate of the embryos.

2 There are no right or wrong answers here, so discussion should be encouraged.

The Diane Blood story

1 and **2** Any sensible reasons for viewpoints.

Who should get IVF on the NHS?

1 Any sensible reasons for viewpoints.

2.1

1 Growth of pubic hair; growth of axillary (underarm) hair; increased activity of sweat glands and sebaceous glands.

2 Visible sign that the male has reached sexual maturity and would therefore increase his chance of mating.

3　**a** Makes it easier to give birth to a big-brained infant.

　b Provides an energy store for mother and fetus should there be a shortage of food during pregnancy.

4 Data combine figures compiled by different researchers, whose methods may not be the same; do not know the sample sizes involved; women who are asked their age at menarche some years after the event may not remember the date accurately.

5 Between dates, there has been a better food supply, children are better-nourished. Social conditions have improved over this period and children are less likely to suffer from serious illness in childhood, which could slow their growth and development. Since 1940, medical care has also improved, children are less likely to suffer from serious infections during childhood.

6　**a** Growth rate is highest at birth, then drops steeply until the child is 2/3 years of age. It continues falling more gradually until the age of puberty. Here it increases sharply again, before falling to 0 when puberty is completed.

　b Growth spurt at puberty starts earlier in girls than boys/starts at about 11 years in girls and 13 years in boys; growth spurt at puberty is greater in boys than girls/ boys stop growing later than girls.

Acne

1　**a** hydrolysis　**b** glycerol

2 Any sensible argument.

2.2

1 $kJ\,kg^{-1}\,hour^{-1}$ (accept any energy unit per mass per time unit)

2 Breathing, pumping blood round the body, moving food along the gut, or active transport across cell membranes.

3 One might be taller and thinner, so would have a larger surface area to volume ratio over which to lose body heat. Alternatively, one man might have a higher proportion of muscle tissue. Muscle tissue is metabolically active and so would use more energy in respiration.

4 Women have a more body fat and less muscle than a man of the same age and body mass, so are likely to have a lower BMR.

Caring grandmas explain evolutionary role of menopause

1 A woman with an allele for menopause would pass the allele on to her daughters. The woman would be able to help her daughter rear her children, as she would no longer have children of her own. These grandchildren would also have the allele for the menopause, and they would be more likely to survive to pass on the allele, since they had a grandmother to care for them.

2 In early human society, it is unlikely that men would have taken much responsibility for child-rearing. Men who survived to a great age might have characteristics that enabled survival, such as great strength, resistance to disease or intelligence. Old men would be more likely to have desirable genetic qualities. The equivalent of the 'menopause' in men would have been a disadvantage for the population.

2.3

Prostate cancer screening

1 Can help scientists to work out the factors that increase the incidence of disease.

2 Screening may worry men who are otherwise well; most men with prostate cancer do not die from it because it is slow-growing; screening might enable men with the fast-growing type of prostate cancer to be identified early enough for treatment to be successful; screening is much less common in the UK than USA yet prostate cancer deaths are not very much higher than the USA.

3 Ultraviolet radiation; chemicals in cigarette smoke; or diesel exhaust.

4 Men have a very small amount of breast tissue behind the nipples.

5 If the cancer is caught early, it is likely not to have spread, or metastasised. Better chance of getting rid of all the cancer cells.

6 Breast cancer is less common in women under 50; may cause anxiety in well women; it is harder to detect; the cost of screening may not be balanced by the number of cases of cancer detected; women over 70 are currently excluded from the screening programme, although the incidence of breast cancer increases with age; screening may save costs because cancer can be cured easily when caught early, whereas long-term treatment is very expensive.

2.4

1 Although dementia can be diagnosed, it is not possible to see the tangles and plaques associated with Alzheimer's disease unless the brain is examined at post-mortem.

2 Loneliness, through being unable to leave the Alzheimer's patient alone; loss of income and pension rights because of taking on a caring role; tiredness, because they have to be constantly on duty; frustration, as the person is less able to communicate; distress, when the patient may have mood swings and become angry with them, or unable to recognise them.

Warning over social care funding

1 Consider whether family members should be expected to offer care free of charge, or whether family members should contribute financially towards their relative's care. Consider paying more money in taxes, or cutting other areas of government expenditure to pay for this care.

2.5

1 Locus, Gene, Genotype, Homozygous, Allele, Phenotype, Heterozygous

2 There is no chance that they will have a child with CF, although all their children will be carriers of CF, i.e. they will be heterozygous. This is shown in the diagram below.

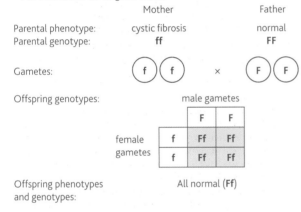

	Mother	Father
Parental phenotype:	cystic fibrosis	normal
Parental genotype:	**ff**	**FF**

Gametes: (f) (f) × (F) (F)

Offspring genotypes: male gametes

		F	F
female gametes	f	Ff	Ff
	f	Ff	Ff

Offspring phenotypes and genotypes: All normal (**Ff**)

3 Yes, he does have the condition. This is because it is a caused by a dominant allele, so it is present in the phenotype even if only one allele is present.

4 Half their children will have cleidocranial dysostosis. This is shown in the diagram below.

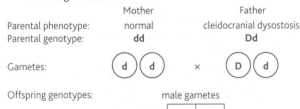

	Mother	Father
Parental phenotype:	normal	cleidocranial dysostosis
Parental genotype:	**dd**	**Dd**

Gametes: (d) (d) × (D) (d)

Offspring genotypes: male gametes

		D	d
female gametes	d	Dd	dd
	d	Dd	dd

Offspring phenotypes and genotypes: Cleidocranial dysostosis 2 (**Dd**) Normal 2 (**dd**)

2.6

1 Sickle-cell haemoglobin has a slightly different sequence of amino acids in two of its polypeptide chains, i.e. a change in the primary structure. Amino acids have different R-groups, so a change in primary structure means that the hydrogen bonds holding the secondary structure together may form in different places. Similarly, the bonds, such as hydrogen bonds and disulfide bridges that hold the tertiary structure together, may be formed in different places as different R-groups will be present.

2 No, they cannot have a child with sickle-cell anaemia, although half their children will have sickle-cell trait.

	Mother	Father
Parental phenotype:	normal	sickle-cell trait
Parental genotype:	$Hb^A Hb^A$	$Hb^A Hb^S$

Gametes: (Hb^A) (Hb^A) × (Hb^A) (Hb^S)

Offspring genotypes: male gametes

		Hb^A	Hb^S
female gametes	Hb^A	$Hb^A Hb^A$	$Hb^A Hb^S$
	Hb^A	$Hb^A Hb^A$	$Hb^A Hb^S$

Offspring phenotypes and genotypes: Normal haemoglobin 2 ($Hb^A Hb^A$) Sickle-cell trait 2 ($Hb^A Hb^S$)

3 Malaria is not present in northern Europe. A child with sickle-cell anaemia would die in early childhood in the days before advanced medical care was available. People with sickle-cell trait would still make some haemoglobin S, so although they do not suffer from sickle-cell anaemia, they would be at a disadvantage compared to people with all normal haemoglobin in a country where malaria was absent, especially when exercising. Therefore these people would be less likely to survive and pass on their alleles.

Cystic fibrosis gene protects against tuberculosis

1 People with two copies of the allele would die before they could reproduce and pass it on. This would reduce the frequency of the allele, as it would only occur in heterozygotes.

2.7

1 A rhesus positive father could have the genotype DD or Dd, but because he has a rhesus negative daughter we know he must be Dd and not DD, as a rhesus negative child must inherit one d allele from each parent.

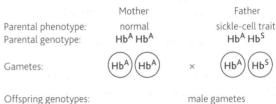

	Mother	Father
Parental phenotype:	Rhesus negative	Rhesus positive
Parental genotype:	**dd**	**Dd**

Gametes: (d) (d) × (D) (d)

Offspring genotypes: male gametes

		D	d
female gametes	d	Dd	dd
	d	Dd	dd

Rhesus positive daughter has this genotype

Rhesus negative daughter has this genotype

2 Red blood cells from the rhesus positive baby, carrying the D antigen, pass into the mother's blood. This antigen causes a specific B cell to become activated. This enlarges and divides to form a clone of plasma cells, which release specific antibodies against the D antigen into the blood.

3 A person with group AB does not have any antibodies in their plasma to agglutinate donor red blood cells, so they can receive blood of any of the ABO blood groups.

4 This blood has no ABO or rhesus antigens on its red blood cells, so it can be given safely to anybody.

5 The parents have a child who is group O, so this child must have inherited the I^O allele from each parent. Therefore we know that both parents are heterozygous for ABO blood group.

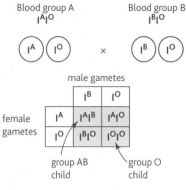

	Mother	Father
Parental phenotype:	Blood group A	Blood group B
Parental genotype:	$I^A I^O$	$I^B I^O$

Gametes: I^A I^O × I^B I^O

Offspring genotypes:

male gametes

		I^B	I^O
female gametes	I^A	$I^A I^B$	$I^A I^O$
	I^O	$I^B I^O$	$I^O I^O$

group AB child — group O child

6 The mother could have genotype $I^A I^A$ or $I^A I^O$, but we know she must be heterozygous as her child has group O, and must have received the I^O allele from each parent. The boyfriend with blood group B was the father. He must have been genotype $I^B I^O$ to have a child with group O.

	Mother	Father
Parental phenotype:	Blood group A	Blood group AB
Parental genotype:	$I^A I^O$	$I^A I^B$

Gametes: I^A I^O × I^A I^B

Offspring genotypes:

male gametes

		I^A	I^B
female gametes	I^A	$I^A I^A$	$I^A I^B$
	I^O	$I^A I^O$	$I^B I^O$

Note: this couple cannot have a child of blood group O

	Mother	Father
Parental phenotype:	Blood group A	Blood group B
Parental genotype:	$I^A I^O$	$I^B I^O$

Gametes: I^A I^O × I^B I^O

Offspring genotypes:

male gametes

		I^B	I^O
female gametes	I^A	$I^A I^B$	$I^A I^O$
	I^O	$I^B I^O$	$I^O I^O$

This child has blood group O

The discovery of ABO blood groups

1 Antigens on the animal's red blood cells cause the human's immune system to make specific antibodies. These antibodies bind specifically to antigens on the animal's red blood cells. One antibody has two binding sites, so one antibody may bind to two different red blood cells. Causes a number of red blood cells to be clumped together by antibodies.

2 I^A and I^B are codominant, while I^O is recessive. A person of group AB cannot have a child of blood group O, because the child must inherit one recessive allele from each parent.

2.8

1 A man passes his Y chromosome, not an X chromosome, on to his sons. DMD is on the X chromosome.

2 This is possible if a woman who carries DMD and a man with DMD have a daughter.

	Mother	Father
Parental phenotype:	normal (carrier)	Duchenne Muscular Dystrophy
Parental genotype:	$X^D X^d$	$X^d Y$

Gametes: X^D X^d × X^d Y

Offspring genotypes:

male gametes

		X^d	Y
female gametes	X^D	$X^D X^d$	$X^D Y$
	X^d	$X^d X^d$	$X^d Y$

Offspring phenotypes and genotypes:
Son with DMD 1 ($X^d Y$)
Normal son 1 ($X^D Y$)
Daughter DMD 1 ($X^d X^d$)
Normal daughter (but carrier of DMD) 1 ($X^D X^d$)

3 **a** Either 4 and 5 do not have colour blindness, but their child does; or 9 and 10 do not have colour blindness, but they have a child who does.

b Only males have the condition.

4 X^b for colour blindness and X^B for normal, 3 is X^B Y, 4 is $X^B X^b$, and 9 is $X^B X^b$.

5 They will not have any children with red-green colour blindness. See diagram. Individual 7 is a male without red-green colour blindness so his genotype is X^B Y and a woman with no family history of red-green colour blindness is likely to have the genotype $X^B X^B$.

	Mother	Father
Parental phenotype:	normal	normal
Parental genotype:	$X^B X^B$	$X^B Y$

Gametes: X^B X^B × X^B Y

Offspring genotypes:

male gametes

		X^B	Y
female gametes	X^B	$X^B X^B$	$X^B Y$
	X^B	$X^B X^B$	$X^B Y$

Offspring phenotypes and genotypes:
All normal

2.9

1 The woman has a 2/3 chance of being heterozygous for CF. We know this because both her parents must be heterozygous, since she has a brother with CF. However, even if she is a carrier of CF, she cannot have a child with CF unless her partner is also a carrier of CF. As CF is quite common in the UK, the couple may be offered genetic screening to see whether they are carriers of CF. If they are both carriers of CF, the genetic counsellor will probably tell them about embryo screening.

2 Different people will have different views about different genetic conditions. Views will depend on factors such as social background, financial status, religious and moral beliefs, other family members and other responsibilities .

3 Some women may feel that termination of a pregnancy is wrong, and although older women have a higher chance of a Down's Syndrome baby than a younger women, they are still very likely to have a normal baby. They may decide that, although Down's Syndrome is a disability, that a child with Down's Syndrome would have a full and satisfying life, and that it would be wrong to deny such a child the right to be born. They may also have had difficulty in conceiving a baby, so the small risk of miscarriage might be unacceptable to them.

4 **a** The individual is a female with Down's Syndrome.

b Prophase or metaphase (since each chromosome consists of two chromatids, and a spindle inhibitor was added so the cell could not progress further into mitosis).

5 **a** Chromosomes are not visible unless a cell is dividing.

b This stops mitosis going to completion, so as many cells as possible will have chromosomes visible inside them.

Saviour siblings

1 The new baby will have different DNA from the sick child's. The proteins on the cell surface that act as antigens will be different from the sick child's, so they would be recognised as foreign by T and B cells. Immune response mounted to the baby's stem cells would result in the stem cells being destroyed. An embryo produced using IVF and embryo screening would have key genes similar to the sick child's, so the antigens are also similar.

2 For: serious life-threatening condition, without the 'saviour sibling' the sick child would have died; the stem cells come from the umbilical cord, so the 'saviour sibling' does not have to undergo painful or life-threatening treatment; the family might want another baby anyway; the 'saviour sibling' is likely to be valued even more by the family because he has offered hope of curing its sibling's disorder. Against: the fate of the unused embryos; it is possible that the 'saviour sibling' was only wanted because they might cure their disorder; this treatment might be used by other people for more trivial reasons, like selecting for children with specific physical features; the cost of such treatment is very expensive and the NHS has a limited budget.

2.10

1 Let **H** be the allele for short hair and **h** be the allele for long hair. The long haired cat must be homozygous recessive, **hh**, and the short-haired cat must be **Hh** because it produces both long- and short-haired offspring. The null hypothesis is that long- and short-haired kittens are produced in a 1:1 ratio. The expected numbers of offspring would be 23 in each category.

Phenotype	O	E	O − E	(O − E)²	$\frac{(O-E)^2}{E}$
Long hair	21	23	2	4	0.075
Short hair	25	23	2	4	0.075
				Σ	0.150

This χ^2 value (with 1 degree of freedom) corresponds to a value of p of 0.7. This means that the observed results would be expected 70% of the time, and so the null hypothesis must be accepted.

2

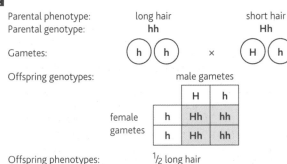

Parental phenotype:	long hair	short hair
Parental genotype:	**hh**	**Hh**

Gametes: (h) (h) × (H) (h)

Offspring genotypes: male gametes

		H	h
female gametes	h	Hh	hh
	h	Hh	hh

Offspring phenotypes: ½ long hair
½ short hair

2.11

1 Scientists are searching for specific DNA sequences. If there is a new mutation they have not met before, or a very rare mutation, this may not be detected by the test.

2 No, this is not a frame-shift mutation. This is because the three bases that code for one amino acid are substituted.

3 Tertiary structure of the protein depends on bonds formed between different R-groups, such as disulfide bridges and hydrogen bonds. Only some R-groups can form these bonds, so substituting a different amino acid with a different R-group means these bonds will form in different places. The tertiary structure of the protein will be altered. Most proteins rely on their specific shape for their function, so a protein with an altered tertiary structure is unlikely to function properly.

4 Cancer is caused when mutations occur in proto-oncogenes or tumour repressor genes. Therefore, anything that causes changes in DNA can also cause cancer.

2.12

1

Mitosis	Meiosis
a one-stage division	**a two-stage division**
produces daughter cells with the same number of chromosomes as the parent cell	produces daughter cells with half the number of chromosomes as the parent cell
homologous chromosomes do not pair – they remain independent of each other	homologous chromosomes pair together forming bivalents in prophase 1
chiasmata are never formed and crossing over does not occur	**chiasmata form and crossing over occurs during prophase 1**
daughter cells are genetically identical to the parent cell	**daughter cells are genetically different from the parent cell and from each other**
two daughter cells are formed	**four daughter cells are formed**

2 Homologous chromosomes have the same genes, but they may have different alleles from each other. Therefore not genetically identical.

2.13

1 continuous variation

2 Caused by several genes/polygenes

3 Children with low birth mass and high birth mass are less likely to survive than babies with an intermediate birth mass. Therefore, these babies of intermediate body mass are more likely to survive, and pass on their alleles in adulthood.

4 mean = 164, median = 175, mode = 175

5 SD = 15.85 or 15.9 to 3 sig figs. (see calculation below).

$$S = \sqrt{\frac{\Sigma(x-\bar{x})^2}{n-1}}$$

$$= \sqrt{\frac{3267}{13}}$$

$$= \sqrt{251.308}$$

$$= 15.85 \ (15.9 \text{ to 3 sig figs})$$

2.14

1 Genetically identical, must be the same sex, as sex is genetically determined.

2 Features such as height or body mass may be influenced by sex. Boys generally grow to be taller than girls.

2.15

1 Inherited a defective chromosome 15 from his father, but the brother and sister have each inherited a copy of their father's other chromosome 15, which is not defective.

2 The counsellor will tell them that individual 11 is a carrier of PWS, because he has the defective chromosome. However, he does not suffer from PWS because he inherited it from his mother. This means that every child born to the couple has a 50% chance of inheriting the defective chromosome 15. As this will be inherited from the father, this will cause PWS. Therefore, they should expect that 50% of their children will have PWS. They will also be told that this is a probability at any pregnancy. The counsellor will also tell the couple about genetic screening tests that may be available, and whether genetic screening of embryos or the fetus might be available to them.

Prader-Willi syndrome and obesity

1 May argue that Hastings Council should modify the house for Mr Leppard, as it is a genetic disorder that is not his fault. Could also offer his mother more support so that he has access to an exercise and strict dieting routine. Others will argue that Mr Leppard should eat less, and that if his mother did not bring unhealthy food into the house he would not be able to eat it.

2 Clearly Mr Leppard is not to blame for his genetic disorder, but if he stopped eating unhealthy, high-energy food, and took more exercise, he would not have such a serious obesity problem.

3.1

1 GCCAUUGAC

2 condensation

3 ACC (note that the tRNA anticodon is complementary to the mRNA codon)

4 DNA replication does not involve RNA; in DNA replication the whole molecule 'unzips', not just a short section.

Genetic code discoveries

1 tRNA, enzymes, ATP

3.2

Men's bad habits could affect the health of their sons

1 Stanozolol, like testosterone, will pass through the phospholipids in the cell membrane and bond to the specific receptor protein in the cytoplasm because it is similar in shape to testosterone. The hormone-receptor complex will then enter the cell nucleus and stimulate transcription of specific genes.

2 The DNA of monozygotic twins will be identical in terms of the DNA base sequence. However, as a result of environmental factors, one twin may have a different pattern of methylation or acetylation of the DNA.

1 Advantages: having large sample sizes and the fact that it allows us to compare the current population with their immediate ancestors. Disadvantages: the data was not collected specifically for the purpose of this study, so the information may not be exact enough; people were asked questions about themselves using a questionnaire, and they might forget details or deliberately mis-report information; the historical records link harvest records to the diet of individuals, and this may not be completely reliable.

2 Arguments for this could include the fact that smoking became much more common in the second half of the 20th century, so could have caused epigenetic effects. On the other hand, growing affluence, lack of exercise and access to 'junk food' could cause the increase in obesity and diabetes without any epigenetic effects.

3.3

1 **a** Herceptin binds to the HER2 receptors, preventing the growth factor from binding. Therefore the growth factor cannot stimulate cell division.

b Herceptin only binds to HER2 receptors, so it will not be effective against any kind of cancer that does not have these receptors on their surface.

Alcohol and breast cancer

1 Very few women develop breast cancer before age 35. After this, the risk increases gradually, rising more steeply after age 45-50 (when the menopause occurs). The risk of developing cancer increases for all women as they get older, but increases more rapidly the more alcoholic drinks consumed.

2 Increase in risk = (13.3 – 8.8)/8.8 x 100% = 51.1%

3 The data show that there is a correlation between drinking alcohol and developing breast cancer. It indicates that drinking alcohol increases the chances of developing breast cancer. However, it does not prove cause and effect. There may be another factor that is causing this increase. Students may suggest another factor, e.g. a dietary factor, or smoking.

3.4

1 Both parents must be heterozygous (Aa) as they do not have PKU but have a child with PKU, who must be aa. Therefore the chance that their next child will also have PKU is 25% or 1 in 4.

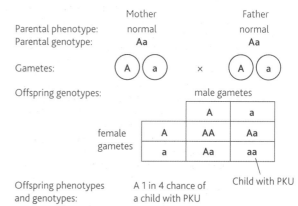

2 This prevents too much of a product being made, so that the cell does not waste energy or raw materials in making a product when it is not needed.

The Guthrie test

1 Any 2 points, e.g. sterilise the agar plate before use; sterilise the instrument used to inoculate the dish; use the lid of the dish as an 'umbrella' while inoculating plate; swabbing bench with disinfectant before and after transfer.

2 B-2-thienylalanine is present in the dish which will inhibit growth of the bacteria if the disc contains blood from a healthy person. However, a person with PKU has extra phenylalanine in it, which means that B-2-thienylalanine no longer inhibits bacterial growth.

4.1

1 Using an enzyme.

2 One for the beginning and one for the end – the base sequence of the DNA where DNA polymerase attached is different on each strand.

3 n cycles produce 2^n copies, so after 8 cycles there will be $2^8 = 256$ (i.e. the number of molecules doubles every cycle). Alternatively, students may work this out the long way:

Cycle number	1	2	3	4	5	6	7	8
Number of DNA molecules	2	4	8	16	32	64	128	256

4 Heating the DNA to 93°C to separate the strands would denature most enzymes.

5 The base sequence of the primers is specific to a base sequence found only in human DNA and not in bacterial DNA, so the bacterial DNA cannot be copied.

6 DNA polymerase from a bacterium is able to replicate DNA from a human (or any other organism) although it comes from a bacterium. This must mean that DNA in a bacterium is the same as DNA in a human or other organism.

Bio-prospecting

1 For: Cetus Corporation has made huge profits from Taq polymerase and can afford to give some money to the National Park which can be used to preserve the biodiversity there. Without the Yellowstone National Park, the Cetus Corporation might not have found this enzyme. We do not yet know what other profitable discoveries will be made from other species found in this, or other National Parks. Against: the idea of using Taq polymerase in PCR is the reason why the profits have been made, and the idea came from an employee of the Cetus Corporation; National Parks are supported by funds from national government for the benefit of everybody, so there should be no need for additional financial support.

2 Microorganisms show variation. Those which, by chance, have a mutation that gives them a more heat-stable enzyme will survive in the extreme environment while other microorganisms will die. The organisms that are best-adapted survive, reproduce and pass on their alleles to their offspring.

3 It is likely to have more disulfide bridges and fewer weak bonds, such as hydrogen bonds, in its tertiary structure.

4 Industrial processes often take place at high temperatures or extreme pH. These extreme organisms may have enzymes that work well in these conditions.

4.2

1 It cuts the DNA from the virus, so that it cannot be inserted into the bacterial DNA.

2 See table below

Name of enzyme	Action
Reverse transcriptase	Makes a single-stranded piece of DNA using an RNA template
DNA ligase	Joins two pieces of DNA together
Restriction enzyme	Cuts DNA whenever it recognises a specific base sequence

3 TAGGCTAC

4 It is slightly different in structure from the human protein, and is therefore seen as 'foreign'. This causes the production of specific T and B cells that will respond to this antigen.

5 TAC-ACC-ATA or TAC–ACC–ATG (note the table gives mRNA codons and you are asked for the DNA sequence)

4.3

1 DNA is invisible on a gel after electrophoresis has taken place. However, fluorescence or radioactivity means that the band of DNA which the probe has bound to can be detected.

2 Pieces of DNA are negatively charged because of the negatively charged phosphate groups.

3 Antibodies bind specifically to a particular protein. This specific antibody will only bind to one specific protein. This protein will only be present if the gene coding for it is present in the bacteria.

4.4

1 Gene technology only transfers one or two genes. Selective breeding may involve more genes than this; gene technology may transfer genes from one species to another, whereas selective breeding transfers genes from one member of the species to another; selective breeding takes much longer, as it relies on sexual reproduction; gene technology is more reliable, as the desired gene is isolated and transferred.

2 BST is a protein, so it will be digested by enzymes in the cow's stomach.

Milk yields

1 A group of similar cattle, of similar age and eating the same food, living in the same conditions, but not receiving BST would have been used. The milk yield for this (control) group would be compared with the milk yield for the group receiving BST.

2 Cattle receiving BST are likely to eat more feed, so it is important to check that the cost of increased food is outweighed by increase in milk yield.

3 There are various suggestions: the breed of cattle at each location might be different, and genetic differences between breeds might lead to different rates of increase in milk production; the type of feed given to each group of cattle might be different, and this could affect milk production; environmental factors might differ in the different locations, such as temperature, type of housing, etc. These environmental factors might affect milk production, as the cattle might use more/less of their energy for heat production, movement, etc.

4.5

1 Any arguments based on scientific fact.

2 The gene needs to be synthesised in a 'gene machine' after looking up the DNA codons required for the antigen's amino acid sequence. Alternatively the DNA sequence may be made from mRNA using reverse transcriptase. This is then placed into a vector such as a plasmid. The plasmid needs to be cut open using a restriction endonuclease, and the gene inserted using DNA ligase. The vector is then mixed with banana cells and taken up by these cells. A marker gene may be used so that banana cells containing the antigen gene can be identified. These modified cells can be used to grow banana plants that produce GM bananas.

Seed patents

1 Students may argue either that this is right or wrong – what is important is that they should be able to find reasons to support their argument.

2 As in Q1, it is the ability to support arguments with reasons that is being tested here.

Media campaigns against GM crops

1 A range of answers is acceptable here.

2 Students may gave a variety of opinions here, but should be able to support their ideas with reasons.

4.6

1 Results are available for everybody, therefore encouraging research that could benefit everybody; it reduces the costs of research, as researchers do not have to pay royalties to investigate patented gene sequences; it allows scientists to work together and share their findings.

2 A lot of the DNA does not code for proteins; we need to find the structure and function of the proteins coded for; some of the DNA is regulatory sequences; there are environmental influences on DNA also (epigenetics).

Low-cost personal DNA readings are on the way

1 There are many possible arguments here. Students may say that it is useful because scientists can learn about individual variation and assess the frequency of certain genetic conditions. It may also help scientists to find out how far lifestyle factors affect conditions for which a person has a genetic predisposition. However, the results may be difficult to interpret and once information is available on the web it is difficult to withdraw it.

2 A variety of responses are possible, but students should give reasons for their views.

4.7

1 Scientists need to identify proteins that act as antigens. Natural selection favours parasites that have antigens that are different from other parasites. This is because antibodies against one antigen will not be effective against a different antigen. This means that genes coding for antigens are likely to show a great deal of variation.

2 Adults who have been exposed to different strains of malaria will have antibodies against several different strains. If they then catch a new strain they will become ill but they are likely to have antibodies that are, at least partly, effective in combating the new strain. This means the new infection will be less severe. Children have a smaller number of different antibodies, so they may not have any antibodies that are effective at all. This means they are likely to have a more severe infection that could kill them.

3 Closely related organisms have DNA sequences that are similar, or show changes in very few nucleotides. The more differences in the DNA sequence, the less closely related the organisms.

4 Advantages: the child will grow up aware of some of the diseases and conditions that they may develop in the future, and may be able to avoid these by making suitable lifestyle decisions; there is early warning of any genetic conditions, which may be less severe if treated early. Disadvantages: the child is too young to give permission, so it is an invasion of privacy; the child may find out they are at risk of serious disease that may not be treatable; the child may become depressed at learning s/he is at risk of serious disease; the child may be unable to gain employment, obtain life insurance or a mortgage if they are at risk of serious disease.

5.1

1 Sensory neurone has one long dendron, while the motor neurone has many dendrites that join directly to the cell body. The cell body of a motor neurone is at one end of the cell, while the cell body of a sensory neurone is not at the end of the cell.

2 **a** eye, ear, **b** leg muscle, pancreas, diaphragm.

The discovery of neurones

1 Purkyne fibres in the heart; the Golgi body in the cell; and the bundle of His in the heart.

2 It stains some parts of cells more than others, allowing structures within the cells to show up.

5.2

1 The channel proteins are a specific shape because of their tertiary structure. Only the 'right' type of ion can fit through it.

2 This uses ATP to transport the ions against their concentration gradient.

3 These are needed for aerobic respiration to provide ATP for active transport.

5.3

1 **a** Muscle weakness/inability to contract muscle/walk.

b Loss of sensation from fingers.

Multiple Sclerosis

1 The myelin sheath on a person's neurones will only have self antigens. Normally, the immune system does not respond to self antigens – it only responds to foreign antigens.

2 This suggests that there may be a genetic factor that pre-disposes people to MS, as the concordance between identical (MZ) twins is much greater than for non-identical twins. However, most MZ twins do not develop MS even if the sibling does, so this suggests that the cause is more likely to be environmental.

5.4

1 Acetylcholine would not be broken down, and sodium ions would continue to enter the postsynaptic neurone. This would cause repeated action potentials to be generated in the postsynaptic neurone.

2 People with Alzheimer's disease do not produce enough acetylcholine, and this causes memory problems. Inhibiting acetylcholinesterase in these patients means that the limited amount of acetylcholine produced stays in place longer before being broken down, thus improving memory.

5.5

1 The transmitter substance is only produced in the presynaptic neurone, and the receptors for it are only found in the postsynaptic membrane.

2 This gives a larger surface area so that there are more receptors for acetylcholine. This allows more sodium ions to enter the muscle so that muscle contraction can occur.

Botox

1 diaphragm and intercostal muscles

2 If these muscles cannot contract and relax, the air in the lungs cannot be ventilated. The lungs will fill with increasing concentrations of carbon dioxide and the blood will not become oxygenated.

3 Yes: this kind of treatment makes people feel better; in the hands of qualified medical practitioners, it need not be dangerous. No: botulinum is a very dangerous toxin so we should not use it to treat 'trivial' conditions; it is very easy to give an excessive dose that can cause severe paralysis.

Poison darts

1 Batrachotoxin causes the neurone membrane to depolarise more easily, so action potentials pass along the neurone constantly.

2 The frog is brightly coloured to warn predators that it produces a poison. Predators will learn to avoid eating these frogs because of their toxic effects.

5.6

1 Nicotine increases blood pressure and heart rate, which are risk factors for heart disease.

2 Although anandamide and THC are similar in shape, they are not the same shape. The enzyme in the synaptic cleft that breaks down anandamide will not break down THC, because THC is not exactly the right shape to fit in the enzyme's active site.

5.7

1 They hold the eyeball in a spherical shape

2 They move the eyes from side to side so that you can look at different things.

Cataracts

1 Weak bonds holding the protein's tertiary structure together, such as hydrogen bonds, have broken. This means that the protein is now a different shape. As a result, the denatured proteins in the lens may tangle together and form clumps.

2 The cornea refracts light, and is still present, so a blurred image can form on the retina.

5.8

1 When you first enter the room, all the rhodopsin in your rods is bleached. However, after a few minutes it has re-formed and you are able to see more clearly.

2 When you look straight at an object, the image forms on the fovea. This part of the eye has very few rod cells, and cones cannot detect the low light intensity from the star. However, if you look at it 'out of the corner of the eye', the image forms elsewhere on the retina, where there are more rod cells, and so you are able to see the dim star.

3 These provide ATP to maintain the resting membrane potential and to re-form rhodopsin after it has been bleached.

4

Rods	Cones
contain the pigment rhodopsin	**contain the pigment iodopsin**
enables monochromatic vision	enable trichromatic vision
low visual acuity	**high visual acuity**
more numerous in the retina	less numerous in the retina
more sensitive to low light intensity	less sensitive to low light intensity

5 a They will see both red and green as similar in colour.

b The woman will carry red-green colourblindness, as she has inherited one of her X chromosomes from her father. Half of her daughters will carry red-green colourblindness, although none will have the condition. However, half her sons will have red-green colourblindness.

X^B = normal vision
X^b = red-green colourblindness

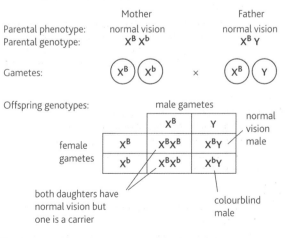

5.9

1 The stronger the stimulus, the greater the frequency of action potentials.

2 They would need accurate judgement of distance when swinging through trees, so that they could make sure they did not fall when leaping for the next branch.

6.1

1 The somatic nervous system carries impulses to skeletal muscles which are under conscious control, while the autonomic nervous system carries impulses to effectors that are not under voluntary control.

2 The ganglion contains the synapses between the pre-ganglionic and the post-ganglionic neurones. Neurones are wider at the synapse, so this causes a bulge in the nerve.

Atropine

1 The parasympathetic nervous system produces acetylcholine at the synapse with the effector. Atropine blocks the action of acetylcholine, therefore blocking the activity of the parasympathetic nervous system. The parasympathetic nerve to the iris would cause the pupil to constrict. Therefore, when atropine is used, the sympathetic nervous system is active. This causes the pupil to dilate.

6.2

1 The nervous system processes information more quickly; blood glucose levels are higher; oxygenated blood is pumped more quickly round the body.

2 Moving food along the gut is mainly a parasympathetic nervous system activity. This is suppressed when the sympathetic nervous system is active.

Khat

1 Cortisol levels in the blood of the baboons dropped when khat was taken, and dropped further afterwards.

2 This shows that khat reduces cortisol levels in the blood of baboons, and cortisol levels are raised when an animal experiences stress. However, this is just one measurement of stress.

3 Many possible comments can be made. Only five animals were used instead of a much larger sample; these were baboons, so the results may not apply to humans; there was no control group – a control group of similar animals should have been given 250 g of a different substance containing no active ingredients; the study only looked at khat use over 8 weeks; khat juice was used, whereas humans usually chew khat.

6.3

1 A temporary, reversible effect might be the effect of adrenaline on the bronchi of the lungs or on heart rate; a permanent and irreversible effect might be the effects of oestrogen or testosterone in developing secondary sex characteristics.

2 It takes time for changes to take place in cells. For example, steroid hormones need time to reach their target cells; then they need to get inside the cells and switch genes on or off; then it takes time for proteins to be produced.

6.4

1 These produce ATP during aerobic respiration. ATP is needed to provide the energy for muscle contraction.

2 This acts an as oxygen store, so it allows the muscle fibre to keep respiring aerobically even when the blood cannot supply enough oxygen directly.

3 A person with a spinal injury is unable to exercise to any great extent, so will carry out very little voluntary movement requiring sustained activity. However, a sprinter will still be carrying out sustained activity at times. Note that this graph does not show the total number of fibres present. A world-class sprinter will have more fast twitch fibres than a person with spinal injury. The graph just shows the proportion of the different kinds of muscle fibre present.

6.5

1 **a** actin, **b** actin and myosin, **c** myosin.

2 **a** gets shorter, **b** gets shorter, **c** stays the same.

3 When a person dies respiration no longer occurs, so calcium ions are no longer pumped out of the sarcoplasm. This allows myosin and actin to bind firmly.

7.1

1 Changes in temperature or pH cause weak bonds in proteins, such as hydrogen bonds, to break. This changes the tertiary structure of the proteins and means that specific protein channels and carriers in cell membranes might no longer be the right shape to transport molecules across membranes.

2

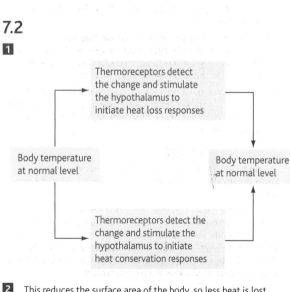

Blood pH falls/carbon dioxide concentration rises above normal level

Chemoreceptors detect change and send impulses to the respiratory centre and cardiovascular centres in the medulla. These send impulses to the intercostal muscles and diaphragm to increase the rate and depth of breathing, and send more impulses along the sympathetic nerve to the SAN to increase the heart rate.

Blood pH/carbon dioxide concentration at normal level

Blood pH/carbon dioxide concentration at normal level

Blood pH rises/carbon dioxide concentration falls below normal level

Chemoreceptors detect change and send impulses to the respiratory centre and cardiovascular centres in the medulla. These send fewer impulses to the intercostal muscles and diaphragm to decrease the rate and depth of breathing, and send more impulses along the parasympathetic nerve to the SAN to decrease the heart rate.

7.2

1

Thermoreceptors detect the change and stimulate the hypothalamus to initiate heat loss responses

Body temperature at normal level

Body temperature at normal level

Thermoreceptors detect the change and stimulate the hypothalamus to initiate heat conservation responses

2 This reduces the surface area of the body, so less heat is lost, especially by radiation.

7.3

1 These people probably died from hypothermia. In advanced hypothermia, the heart rate falls so low that blood flow is reduced and the person feels warm. The person also develops confusion. These people have felt too warm and removed their clothes. They were so confused from hypothermia that they did not realise this was inappropriate in the conditions.

2 This traps an insulating layer of air around the body. It also reduces air movement, cutting down heat loss from the body.

Fever

1 This may increase the rate at which the immune system responds to infection, by stimulating enzyme activity. It may also increase the rate of mitosis so that tissue repairs more quickly.

First aid treatment for hypothermia

1 This insulates the body, keeping in any heat generated in metabolism. This means that the person should warm up all the way through the body. Warming a person up quickly, e.g. by putting them in front of a gas fire, will warm up the outer part of the body quickly. This could result in heat loss responses, such as vasodilation and sweating, while the body's internal temperature is still very low.

2 The head has a large surface area so loses a lot of heat by radiation.

3 The warm drinks will help to warm up the gut and the blood supply to the gut. This helps to keep internal organs warm.

7.4

1

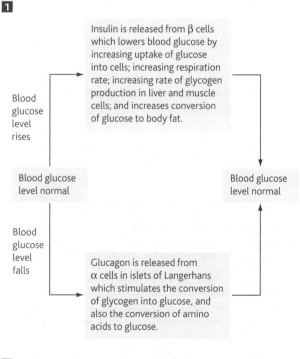

Blood glucose level rises

Insulin is released from β cells which lowers blood glucose by increasing uptake of glucose into cells; increasing respiration rate; increasing rate of glycogen production in liver and muscle cells; and increases conversion of glucose to body fat.

Blood glucose level normal

Blood glucose level normal

Blood glucose level falls

Glucagon is released from α cells in islets of Langerhans which stimulates the conversion of glycogen into glucose, and also the conversion of amino acids to glucose.

2 When a person is active their muscles have a higher rate of respiration. This uses up glucose.

3 It is insoluble so it does not affect osmosis. Glucose is soluble, so it lowers the water potential of the cell. This would cause water to enter the cell, and if enough glucose entered a cell, it could burst. Glycogen is insoluble so does not have this effect.

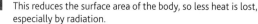

7.5

1 It produces human insulin, rather than pig (or other animal) insulin, so it is identical with human insulin; also human insulin is less likely to produce an immune response in the person being injected; there is less risk of disease as recombinant insulin is not taken from an animal; avoids ethical issues involved in using animals for insulin production.

2 a Exercise increases metabolic rate and rates of respiration in cells, lowering blood glucose levels.

  b Foods with a low glycaemic index release glucose slowly, avoiding a rapid rise in blood glucose level following a meal.

3 a The graph shows that people with a higher percentage of central abdominal fat tend to have a lower sensitivity to insulin than those with a lower percentage of central abdominal fat (although there is some variation).

  b There are various answers here and the following are points that might be valid. There is evidence here that people with more central abdominal fat have lower sensitivity to insulin, and this is a factor that can lead to type 2 diabetes. However, some people with 40-45% central abdominal fat have very low sensitivity to insulin, while others with similar abdominal fat have much greater sensitivity. This shows that another factor may well be involved. The graph shows a correlation, not cause and effect. The data relate to insulin sensitivity, not diabetes.

Testing for diabetes

1 So that the results are not affected by glucose in food or drink that has been consumed recently.

2 A non-diabetic person will produce insulin when the drink is consumed. This will cause liver and muscle cells to take up more glucose, lowering the blood glucose level.

Monitoring blood glucose levels

1 Only glucose will fit into the active site of the enzyme, glucose dehydrogenase.

2 It allows people to measure their blood glucose level several times a day, and at key times, e.g. after meals. This means that they can inject the right amount of insulin, and/or control their food intake to keep their blood glucose level within normal limits.

8.1

1 Population; all the organisms of the same species that occupy the same place at the same time and that have the chance to interbreed with one another. Phenotypic variation: variation in the observable characteristics of an organism. Mutation: any change to the quantity or structure of the DNA. Allele frequency: the frequency of an allele within a population. Evolution: a change in the allele frequency of a population over time, a gradual process in which species develop from pre-existing ones.

2 Genetic factors: crossing over and independent assortment during meiosis, random fertilisation. Environmental factors: competition for food/mates; disease.

3 A mutation that produced warfarin resistance in an individual allowed it to survive in areas where warfarin was used, breed and pass the allele on to its offspring. The offspring would also survive, increasing the frequency of the allele within the population.

Head lice forming a resistance movement

1 An organism that lives on or in a host organism and gains an advantage while causing harm to the host.

2 $\dfrac{8 \times 3000}{100} = 240$

3 to remove sampling bias

4 mutation

5 Pyrethrin is still being used in southern Wales so resistant head lice are surviving the treatment and susceptible head lice are being killed. This means there is an increased likelihood that resistant head lice will interbreed to produce resistant offspring. This will increase the incidence of head lice that are homozygous for the pyrethrin resistance allele (i.e. head lice homozygous for the pyrethrin resistance allele are selected for). In northern Wales, where the use of pyrethrin has been discontinued for 5 years preceding the study, head lice without any alleles for pyrethrin resistance are just as likely to survive and reproduce as the head lice that are homozygous for the pyrethrin resistance allele, so there is no selective advantage to the lice in being homozygous for the pyrethrin resistance allele in north Wales.

6 Recommend different lotions in different parts of the health authority and at different times; use non-chemical methods of treatment.

8.2

1 A small, circular, independently replicating piece of extra-chromosomal cytoplasmic DNA found in prokaryotic cells. Capable of moving from one one cell to another. Carry genes that are additional to those on the bacterial chromosome. E.g. antibiotic resistance genes.

2 There is little chance that any bacteria will survive. (The antibiotic will reduce the size of the bacterial population and the patient's immune system can destroy the remaining cells.)

3 It increases the likelihood of selection of resistant strains over ones that are more susceptible.

4 Resistant bacteria are at a disadvantage relative to those bacteria that are not resistant. Therefore, the resistant bacteria are out-numbered/ out-competed for nutrients/can't reproduce as quickly as the susceptible bacteria so the proportion of resistant bacteria (and hence genes for resistance) will decrease within the population.

Antibiotics prescriptions

1 Antibiotics kill bacteria by interfering with their metabolism. Viruses do not have any metabolism of their own so antibiotics will have no effect.

2 Antibiotics have provided a selection pressure. Any bacteria with resistance to antibiotics are more likely to survive, reproduce, and pass on their alleles to their offspring. Eventually, the bacteria accumulate so many alleles for resistance to different antibiotics that there are very few antibiotics that can kill them. This is how MRSA has developed.

3 The person will mount an immune response. Plasma cells will produce specific antibodies against the antigens on the surface of the bacteria causing the sore throat, and killer T cells will destroy bacteria with the specific antigen on their surface.

4 The very young or elderly may have weaker immune systems, that are less able to fight the infection. Also, they may have other health problems that might get worse if the respiratory infection is not treated as quickly as possible.

Birth weight in humans

1 This is the time period during which humans have had expanded brains. A baby with a large cranium may become stuck in the birth canal.

2 Antenatal care means that more low birth weight babies survive. Caesarean sections are readily available for mothers whose babies are difficult to deliver if, for e.g. a vaginal delivery (of a large baby) would put the health of mother and/or baby at risk. Today, the chances of baby surviving in a MEDC are high, regardless of birth weight, because of readily available medical care.

8.3

1. Gene pool: all the genetic information (genes) present within a population at a given time. Geographical isolation: the isolation of populations of a species by a geographical barrier, such as a river, so that they no longer interbreed. Prezygotic isolation mechanism: an isolation mechanism that prevents the formation of hybrid zygotes. Postzygotic isolation mechanism: an isolation mechanism that reduces the viability or fertility of hybrid zygotes.

2. Hybrid animals, like the mule, inherit one set of chromosomes from the gamete of one species and another set from the gamete of the second species. These two sets of chromosomes do not pair up or segregate regularly at meiosis, so viable gametes cannot be formed.

Spartina alterniflora

1. Spartina x townsendii is sterile and can only reproduce by vegetative reproduction.

2. Individuals of this species can breed with each other to produce fertile offspring.

3. When pollinated with pollen from another Spartina anglica plant, seeds are produced that will germinate to produce fertile offspring. When S. anglica plants are pollinated by S. alterniflora or S. maritima, no fertile offspring are produced. Spartina x townsendii plants are male and sterile, so will not produce seeds. Chromosome analysis to look for differences arising from a chromosome mutation.

Butterflies and evolution

1. sympatric speciation

2. prezygotic, behavioural

3. **Phenotypic variation** in wing colouration exists within the population. Wing colouration stimulates breeding behaviour. Within a population, butterflies with distinctive wing colouration will be stimulated to breed together successfully to produce offspring that carry alleles for this distinctive wing pattern. The offspring survive to reproduce and the **allele frequency** for distinctive wing colouration increases within the population. Butterflies in the same population with wing colouration that resembles that of a neighbouring population of a different species will be stimulated to breed with their neighbours to produce offspring that are hybrid and do not survive to reproduce. This results in **genetic isolation** of the two neighbouring groups and species divergence.

8.4

1. Yes. Evolution is defined as a change in allele frequency within the gene pool of a population. The process doesn't have to result in speciation. It can cause, for example, polymorphism within the population.

2. Many pest species have a very rapid rate of reproduction and their allele frequencies within their gene pools will therefore alter rapidly in response to changed selection pressures. Tigers have much slower rates of reproduction and produce far fewer offspring in a given time than many pest species. Changes in the frequency of alleles that may allow individuals to survive altered selection pressures are not great enough to prevent the their survival from being threatened.

3. Milder winters and warmer springs associated with climate change may be responsible. **Human impact** affects levels of CO_2 in the atmosphere, and the altered climate patterns resulting from this may be responsible for earlier breeding. **Phenotypic variation** within the populations of these organisms will mean that the individuals that are able to complete their reproductive cycles earlier in the year, coinciding with an earlier abundance of suitable food sources, will successfully raise more young. As a result of **natural selection**, the offspring of these individuals are more likely to survive than the offspring of individuals that have bred later in the spring, when food supplies may have become depleted. If the timing of reproduction is genetically controlled, the successful, early-breeders will pass on the alleles that allow earlier reproduction to their offspring. The **frequency of these alleles** will therefore increase within the **gene pool** of the population.

City song birds

1. Low frequency sounds, such as traffic noise in the city may prevent potential mates from hearing a lower frequency song. Higher frequency songs will be heard above the sound of the traffic and so potential mates will hear it and be attracted by it.

2. In an urban environment great tits with higher frequency songs are able to attract mates, breed and produce offspring. Birds with a lower frequency song will not be as successful at attracting mates in an urban environment and so will produce fewer offspring (there is a selection pressure for birds whose song can be heard above the traffic noise). The frequency of alleles allowing the birds to produce higher-pitched songs will increase in the urban environment.

3. Pre-zygotic, behavioural isolation, which could lead to sympatric speciation.

4. Students may devise many different investigations. Here is one idea: Hatch great tit eggs from birds that have bred in the city for several generations and birds that have bred in the forest for many generations in isolation in the lab. Measure the frequencies that these birds sing at. If the change in frequency of the city bird song is the result of altered allele frequencies, the isolated city great tits should sing at a higher frequency than the isolated forest great tits. If both types of isolated bird sing at the same frequency, measure how their song changes when exposed to the higher frequency song. It may be that the city birds inherit alleles giving them the ability to learn quickly, in which case it would be expected that their song would alter to match the high frequency song faster than the forest birds (genes and environment interacting). If both groups of birds begin to sing at the higher frequency at the same rate, then the effect observed in the city may simply be that the young birds learn form the birds around them, which mostly sing at the higher frequency (environment).

9.1

1. They pupate as chrysalises which don't feed, and emerge as adults in the summer.

2. Woodlice feed on decaying vegetation and wood, so they have plenty of food all year round. This could be investigated by measuring the size of a woodlouse population under a log, for example, at regular intervals throughout the year. The numbers should be relatively constant, as food supply and environmental conditions are fairly stable throughout the year.

3. Biodiversity refers to the variety of living things and all the places where they are found. It is the diversity of life, from genes to ecosystems.

4. An oak woodland contains many different kinds of habitat. This means that many different species of organism live there, each species having its own ecological niche.

The harlequin ladybird

1. Harlequin ladybirds are very effective predators of aphids and have a wider food range and habitat than the 7-spot ladybird.

2. Wider food range and habitat; have a longer reproductive period than most British ladybirds so their populations can increase in size relatively quickly; can disperse over long distances.

3. Two possible reasons for eradication (there are others): the harlequin ladybirds are out-competing native British ladybirds and therefore disrupting the ecosystems where native British ladybirds are found; harlequin ladybirds will also eat the eggs, larvae and pupae of native ladybird species, and the eggs and caterpillars of butterflies and moths and are therefore threatening the survival of populations of other species. Two possible reasons against eradication (there are others): the chemicals used to kill the harlequin ladybirds may also kill native ladybirds and other insects; eradication would cost a great deal.

9.2

1 Predators have a density-dependent effect on the population size. When the prey population is large, there is plenty of food for predators, and the size of the predator population increases. The increase in predation leads to a decrease in the size of the prey population. There is less food available for the predators so the size of the predator population decreases. The decrease in predation allows the population of the prey to increase in size and so on.

2 **Tree-felling** - the habitat of many insects, such as ladybirds and birds, such as owls will be lost, leading to increased competition (inter- and intraspecific) for nesting sites, food sources, shelter. This may reduce the size of the populations of these organisms. **Climate change** – patterns of bird migration may change, leading to changes in inter- and intraspecific competition for territories, mates and food. This may cause the populations of some organisms in the community to increase and others to decrease. **Introduction of alien species** – interspecific competition with existing community members will probably lead to the reduction in the size of the native population and an increase in the size of the population of the introduced species. This will disturb the dynamic equilibrium of the woodland community.

Chiffchaffs change their migration pattern

1 The Chiffchaff has to store up enough body fat to make the journey, and inter- and intraspecific competition for food may mean they do not store enough to cover the entire distance. They also have to avoid predators and adverse weather conditions.

2 a Chiffchaffs that have remained in Britain will be able to feed on the adult insects as soon as they emerge and will therefore be able to breed and feed their young. Birds returning to Britain may miss this food surplus all together, or may have to compete (intraspecific compeition) with other Chiffchaffs for a limited food supply. The migrating Chiffchaffs will have used up all their body fat reserves on their long journey. This means that they will be physically weak and so less able to compete with the Chiffchaffs that spend winter in Britain. This intraspecific competition will make it difficult for the migrating birds to establish territories, attract a mate, breed and successfully raise young.

b There will be greater interspecific competition for insect food sources in these ecosystems, which could threaten the survival of other species and make the ecosystem unstable.

3 Yes. Chiffchaffs that don't migrate are at a selective advantage over those that do, so their offspring are more likely to survive and pass on their alleles to their offspring, altering the allele frequency within the population.

9.3

1 **Predators (carnivores)** feeding on **prey**, e.g. barn owls eating voles. **Scavengers (carnivores)** feeding on animal remains, e.g. crows. **Parasites** feeding on/in a **host**, e.g. lice on hedgehogs. **Decomposers** (saprophytes), e.g. fungi and bacteria breaking down non-living organic material such as herbivore dung. **Detrivores**, e.g. earthworms feeding on detritus such as dead leaves.

2 Its ecological niche does not overlap with those of species that are already established within the ecosystem.

Reds versus greys

1 Interspecific competition for food between the grey and red squirrels.

2 To prevent them from becoming established in areas that are occupied by red squirrels because research has shown that the grey squirrels will displace the red squirrel population in an area within 15 years. Grey and red squirrels cannot co-exist because their ecological niches overlap.

3 If the threatened species is rare or has economic value some people may justify the killing of a species that poses a threat to it. Others may not be able to justify killing another living thing for any reason.

4 If grey squirrels can be repelled from red squirrel habitats, it may not be necessary to kill the grey squirrels.

5 A large population of grey squirrels will compete with other animals that eat nuts and seeds, reducing the size of their populations. The grey squirrels may prevent re-generation of woodland plant species if they eat large quantities of fruits, nuts and seeds. The grey squirrels may provide food for large predators such as foxes and birds of prey so their populations may increase in size.

A growing concern

1 Any reasonable answers: Yes, because this plant threatens native wildlife and buildings/surfaces. No, because there are more important things to spend money on such as alleviating poverty, improving health services and education.

2 Yes, because the stable relationships that have developed among native species over thousands of years are threatened. Ecosystem stability is important if biodiversity is to be maintained.

9.4

1 They have been able to occupy an ecological niche that is created by the urban environment. They can find plenty of food. They are safe from predators.

2 Habitat destruction; more people encouraging wildlife into their gardens in towns and cities so more food available; more homes built on the urban fringe, so the urban landscape is encroaching into the countryside so the wood pigeons don't have far to travel.

Felis catus on the prowl!

1 Mammals = 9915; Birds = 3449; Amphibians = 575; Reptiles = 144.

Calculations made as follows: $\dfrac{\% \times \text{total}}{100}$

e.g. for mammals $= \dfrac{69 \times 14370}{100}$

2

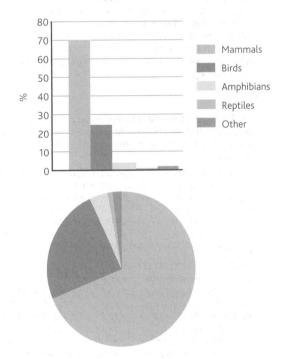

3 986 cats brought home 14370 prey items. 9 000 000 cats would therefore have brought home 14370 × 9 000 000 / 986 prey items = 131 160 000 prey items during the study period.

4. An underestimate. The results depend on the goodwill of the cat owners to respond to the questionnaire. There may be other cats that caught many prey items during the survey period but their owners did not respond to the questionnaire .Cats hunt at night and probably don't bring their night-time prey home because they are locked out. However, the survey was carried out in the spring and summer when cats are outside a lot and prey populations are high, so more prey are more likely to be caught than in the winter months.

5. Direct observation. This is more valid that a questionnaire because observations could be made around the clock and throughout the year, so night time and winter prey items could be included. However, the questionnaire did gather a lot of data. Direct observation could not produce results from so many different sources unless a lot of money was spent on the research project.

Flock evicted from favourite feeding ground

1. Public health issues; health and safety issues such as slippery surfaces; cost of cleaning up after them; effect of their droppings on historic buildings such as Nelson's Column. There are many more!

2. No, the birds will move on and find food elsewhere. Yes, these are domesticated animals and Trafalgar Square is their home.

3. Yes, the bird of prey is only scaring the pigeons. No, the bird of prey is likely to catch and kill some of the pigeons.

9.5

1. If a random **mutation** in the DNA of a plant pest species allows it to survive the Bt toxin, the pest may reproduce and pass the allele for resistance to its offspring. The offspring will also survive, the **frequency of the resistance allele** will increase within the population and the GM crop will be eaten by the growing Bt-resistant pest population. The Bt toxin is exerting a **selection pressure**. Populations of the pest living in the conventional crop will not be exposed to this selection pressure, so the frequency of the allele for Bt resistance in this population will probably remain low. Individuals of the two populations will be able to interbreed, preventing the production of a population that is homozygous for resistance to the Bt toxin.

3. The GM bacteria are contained within a sterile lab and are unlikely to survive outside that environment. Mosquitoes will be released into the environment, where their effects on populations of other organisms within the ecosystem can only be forecast and are not known for sure.

Farm-scale evaluations

1. The crop plants need 12 months to complete their lifecycles. Replicates need to be carried out to increase the reliability of the data.

2. For comparison. To allow the effects of the herbicide used to control weeds in the GM crops to be compared with the effects of the herbicide used to control weeds in the non-GM crops.

3. Based on the results of this trial, if a farmer is going to use herbicides on his crops, the best combination for wildlife would be: GM maize, non-GM beet and non-GM spring rape. However, the study did not allow a comparison of the effects of farming using chemical weed control with the effects of organic farming on wildlife.

Security for GM crop trials?

1. To evaluate whether GM crops can reduce the cost and environmental impact of farming, and whether GM variants will grow better in harsh environments where droughts have devastated harvests.

2. They do not want GM crops to be developed.

3. So that other farmers in the area are aware of the trial. If, for example, they are an organic farm, it is very important that their crop is not genetically polluted by genes (from pollen) from the GM crop, so they can lodge an objection to the trial, or take measures to try to minimise the risk of genetic contamination.

4. Public pressure, government pressure, lack of reliable field trial data because trials are invalidated by vandalism.

Genetically modified mosquitoes

1. The GM mosquitoes would interbreed with wild ones and over several generations, the frequency of resistance to *Plasmodium* within the population would increase. This effect would only be seen in areas where malaria is common because the GM mosquitoes only have a selective advantage over the wild type mosquitoes in these areas.

2. So that they can estimate the frequency of resistant individuals within the population.

3. *P. falciparum* infects humans and it would be unethical if humans were deliberately infected with malaria during the course of this research. It does mean that another organism has been infected with malaria during the course of this research and this also raises ethical issues.

4. The sterile mosquitoes could not transfer their genes for green testicles to their offspring, so no GM genes would be released into the wild. The resistant mosquitoes breed and pass their GM genes onto their offspring. The American research could lead to the replacement of a naturally occurring species with a genetically modified variant.

10.1

1. A substance, usually a protein, that produces an allergic reaction.

2. Hay fever: sneezing, a blocked or runny nose, itchy eyes, nose and throat, headaches. Asthma: coughing, wheezing, shortness of breath, tightness in the chest.

3. Allergies result from the interaction of genetic and environmental factors. The genetic make-up of the UK population has not changed significantly enough to account for the increase in the incidence of allergy in the last 50 years, which means that the change must be the result in an increase in environmental factors that trigger allergic responses.

Peanut allergy trials

1. Scientists already know that these babies have an increased risk of becoming allergic to peanuts at school age.

2. So that the data collected is reliable.

3. Children will be tested when they are five years old, so the trial needs to be at least 5 years long. The research team are unlikely to find 480 babies straight away, so there is a two-year recruitment period.

4. The babies who do not receive peanuts will be the control group, set up for comparison with the experimental group (the babies who are given peanuts). Scientists can find out which potentially allergic children in the trial actually develop a peanut allergy at the age of five. They can then look for patterns in the data which may help them to decide whether the children in the trial that were exposed to peanut snacks early in life are likely to develop peanut allergy or whether the children in the trial that were not allowed any peanut products are likely to develop peanut allergy.

5. Reasons for involvement could include: the advice given to parents about how to avoid peanut allergy is conflicting, so parents really don't know what to do; parents may feel their child has nothing to lose; parents may wish to contribute to a study that will provide valid scientific data on which to base sound reliable and helpful advice; their child is already known to be allergic. Being involved in this trial will give the families access to specialist medical support throughout the trial. Reasons against involvement could include: if the babies are randomly assigned to the two halves of the trial, parents may feel that the treatment goes against what their parenting instinct tell them is right for their baby; parents may not wish to/be able to regulate their child's diet according to the trial procedure.

6 Formula infant milk is a modern invention. The human immune system develops during early infancy. During early human evolutionary history, breast milk was the only form of nutrition available for human babies. Natural selection will therefore have selected for individuals that thrive on breast milk. Genes for tolerance of breast milk would therefore be widespread within human populations. Formula milk contains many ingredients that are 'new' to the human immune system, such as those in cows milk, and these may act as allergens.

7 Survey large numbers of children to gather data on how they were fed as babies and whether they have or have ever had food allergies or eczema. Carry out a statistical test for association. Students may also suggest a clinical trial, e.g. breast fed for 4 months/no peanuts between the ages of 4 months and three years; breast fed for 4 months/regular peanuts between the ages of 4 months and three years; formula fed for 4 months/no peanuts between the ages of 4 months and three years; formula fed for 4 months/regular peanuts between the ages of 4 months and three years.

Oilseed rape

1 Oilseed rape is a profitable crop because there is demand for its oil. Some is used to make extra virgin rapeseed oil, the rest is used to produce bio-diesel.

2 Oilseed rape flowers are insect-pollinated. Oilseed rape pollen does not get blown over large areas by the wind so it is unlikely to be the cause of hay fever. The types of pollen that are major causes of hay fever are from wind-pollinated plants, like grass.

10.2

1 A type of antigen that produces an abnormal immune response.

2 There is an over-reaction of their immune system when they are exposed to an allergen.

3 **In the nucleus of the B cell, mRNA is transcribed from the gene for the polypeptide chain**: A section of DNA in the nucleus of the cell 'unzips'. The hydrogen bonds between the DNA bases are broken by an enzyme. The two DNA strands separate and one strand becomes a template. Nucleotides of RNA, which are present in the nucleus, attach to the exposed bases of the template strand, by complementary base-pairing. The enzyme RNA polymerase joins the RNA nucleotides together to form a single strand of messenger RNA (mRNA).Once the mRNA has been formed, an enzyme 'zips up' the DNA molecule again. Pieces of the mRNA, called introns, are spliced out of the mRNA. These sections are not required to make the final protein. The mRNA leaves the nucleus via a pore in the nuclear envelope. **Translation of the mRNA molecule occurs in a ribosome on the rough endoplasmic reticulum. Codons in the mRNA molecule specify the sequence of amino acids in the polypeptide chain**: A ribosome attaches to the start of the mRNA molecule. Three bases on the mRNA form a codon that codes for one amino acid. Two codons fit into the ribosome at any one time. The tRNA molecules with the complementary anticodons to the mRNA codons enter the ribosome, and bind by complementary base-pairing. They bring their specific amino acids. The two amino acids carried by the tRNA molecules come very close to each other. They join to each other by a peptide bond. The first tRNA is now released from the amino acid. It leaves the ribosome and can now bring another amino acid of the same kind to the ribosome. The ribosome moves along the mRNA and a new tRNA enters the ribosome, bringing its specific amino acid. This continues until the ribosome reaches a stop or non-sense codon. This does not code for any amino acid. It acts like a 'full stop'. The ribosome, the last tRNA and the polypeptide chain all separate from each other.

4 Antigen binds to specific receptors on B cell ⟶ B lymphocyte releases IgE antibodies that are specific and complementary to the allergen ⟶ Antibodies bind to specific receptors on mast cells ⟶ Allergen attaches to specific antibodies on mast cells ⟶ Allergen-antibody complex results in the mast cell releasing histamine ⟶ Histamine produces the symptoms of the allergy

5 Anaphylaxis can involve oedema (swelling) in the airways leading to the lungs, or a large and sudden fall in blood pressure. The onset can be so rapid and severe that it can result in death by asphyxiation and/or lack of adequate blood circulation.

Anaphylactic shock

1 Exposure to wider range of foods in the diet/ medications/ treatments/ Increase in the number of environmental triggers to allergy during this period.

2

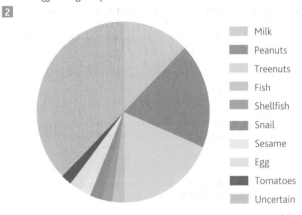

Legend: Milk, Peanuts, Treenuts, Fish, Shellfish, Snail, Sesame, Egg, Tomatoes, Uncertain

Reported cases of fatal anaphylaxis to food from 1999-2006

3 nuts (peanuts and tree nuts)

Anaphylaxis has resulted from an ingredient in the food that has not been identified. The food may have been home-made and not pre-packed, or the food item could be packaged but may not have been labeled or may have been incorrectly labeled.

10.3

1 If the patient has a risk of an anaphylactic (shock) reaction, skin prick testing may trigger anaphylaxis and would be considered dangerous. If a patient has extensive eczema, skin prick testing may be impractical. If antihistamine medication cannot be stopped because of the severity of the symptoms. If unusual and rare allergens are suspected

2 Histamine released from mast cells produces the symptoms of allergy during an allergic response. The histamine applied to the skin should also produce an allergic response.

3 The medication may interfere with the release of histamine which is needed to indicate an allergic response.

4 Adrenaline causes widening of the airways (bronchodilation) and raising of blood pressure through narrowing of blood vessels (vasoconstriction). Neither of these effects will relieve the symptoms of hay fever, but antihistamines will.

5 Histamine receptors are (glyco)proteins with specific shapes into which only histamine molecules fit.

Allergy advice for schools

1 Yes, because if an allergic child comes into contact with nuts and seeds it could be fatal. No, because children with nut allergies must learn to be vigilant about what they eat and they can only develop this behavior if there is the potential for them to be among people who are consuming nuts and seed. No, because people should have the right to choose what they eat. No, because you cannot guarantee a nut-free environment. Children with nut allergies may not be vigilant because they don't realise they may be at risk from nut-containing foods that could have been sneaked in to school against the rules.

2 Yes, because their allergy is life threatening and the onset of symptoms can be very rapid and severe. No, because they may use the epipen even if they experience mild symptoms. No, because there should be trained adults at home and at school who can respond as soon as anaphylaxis is suspected.

10.4

1 alpha-1-antitrypsin

2 Levels of ground level ozone may be high and if inhaled this can cause irritation of the walls of the bronchi and bronchioles, leading to a worsening of symptoms.

3 Even though a correlation may exist between these two variables, it cannot be concluded that vehicle emissions cause asthma. There are many risk factors associated with the development of asthma, and it is possible that high local traffic density exposes people to other factors that could increase the risk of developing asthma. For example, if traffic volume is high, windows in the home could be kept shut to keep down noise/keep out vehicle emissions, and levels of potential allergens in the house could become very high because of lack of ventilation.

Traffic pollution

1 chronic bronchitis, emphysema

2 Factors such as income could affect the health of the individual. For example, children from lower socio-economic groups may be more prone to lung infection during childhood than children from higher socio-economic groups because they may not receive a healthy balanced diet. It may be this difference that is causing the reduction in lung function in a child living close to a road rather than the distance from the major highway. The researchers have tried to control this variable by, for example, selecting children from similar socio-economic backgrounds for their research.

3 It may be that respiratory infections between the age of 10 and 18 (when the lungs are undergoing development) have a more serious effect upon lung development than those suffered up to the age of 10. Alternatively, it may be that up to the age of 10 both groups suffer similar numbers of respiratory infections (and so have similar reduction in lung function), but by the time the children have reached the age of 18, children living close to major highways will have suffered more lung infections than their counterparts, and so have suffered greater reduction in lung function.

10.5

1 Gastroenteritis and ear and respiratory tract infections.

2 Coliform bacteria are found in the intestine of humans and other endothermic animals, and their presence in bathing water indicates that the water is polluted by faeces.

3 To provide an accurate, up to date assessment of the quality of the coastal destination and to encourage management practices to be appraised annually.

Beach award schemes

1 Total coliform 500/100 cm³, Faecal coliform 100/100 cm³, Faecal streptococci 100/100 cm³.

2 MCS recommended beach

3 Some schemes, e.g. Blue Flag, may be very expensive to implement so only large coastal destinations may be able to apply. The authorities operating the resorts may be more encouraged to improve the quality of their coast if there are several schemes that they can choose from, rather than just one.

4 Resort too small to be able to meet some or all of the criteria, beach too remote to be called a resort, beach lacks facilities, coastline unmanaged.

10.6

1 Epithelial cells from the small intestine

2 Vomiting/diarrhoea may cause dehydration. Insufficient T helper lymphocytes to bring about destruction of the parasite.

3 Oocysts leave the gut of an infected host in the faeces. The *Cryptosporidium* causes severe diarrhoea, increasing the likelihood that oocysts will be released from the gut of one host to infect another.

4 They are very small and resistant to the levels of chlorination used in water treatment.

5 From the faeces of hill sheep and other mammals such as rabbits.

6 In healthy people infectd with *Cryptosporidium*, cells of their immune system help to destroy the parasites.

Rabbit threatens the health of 250 000 people

1 To check that the water entering the water supply network are below the *Cryptosporidium* alert level of 0.1 oocysts in 10 litres.

2 To destroy the oocysts.

3 There may be residual water that is still contaminated with oocysts in the system and the 8 days should be sufficient time for it to be flushed out.

4 To identify and eliminate the source. If the patients were infected with a strain of *Cryptosporidium* that was different from the strain found in the water and carried by the rabbit, then the rabbit and the drinking water could not be the source of the infection, and an alternative would have to be tracked down.

5 Cleaning out the water distribution network/ Alerting the public to the risk/ Alerting Health Authority to the risk/ Testing water and faecal samples for *Cryptosporidium*/ Genetic analysis of *Cryptosporidium* from water and faecal samples/ Treating patients who develop severe symptoms/ Writing reports/ Rabbit-proofing the water distribution network!

11.1

1 pioneer community

2 The pioneers decay, forming shallow soil in which other colonising plants can take root. This increases the number of ecological niches that are available.

3 The total number of species increases, then declines slightly. The gradual increase occurs because the biomass (living material) that accumulates throughout the process of succession can increase the numbers of niches available for plants and animals and can increase the depth and nutrient content of the soil. The slight decrease occurs because the dominant plant species in the climax community can out-compete other plant species for light and water, so species diversity decreases slightly.

4 At the beginning of the Neolithic period, most of Britain was covered by deciduous woodland. About 5,000 years ago, early farmers began to clear areas of woodland for the cultivation of crops and for pasture for domesticated grazing animals such as cattle. The wood would be used for fuel and the construction of buildings and barriers and the leaves would be used for animal feed. The clearance of woodland for agriculture has continued, and grazing and ploughing prevent the climax community developing on agricultural land. Woodlands are also cleared to make room for housing, roads and commercial/industrial developments.

11.2

1 Wasteland can provide a mosaic of different, often unusual habitats in the early stages of succession, helping to increase biodiversity. Wasteland sites can become colonised by rare or exotic species that are not usually found in the geographic area.

2 An area that has previously been developed for human use, e.g. a former factory site.

3 They offer a wide variety of habitats, important to the survival of many different species. The habitats they contain may be rare or threatened by urbanisation or intensive agriculture.

4 Flower-rich grasslands may develop on a brown-field site. Intensive farming has led to the loss of these biodiverse grasslands from the countryside and many of the species that depend on this type of habitat for survival have found refuge in brownfield sites.

5 Lots of ideas here: recreate the grassland on the roof of the building (a brown roof); retain some of the brownfield site at the development and work with local wildlife experts to manage it for wildlife; ask the developers to find another brownfield site in the area that contains suitable habitat and can be managed for wildlife. These sites could be made into nature reserves so that they are carefully managed and well-used by wildlife and people.

Butterflies use the railway

1 They link habitats together, allowing wildlife to reach new sites.

2 Their existing habitats may be destroyed by human activity, or may become unsuitable as the climate changes.

3 If left unmanaged the embankment would undergo succession, the soil would become deeper and the wild thyme would be out-competed for light by taller plants that could colonise the area. This would mean that the food plant of the Large Blue caterpillars would become scarce and the long-term survival of the species at this site would be threatened.

Brown-field sites and development

1 To reduce urban sprawl by in-filling previously developed land, rather than building on greenspace land beyond the urban fringe.

2 Many brown-field sites are associated with antisocial behaviour, such as fly-tipping, drug-abuse, motor-cycling and joy-riding. They are also considered by some to be eyesores.

3 Brown-field sites contain many different types of habitat, many of them in the early stages of succession. Many greenspaces contain relatively few types of habitat.

4 Brown-field sites are easy to identify from records of land-use. If a 'biodiversity-first' approach was to be adopted, the biodiversity of all potential development sites would have to be accurately measured to assess their biodiversity value, and this may make the planning application very expensive. A biodiversity first approach may also meet with public opposition as it could lead to the spread of towns and cities into 'greenspace', which may be less diverse than a brownfiled site, but may be more visually appealing.

11.3

1 The number of species present approximately doubles. This is because the numbers of habitats and niches within the ecosystem will increase as the area increases.

2 Increase the area of the nature reserve and work with other organizations to provide wildlife corridors that will allow the plants and animals in the reserve to migrate into new areas in response to climate change. The wildlife corridors will, in effect, produce one large habitat for wildlife. Scientific evidence suggests that the numbers of different species present in an ecosystem doubles for a ten-fold increase in area, so this strategy will increase biodiversity and improve the prospects for the long-term survival of the wildlife in the reserve.

Living landscapes

1 Climate change may alter a species' habitat and it will have to migrate to find a suitable habitat elsewhere.

2 Core areas of high quality habitat/ Inter-linkages between core areas/ Land between core areas and its connections that allows wildlife to move through it.

3 a green space to enjoy with wildlife on the doorstep, b migratory routes to suitable habitats that can be colonised, providing a larger area for wildlife in general.

11.4

1 Succession will change their habitats and they many no longer be suitable for an organism that requires an open habitat. For example, species-rich grassland may become colonised by shrubs, and the grassland plants may not receive sufficient sunlight to survive. The site may be managed by clearing the shrubs from some or all of the area.

2 Samples are taken because it is impossible to identify and count every organism present within a particular habitat, so instead, samples are taken which are thought to be representative of the whole habitat. The samples are random because this ensures that every organism in the sampling grid has an equal chance of being sampled, helping to remove sampling bias.

3 a Divide the study area up into grids with co-ordinates. Use a random number table to generate co-ordinates. Place a frame quadrat at the intersection of the co-ordinates and count the total number of bee orchid plants within the quadrat. Repeat this procedure many times so that a mean number of bee orchids per unit area can be calculated. Estimate the area of the site and multiply this by the numbers of bee orchids per square metre to give a population estimate for the whole site.

b Repeat the population sampling every year for five years.

c Abiotic factors affect the growth of the plants. Any change in the size of the population of bee orchids could be associated with a change in abiotic factors. This may mean that the management of the site needs to be altered.

11.5

1 Pitfall traps placed on a grid system using random numbers. At least 10 traps would be set up in each area to increase the reliability of the data.

2 Repeat the sampling exercise many more times in different parts of the same woodland and in other woodlands to gather as much data as possible. At present, any conclusions they draw are based on one survey only.

3 Chi squared, because the sampling produced frequency values, not measurements.

Sampling site	O	E	$O-E$	$(O-E)^2$	$\dfrac{(O-E)^2}{E}$
A	32	22	10	100	4.55
B	26	22	4	16	0.73
C	21	22	−1	1	0.045
D	9	22	−13	169	7.68
Σ 88		Σ			13.01

With 3 degrees of freedom, the calculated value of χ^2 (13.01) is larger than the critical value of 7.82 given in the table. The calculated value of χ^2 is greater than the critical value at $p = 0.05$ so the null hypothesis is rejected and we can conclude that there is a significant difference between the observed and expected values, i.e. there is a significant difference between the numbers of moths trapped in the three sites.

11.6

1 People who are more prosperous tend to have more 'stuff', make more waste and may value things less.

2 The BPEO is the option which provides the most benefit or least damage to the environment as a whole, at an acceptable cost in both the long and short term.

3 The waste hierarchy sets out the main options for the management of waste. It is the primary tool for assessing the 'Best Practical Environmental Option' (BPEO) for waste management. The options within the waste hierarchy are presented on a sliding scale with the most sustainable option first (reduction) and the least sustainable option last (disposal).

4 Incentives: more collections of recyclable waste; rewards for recycling, e.g.vouchers or a reduction in council tax payments for residents of authorities that meet their recycling target. Penalties: introduce a threshold for non-recyclable waste collection and charge households that go above the threshold (this would require some sort of waste metering).

Recycling in the UK

1 There was more recycling during this period.

2 East England (34%)

3 Northumberland and London

4 It may depend upon how active the local authority is in promoting recycling by providing recycling facilities and schemes, such as kerbside collection of recyclable waste.

11.7

1 The Polluter Pays Principle states that the polluter pays for the direct and indirect environmental consequences of their actions.

2 Landfill tax is an environmental tax paid on top of normal landfill rates by any company, local authority or other organisation that wishes to dispose of waste in landfill.

3 The EU Landfill directive requires biodegradable municipal waste in England be reduced to 5.2 million tonnes in 2020. This means that less organic waste will be sent to landfill, and so the amount of landfill gas and hence electricity produced from it will also reduce.

4 Incineration (burning the waste): advantage – can produce heat or use the burning waste as a fuel for electricity generation; disadvantage – hazardous air pollutants can be produced. Anaerobic digestion (waste is decomposed by microorganisms in an enclosed oxygen-free atmosphere): advantage – methane gas is produced which can be used as a fuel, a small volume of solid waste is produced that can be used as fertiliser; disadvantage – UK can only deal with a small amount of waste by this method at present, but it is set to increase. Composting (vegetable and garden waste is broken down aerobically to form compost. Has been carried out on domestic scale for many years. UK government want larger scale composting to go ahead): advantage – kerbside collection of garden waste already widespread in the UK; disadvantage – probably none.

Taxing waste disposal

1 **Advantages**: The polluters pay i.e. the people who produce most waste pay for their own waste disposal, rather than everyone sharing the cost; people don't like being charged again for something they already pay tax on, so they may be encouraged to seek an alternative, like recycling. **Disadvantages**: people don't like being charged again for something they already pay tax on, so they may be encouraged to seek an alternative, like fly-tipping; it would be very difficult to police, especially if there was nothing to stop one household putting their waste in a neighour's bin.

2 Reasons to support answers could be those given in answer to Q1. Another good reason for supporting pay as you throw may be the evidence of its success in Europe, as given in the passage.

12.1

1 The carbon footprint is a measure of the impact that human activities have on the amount of greenhouse gases produced, measured in terms of kilograms of carbon dioxide produced per year.

2 Any reasonable answer: using a computer; using public transport.

3 The UK hopes to achieve an 80% reduction in carbon emissions by 2050. A zero carbon building will pay back the carbon invested in its construction through exporting zero carbon energy back into the national grid, helping to reduce emissions of CO_2.

4 Carbon-offsetting means paying someone to make a greenhouse gas saving on your behalf. It may not be practical or possible to reduce the size of the carbon footprint associated with a particular process, so carbon-offsetting attempts to compensate for these unavoidable emissions.

5 Any reasonable answer: **Primary**: transport to the store or energy consumption by the computer if bought on-line. **Secondary**: Extraction, processing and transportation of raw materials for production of disc and case to the manufacturer; manufacture of blank disc, case and label (lighting and heating the factory, manufacturing the disc, case and label, transportation of all personnel to and from production site); production of track/film for DVD (lighting and heating the recording studio, recording the disc, transportation of all personnel to and from production site); advertising the disc, distributing the disc, retailing the disc.

12.2

1 Ecological succession is the gradual process by which ecosystems change and develop over time. Nearly all ecosystems on earth are influenced by climate change.

2 **a** Over the last 60 years, the comma butterfly has expanded its range northwards in the UK; Little Egrets are now seen frequently in southern Britain, since their arrival from the Mediterranean around 20 years ago.

b Many chiffchaffs still make the long journey south for the winter, however observation show that some are staying in Britain and not migrating at all. Milder winters in Britain may mean that there is plenty of food to support some chiffchaff populations, without the need to make the dangerous journey south for the winter.

Climate change and succession

1 Dormice need linked woodland to allow them to move between habitats. Many of these links have been lost as land is used for agriculture and developed.

2 If there is a decrease in the amount of snow on the ground in winter, the ptarmigan will not be camouflaged and will be easily spotted by predators. This will decrease the size of the breeding population in the spring, leading to a gradual decline in numbers./ Suitable habitat will disappear and may only be present at much higher latitudes.

12.3

1 Both have an outer and an inner membrane. (There are other correct answers.)

2 In respiration, glucose is broken down to make ATP. This glucose is made by photosynthesis. So this would not work – without another source of ATP, there would be no glucose for use in respiration.

3 It does not occur when there is no light because it needs ATP and reduced NADP from the light dependent stage.

12.4

1 The animals lose less energy in movement and heat, so they convert more of the food they eat into body tissue.

2 If you eat a vegetarian diet, you are a primary consumer. If you eat meat, you are a secondary consumer and there are losses of energy from two trophic levels instead of just one.

3 There is not enough energy left to sustain another trophic level.

12.5

1 A source of energy that is not depleted because it is sustainably replaced.

2 To produce significant reductions in the use of fossil fuels, biofuel crops have to be grown on a very large scale. This will have impacts on the environment, reducing the availability of land used for the production of food for human consumption. Development of biofuels could divert land, water and other resources away from food production, reducing the production of badly-needed basic foodstuffs and driving up their price. In parts of the world where food is in shortest supply and many people are affected by poverty, food shortages and inflated prices could lead to food riots.

3 Environmental advantages: renewable, therefore sustainably replaced; reduce the use of fossil fuels; reduce the emission of greenhouse gases, compared to fossil fuels. Environmental disadvantages: destruction of habitats and loss of biodiversity as land is cleared for biofuel crop cultivation; reduced availability of food/water for human consumption, leading to starvation/decline in health of human population,/civil unrest/political instability and subsequent possible environmental destruction.

Just how sustainable is the production of biofuels?

1 So that they can include indirect greenhouse gas costs in their calculations of overall greenhouse gas savings. Consumers have the right to know this information so that they can make informed decisions about their choice of transport fuel.

2 Aerobic respiration by microorganisms in the soil increases as the soil dries out, releasing carbon dioxide.

3 Greenhouse gas emissions from the fossil fuels used: by the machinery needed to clear the area and plant the biofuel crop; in the manufacture of the fertilisers and pesticides applied to the biofuel crop to increase the yield; in the processing of the biofuel; in the transportation of the biofuel from the site of manufacture to the point of use.

12.6

1 If oxygen is not present, the reduced coenzymes cannot 'lose' their hydrogen atoms, so they remain reduced. This means there are no coenzymes to pick up hydrogens in the link reaction and Krebs cycle, so these processes stop.

2 Any **3** suitable suggestions such as active transport, protein synthesis, movement of sperm, replication of DNA, muscle contraction, and movement of chromosomes in mitosis and meiosis.

13.1

1

Prokaryotic cell	Eukaryotic cell
No true nucleus with a nuclear envelope	True nucleus with nuclear envelope
DNA in the form of a long, circular DNA molecule with no protein attached to it	DNA in the form of linear chromosomes inside the nucleus
No membrane bound organelles	Membrane bound organelles
Small ribosomes	Larger ribosomes
May have a capsule	Does not have a capsule
Cell wall made of peptidoglycan	Cell wall, if present, made of cellulose
Plasmids	No plasmids

2 Because the uterus/ amniotic sac is not open to the external environment, and substances from the mother's blood can only reach the blood of the fetus if they can cross the placenta. Virus particles in the maternal blood plasma may cross the placenta and enter the blood of the fetus, but bacterial cells, protoctists and fungi cannot.

3 A baby that is delivered vaginally will first encounter the flora that occupy its mothers vagina and perineum (the tissue between the vagina and the anus). It is the microbes from these habitats that will gain access to the fetal gut first of all and will be joined by those from the skin of the mother and anyone else who handles the baby soon afterwards. A baby that is born by caesarean section will encounter the microorganisms that live on the skin of its mother's abdomen, the bacteria that may be on the surgeon's gloves and in the operating theatre first of all. It too will then be exposed to those from the skin of the mother and anyone else who handles the baby.

4 The human ecosystem is very diverse, containing many different habitats and thus niches that can be occupied by a great diversity of microorganisms.

13.2

1 They live harmlessly on the skin, helping to prevent infections by out-competing potentially pathogenic microorgansims.

2 The sebaceous glands of people with acne are especially sensitive to normal blood levels of the hormone, testosterone, found naturally in both men and women. The levels of testosterone in the blood begin to increase during puberty, and this triggers the sebaceous glands on the face, chest and back to produce excess sebum. If this can't leave the hair follicles because they are blocked by dead skin cells, the sebum builds up inside the follicles, causing blackheads and whiteheads (spots) to form.

3 The antiseptics may kill the bacteria (bacteriocidal) or they may simply inhibit further growth of the bacterial population (bacteriostatic).

4

Treatment	Bacteriocidal/bacteriostatic effect on bacterial populations	Encourages shedding of surface layer of skin cells	Reduces production of sebum
Antiseptic cleanser	✓		
Benzoyl peroxide		✓	
Retinoid cream		✓	
Antibiotic tablets e.g tetracycline	✓		
Oral contraceptive			✓
Isoretinoin tablets			✓

Can over-the-counter acne treatments be as effective as antibiotics?

1 Effectiveness (of a treatment) relative to its cost.

2 Populations of bacteria with genes for antibiotic resistance are selected-for by the use of antibiotics. Some populations of bacteria that cause acne have evolved resistance to the antibiotic tetracycline and do not respond to treatment with this antibiotic.

3 They may produce a desired effect in less than 18 weeks (the length of this trial), and if the case of acne is severe, this faster rate of treatment may be desirable. They may be easier to use – one or two tablets swallowed each day may be more convenient than applying cream to cleansed skin twice a day. Their effect may be more long-lasting than the effect of a cream.

Comparing the effectiveness of face washes

1 A technique used in microbiology that minimizes the risk of cross-contamination.

2 The experiment was being carried out using P.acnes. This is a potentially pathogenic bacterium and the risk of contamination to people and the environment must be minimised. It is also important to prevent contamination of the pure culture of P.acnes with other bacteria. Contamination would mean that any results obtained could be due to the contaminating organism.

3 Mean diameter of zone of inhibition/mm

Face wash 1 Concentration/%				Face wash 2 Concentration/%			
0	2.5	5	10	0	2.5	5	10
5	9	14	21	5	8	10	15

4 To act as a control. The sterile paper discs may have an antibacterial effect without being impregnated with face-wash. The sterile paper discs do not have an antibacterial effect as no zone of inhibition was produced around them (the 5mm result is simply the diameter of the filter paper disc).

5 face-wash one

6 t-test

7 H_0: there is no difference between the means of the two samples and any difference is due to chance.

8 $t = 6.410$

Critical value from t-distribution table = 2.048

If the null hypothesis were true, the probability of obtaining a value for t of 6.410 with 28 degrees of freedom is less than 5%. Therefore the null hypothesis is rejected and it is concluded that there is evidence at the 5% level of significance to suggest that there is a difference in the means.

Sample no	value	x-mean	$(x\text{-mean})^2$		Sample no	value	x-mean	$(x\text{-mean})^2$
1	20	−1.000	1.000		1	14	−1.000	1.000
2	24	3.000	9.000		2	15	0.000	0.000
3	21	0.000	0.000		3	21	6.000	36.000
4	18	−3.000	9.000		4	16	1.000	1.000
5	19	−2.000	4.000		5	15	0.000	0.000
6	18	−3.000	9.000		6	18	3.000	9.000
7	22	1.000	1.000		7	11	−4.000	16.000
8	20	−1.000	1.000		8	15	0.000	0.000
9	18	−3.000	9.000		9	13	−2.000	4.000
10	21	0.000	0.000		10	14	−1.000	1.000
11	22	1.000	1.000		11	16	1.000	1.000
12	25	4.000	16.000		12	16	1.000	1.000
13	21	0.000	0.000		13	12	−3.000	9.000
14	19	−2.000	4.000		14	13	−2.000	4.000
15	27	6.000	36.000		15	16	1.000	1.000
Totals	315		100.000		Totals	225		84.000

Mean	21.00		Mean =	15.00
Variance =	6.67		Variance =	5.60

H_0: That there is no difference between the means of the two samples

H_1: That there is a difference between the means of the two samples

test statistic $t =$

degrees of freedom $= n_1 + n_2 - 2$ 28

two tailed test at 5% level of significance .... $t = 6.410$

critical value t from statistical tables $= 2.048$

9 The growth of *P.acnes* on agar plates will be different to its growth in the sebaceous glands of human skin. For example, the bacteria may not be provided with the same nutrient sources; in the experiment, the face-wash is applied to paper discs, not to the skin, so the technique does not reproduce the way in which the face-wash is applied to the skin; in the experiment, the face-wash is applied and left in place for 24 hours, whereas most face-washes are applied to the face and are then rinsed off; all the tests used diluted facewash, the facewash may be applied to the skin without dilution; the dishes were incubated at 30 °C, not 37 °C and this does not recreate the temperature at which the P.acnes bacteria will grow on the skin.

13.3

1 A food that contains populations of *Lactobacillus* and *Bifidobacteia*.

2 They compete with potential pathogens, preventing the growth of populations of pathogenic bacteria.

3 The baby's gut flora receives *Bifidobacteria* from the outset; breast milk contains the specific proteins needed for the growth of the *Bifidobacteria*; both of these responses mean that pathogenic bacteria are unlikely to become established in the gut of a breast-fed baby; breast milk contains antibodies.

4 The normal gut bacteria have to increase in number sufficiently to out-compete the pathogenic bacteria. This may be difficult if they are unable to obtain nutrients because of, for example, diarrhoea, which will reduce the amount of time that the undigested food spends in the large intestine. The populations of normal gut bacteria will be ingested with food, and until normal eating habits are re-established and a healthy balanced diet is consumed, the population balance may be slow to change.

Probiotics and health

1 To reduce the risk of developing infections.

2 The results do support the hypothesis, because patients given antibiotics developed infections (suggesting that their immune systems were not functioning effectively), whereas patients not given antibiotics did not develop infections (suggesting that they produced an immune response and prevented infection by any potential pathogens). However, only thirty-three patients were involved in the trial so the validity of the results is questionable.

3 Seriously ill people may be vulnerable to infection by potential pathogens via their deep surgical wounds. If probiotics help to maintain a diverse gut flora and a healthy immune system, populations of these potential pathogens are more likely to be destroyed by an appropriate immune response.

4 Healthy people probably have a healthy and diverse gut flora, and a healthy balanced diet. If they consume plenty of fresh fruit and vegetables and non-processed food they will probably receive adequate doses of probiotic bacteria that can 'top up' the populations of their 'good' gut bacteria. If they are healthy but do not have a balanced diet, probiotic supplements may be useful, but only if the bacteria that they contain are specific and can survive the conditions experienced in the stomach and upper part of the small intestine.

Index